RECENT ADVANCES IN SUSTAINABLE AQUACULTURE AND FISHERIES

RECENT ADVANCES IN SUSTAINABLE AQUACULTURE AND FISHERIES

Shabir Ahmad Dar, PhD
Imtiyaz Qayoom, PhD
Gohar Bilal Wani, PhD
Adnan Abubakr, PhD
Seemab Zehra, PhD
Editors

AAP | APPLE ACADEMIC PRESS

First edition published 2026

Apple Academic Press Inc.
1265 Goldenrod Circle, NE,
Palm Bay, FL 32905 USA

760 Laurentian Drive, Unit 19,
Burlington, ON L7N 0A4, Canada

CRC Press
2385 NW Executive Center Drive,
Suite 320, Boca Raton FL 33431

4 Park Square, Milton Park,
Abingdon, Oxon, OX14 4RN, UK

Library and Archives Canada Cataloguing in Publication

Title: Recent advances in sustainable aquaculture and fisheries / Shabir Ahmad Dar, PhD, Imtiaz Oayoom, PhD, Gohar Bilal Wani, PhD, Adnan Abubakr, PhD, Seemab Zehra, PhD, editors.
Names: Dar, Shabir Ahmad, editor.
Description: First edition. | Includes bibliographical references and index.
Identifiers: Canadiana (print) 20250181304 | Canadiana (ebook) 20250181339 | ISBN 9781998511365 (hardcover) | ISBN 9781998511372 (ebook)
Subjects: LCSH: Sustainable aquaculture—Management. | LCSH: Fishery management.
Classification: LCC SH136.S88 R43 2026 | DDC 639.8028/6—dc23

Library of Congress Cataloging-in-Publication Data

Title: Recent advances in sustainable aquaculture and fisheries / Shabir Ahmad Dar, Imtiyaz Qayoom, Gohar Bilal Wani, Adnan Abubakr, Seemab Zehra, editors.

Description: First edition. | Palm Bay, FL : Apple Academic Press, 2026. | Includes bibliographical references and index. | Summary: "This comprehensive new book, Recent Advances in Sustainable Aquaculture and Fisheries, delves into the latest developments and innovations in the field of aquaculture and fisheries management. Authored by experts in the field, the book presents a thorough exploration of sustainable practices, technological advancements, and ecological considerations that shape the future of aquatic resource management. The book discusses cutting-edge techniques and methodologies for sustainable aquaculture, including recirculating aquaculture systems (RAS), integrated multitrophic aquaculture (IMTA), and biofloc technology. It emphasizes environmentally friendly approaches that minimize resource depletion and pollution. Readers are introduced to modern fisheries management strategies aimed at ensuring the long-term viability of fish stocks. Topics include stock assessment methods, ecosystem-based management, and the role of marine protected areas (MPAs) in conservation efforts. The book explores advancements in aquafeed formulations, including the use of alternative protein sources, microalgae, and probiotics. Also discussed are sustainable feed practices that play a crucial role in reducing dependency on wild fish stocks and minimizing environmental impacts. A significant focus of the book is on disease prevention and management in aquaculture systems. The book covers topics such as biosecurity measures, vaccination strategies, and emerging technologies for disease diagnosis and treatment. Recognizing the challenges posed by climate change, the book discusses adaptive strategies for aquaculture and fisheries resilience. It addresses issues such as water quality management, species diversification, and climate-smart aquaculture practices. Bringing together a wealth of research-based insights, case studies, and practical recommendations for industry professionals, researchers, policymakers, and students, this volume bridges the gap between theory and practice, highlighting successful sustainable aquaculture and fisheries initiatives from around the globe"-- Provided by publisher.

Identifiers: LCCN 2025013749 (print) | LCCN 2025013750 (ebook) | ISBN 9781998511365 hardback | ISBN 9781998511372 ebook

Subjects: LCSH: Sustainable aquaculture--Management | Sustainable fisheries http://id.loc.gov/authorities/subjects/sh94008175 | Fishery management http://id.loc.gov/authorities/subjects/sh85048696

Classification: LCC SH136.S88 R43 2026 (print) | LCC SH136.S88 (ebook) | DDC 639.8068--dc23/eng/20250616

LC record available at https://lccn.loc.gov/2025013749

LC ebook record available at https://lccn.loc.gov/2025013750

ISBN: 978-1-998511-36-5 (hbk)
ISBN: 978-1-998511-37-2 (ebk)
DOI: 10.1201/9781998511372

About the Editors

Shabir Ahmad Dar, PhD
Assistant Professor (Gear & Craft), Division of Fishery Engineering, Faculty of Fisheries, Sher-e-Kashmir University of Agricultural Sciences and Technology, Kashmir (SKUAST-K), India

Shabir Ahmad Dar, PhD, is working as Assistant Professor (Gear & Craft) in the Division of Fishery Engineering, Faculty of Fisheries, Sher-e-Kashmir University of Agricultural Sciences and Technology Kashmir (SKUAST-K), India. He has more than 10 years of experience in teaching, research, and extension in the field of fishing technology. He has a total of 45 publications to his credit in reputed national and international journals, which includes 25 research articles, 13 popular articles, and seven book chapters. He has organized and coordinated several training programs in fisheries for the benefit of fish farmers, rural folks/fishers, and entrepreneurs. He did his PhD in Fisheries Resource Management (FRM) with specialization in Fishing Technology from Central Institute Fisheries Education (CIFE) Mumbai, India.

Imtiyaz Qayoom, PhD
Assistant Professor/Junior Scientist Head, Division of Aquatic Environmental Management, Faculty of Fisheries, Sher-e-Kashmir University of Agricultural Sciences and Technology, Kashmir, SKUAST-K, India

Imtiyaz Qayoom, PhD, works as Assistant Professor/Junior Scientist Head in the Division of Aquatic Environmental Management, Faculty of Fisheries, Sher-e-Kashmir University of Agricultural Sciences and Technology, Kashmir, SKUAST-K, India. He is working in the field of aquatic toxicology, fish toxicity testing, environmental monitoring, residue analysis, and *in vitro* toxicity assessment in fish models. He is involved in various teaching, research, extension, and administrative activities of the university, where he has more than five years of experience. Dr. Imtiyaz has been instrumental in conducting various workshops and trainings to researchers, fisher communities, and stakeholders as well as awareness campaigns on lake de-contamination, restoration, and management. In addition to having several externally and university-funded projects, Dr. Imtiyaz has had the privilege to publish research papers and review articles in high-impact factor journals of repute and to guide master's and PhD scholars for their degrees related to aquatic/fish toxicology.

Gohar Bilal Wani, PhD
Professor/Chief Scientist (Aquaculture Engineering), Head, Division of Fishery Engineering, Faculty of Fisheries, Sher-e-Kashmir University of Agricultural Sciences & Technology of Kashmir, India

Gohar Bilal Wani, PhD, is working as Professor/Chief Scientist (Aquaculture Engineering) and Head of the Division of Fishery Engineering, Faculty of Fisheries, Sher-e-Kashmir University of Agricultural Sciences & Technology of Kashmir (SKUAST-K), India. He is a fellow of various societies and academies, including Fisheries and Life Science of India, Zoological Society of India (ZSI), Coldwater Fisheries Society of India, and The Academy of Environmental Biology (AEB). He has standardized the RAS-based trout hatchery technology for the cold arid region of Ladakh, India, and has worked on seed production of snow trout fish. In addition this, he has been involved in teaching, research, and extension for more than 16 years in the field of fisheries sciences. He has been instrumental in instituting a master's degree program in Fisheries Engineering and Aquaculture. He has also instituted a certificate course in Aqua-farming, Re-circulatory Aquaculture System, and Aquaponic System in Faculty of Fisheries, SKUAST-K., India. He has been involved in several national and international funded projected worth over $3 million USD. He has also organized international and national training programs.

Adnan Abubakr, PhD
Professor of Limnology and Head, Division of Aquatic Environmental Management, Faculty of Fisheries, Sher-e-Kashmir University of Agricultural Sciences and Technology, Kashmir, SKUAST-K, India

Adnan Abubakr, PhD, is working as Professor of Limnology and Head of the Division of Aquatic Environmental Management, Faculty of Fisheries, Sher-e-Kashmir University of Agricultural Sciences and Technology, Kashmir, SKUAST-K, India. He is specialized in designing artificial wetland systems for treatment of wastewaters. He has been involved in teaching, research, and extension for more than 15 years in the field of aquatic ecology. He has been instrumental in instituting a master's program in Fisheries Sciences (MFSc) and PhD degree program in Aquatic Environment Management (AEM) in the Division of AEM. He has more than 60 research publications in peer-reviewed journals and has guided more than 20 students for their master's and PhD degrees in Fisheries Sciences. He has also completed two externally funded and three university-funded research projects. He earned MSc and PhD degrees in Limnology from the Department of Limnology, Barkatullah University Bhopal, India, and has specialized in water and wastewater examination, analytical and bacteriological techniques related to water, freshwater, and brackish water.

Seemab Zehra, PhD
Senior Aquaculture Specialist, King Abdullah University of Science and Technology, Saudi Arabia

Seemab Zehra, PhD, is a Senior Aquaculture Specialist at King Abdullah University of Science and Technology, Saudi Arabia, who is an internationally acclaimed fish nutritionist with more than 15 years of research experience. She has been awarded the prestigious national UGC postdoctoral fellowship as a woman scientist. She was an Assistant Professor on ad-hoc basis from 2015 to 2020. She has attended an advanced training programs on Nutritional Strategies and Feeding Management in Finfish and Shellfish at the Central Institute of Fisheries education [CIFE], Mumbai, India, held in 2007. She has participated in several national and international conferences. She has made remarkable contributions in the field of fish nutrition. Her pioneering research contribution is the nutritional efficacy and digestibility of alternative fed ingredients to fish meal, the feasibility of replacing fishmeal, and quantitative amino acid, protein, mineral, and vitamin requirements of the different cultivable fishes. She has published more than 30 research papers in journals of international repute such as *Aquaculture*, *Aquaculture Research, Aquaculture Nutrition, Journal of the World Aquaculture Society, Journal of Applied Ichthyology*, *Journal of Applied Aquaculture*, *Fish Physiology and Biochemistry,* and *Aquaculture International*, etc.

Contents

Contributors

Adnan Abubakr
Division of Aquatic Environmental Management, Faculty of Fisheries, Sher-e-Kashmir University of Agricultural Sciences and Technology, Kashmir, SKUAST-K, India

Irfan Ahmad
Division of Fish Genetics and Biotechnology, Faculty of Fisheries, Sher-e-Kashmir University of Agricultural Sciences and Technology of Kashmir, Rangil, Ganderbal, Jammu and Kashmir, India

Imtiaz Ahmed
Fish Nutrition Research Laboratory, Department of Zoology, University of Kashmir, Srinagar, Jammu and Kashmir, India

Ishtiyaq Ahmad
Division of Fish Genetics and Biotechnology, Rangil Sher-e-Kashmir University of Agricultural Sciences and Technology, Jammu and Kashmir, India

Kamil Akamad
ICAR-Central Institute of Fisheries Education, Mumbai, Maharashtra, India

Chalungal Thampi Amal
ICAR-Central Institute of Fisheries Education, Mumbai, Maharashtra, India

Adnan Amin
Division of Aquatic Environmental Management, Faculty of Fisheries, Sher-e-Kashmir University of Agricultural Sciences and Technology, Rangil, Ganderbal, Kashmir, India

P. S. Ananthan
ICAR-Central Institute of Fisheries Education, Mumbai, Maharashtra, India

Vidya Shree Bharti
ICAR-Central Institute of Fisheries Education, Mumbai, Maharashtra, India

Farooz Ahmad Bhat
Division of Fisheries Resource Management, Faculty of Fisheries, Skuast-K, Rangil, Ganderbal, Jammu and Kashmir, India

K. Anantharaja
RRC of ICAR-Central Institute of Freshwater Aquaculture, Bangalore, Karnataka, India

Shabir A. Dar
Division of Fishery Engineering, Faculty of Fisheries-Rangil, Sher-e-Kashmir University of Agricultural Sciences and Technology, Rangil, Jammu and Kashmir, India

Leela Edwin
Fishing Technology Division, ICAR-Central Institute of Fisheries Technology, Cochin, India

Burhan Ellahi
Division of Fish Nutrition and Biochemistry, Faculty of Fisheries, Sher-e-Kashmir University of Agricultural Sciences and Technology, Ganderbal, Jammu and Kashmir, India

Aimen Firdous
Division of Post-Harvest Technology, Faculty of Fisheries, Sher-e-Kashmir University of Agricultural Sciences and Technology, Ganderbal, Jammu and Kashmir, India

Yassine Mubark Ali Gameredinn
Agricultural Engineer, Aquaculture Expert and Specialist, Secretary General of the Sudanese Fisheries Sciences Society, Sudan, Africa

Ediga Arun Goud
Dr. Balasaheb Sawant Konkan Krishi Vidyapeeth, Dapoli, Department of Aquaculture, College of Fisheries, Shirgaon, Ratnagiri, Maharashtra, India

Sipaee Venkatesh Goud
Department of Fisheries Resource Management, Faculty of Fishery Science, West Bengal University of Animal and Fishery Sciences, Kolkata, India

Mehak Hafeez
Division of Fish Genetics and Biotechnology, Faculty of Fisheries, Sher-e-Kashmir University of Agricultural Sciences and Technology of Kashmir, Rangil, Ganderbal, Jammu and Kashmir, India

Farah Hanief
Division of Aquatic Environmental Management, Faculty of Fisheries, Sher-e-Kashmir University of Agricultural Sciences and Technology, Rangil, Ganderbal, Jammu and Kashmir, India

S. Abuthagir Iburahim
ICAR-Central Institute of Fisheries Education, Mumbai, Maharashtra, India

Inain Jaies
Division of Aquatic Animal Health Management, Faculty of Fisheries, SKUAST-K, Rangil, Ganderbal, Jammu and Kashmir, India

Paras Nath Jha
Fishing Technology Division, ICAR-Central Institute of Fisheries Technology, Cochin, India
Faculty of Marine Sciences, Cochin University of Science and Technology, Cochin, India

G. Kantharajan
ICAR-National Bureau of Fish Genetic Resources, Lucknow, Uttar Pradesh, India

Tao Kara
ICAR-Central Institute of Fisheries Education, Mumbai, Maharashtra, India

Aboobucker Siddik Abul Kasim
Beacon Development, King Abdullah University of Science and Technology, Thuwal, Jeddah, Saudi Arabia

Jahanzaib Khan
Division of Fish Nutrition and Biochemistry, Faculty of Fisheries, Sher-e-Kashmir University of Agricultural Sciences and Technology, Ganderbal, Jammu and Kashmir, India

Y. M. Khan
Cluster University of Srinagar, Srinagar, Jammu and Kashmir, India

Keezia Khurshid
Division of Aquatic Animal Health Management, Faculty of Fisheries, Sher-e-Kashmir University of Agricultural Sciences and Technology of Kashmir (SKUAST-K), Rangil, Ganderbal, Jammu and Kashmir, India

P. Krishnan
Bay of Bengal Programme Inter-Governmental Organisation (BOBP-IGO), Abhiramapuram, Chennai, Tamil Nadu, India

Joseph Leopoldo Q. Laranja
Beacon Development, King Abdullah University of Science and Technology, Thuwal, Jeddah, Saudi Arabia

Madhulika
ICAR-Central Institute of Fisheries Education, Mumbai, Maharashtra, India

Monisa Malik
Division of Aquatic Environmental Management, Faculty of Fisheries, Sher-e-Kashmir University of Agricultural Sciences and Technology, Rangil, Ganderbal, Kashmir, India

Sandhya Kavitha Mandhir
Fishing Technology Division, ICAR-Central Institute of Fisheries Technology, Kochi, Kerala, India

Asaad H. Mohamed
Beacon Development, King Abdullah University of Science and Technology, Thuwal, Jeddah, Saudi Arabia

Ishrat Mohd
Division of Fisheries Resource Management, Faculty of Fisheries-Rangil, Sher-e-Kashmir University of Agricultural Sciences and Technology, Rangli, Jammu and Kashmir, India

Naveed Nabi
Fish Nutrition Research Laboratory, Department of Zoology, University of Kashmir, Srinagar, Jammu and Kashmir, India

T. Paul Nathaniel
ICAR-Central Institute of Fisheries Education, Mumbai, Maharashtra, India

Sanju Nehra
ICAR-Central Institute of Fisheries Education, Mumbai, Maharashtra, India

B. Chanikya Naidu
ICAR-Central Institute of Fisheries Education, Mumbai, Maharashtra, India

Soibam Ngasotter
ICAR-Central Institute of Fisheries Education, Mumbai, Maharashtra, India

Edoardo Pantanella
Beacon Development, King Abdullah University of Science and Technology, Thuwal, Jeddah, Saudi Arabia

Remesan Meleth Parambil
Fishing Technology Division, ICAR-Central Institute of Fisheries Technology, Kochi, Kerala, India

Hema Prasanth
RRC of ICAR-Central Institute of Freshwater Aquaculture, Bangalore, Karnataka, India

Imtiyaz Qayoom
Division of Aquatic Environmental Management, Faculty of Fisheries, SKUAST-K, Rangil, Ganderbal, Jammu and Kashmir, India

Karankumar Ramteke
ICAR-Central Institute of Fisheries Education, Mumbai, Maharashtra, India

R. Bharathi Rathinam
ICAR, Central Institute of Fisheries Education, Versova, Mumbai, India

Quseen Mushtaq Reshi
Fish Nutrition Research Laboratory, Department of Zoology, University of Kashmir, Srinagar, Jammu and Kashmir, India

P. Routray
ICAR-Central Institute of Freshwater Aquaculture, Kausalyaganga, Bhubaneswar, Odisha, India

Tabinda Shabir
Division of Post-Harvest Technology, Faculty of Fisheries, Sher-e-Kashmir University of Agricultural Sciences and Technology, Ganderbal, Jammu and Kashmir, India

Feroz Ahmad Shah
Division of Aquatic Animal Health Management, Faculty of Fisheries, Sher-e-Kashmir University of Agricultural Sciences and Technology of Kashmir (SKUAST-K), Rangil, Ganderbal, India

Shahziya Shah
Division of Fisheries Resource Management, Faculty of Fisheries, Sher-e-Kashmir University of Agricultural Sciences and Technology of Kashmir, Rangil, Ganderbal, Jammu and Kashmir, India

Tasaduq Hussain Shah
Division of Fisheries Resource Management, Faculty of Fisheries-Rangil, Sher-e-Kashmir University of Agricultural Sciences and Technology, Jammu and Kashmir, India

A. Al Shaikhi
Ministry of Environment, Water and Agriculture, Riyadh, Saudi Arabia

Latha Shenoy
ICAR-Central Institute of Fisheries Education, Mumbai, Maharashtra, India

Soibam Khogen Singh
College of Fisheries, Central Agricultural University (Imphal), Tripura, India

Saroj Kumar Swain
ICAR-Central Institute of Freshwater Aquaculture, Kausalyaganga, Bhubaneswar, Odisha, India

Amal C. Thampi
ICAR-Central Institute of Fisheries Education, Mumbai, Maharashtra, India

D. M. Affarin Tinku
ICAR-Central Institute of Fisheries Education, Mumbai, Maharashtra, India

Suryawanshi Umesh
Karnataka Veterinary, Animal and Fisheries Sciences University, Bidar, Department of Fisheries Resource Management, College of Fisheries, Mangaluru, Karnataka, India

Aitwar Vaijnath
Karnataka Veterinary, Animal and Fisheries Sciences University, Bidar, Department of Aquatic Environment Management, College of Fisheries, Mangaluru, Karnataka, India

Sandhiya Venkatesh
ICAR-Central Institute of Fisheries Education, Mumbai, Maharashtra, India

A. K. Verma
ICAR-Central Institute of Fisheries Education, Versova, Mumbai, Maharashtra, India

Asifa Wali
Division of Fisheries Resource Management, Faculty of Fisheries-Rangil, Sher-e-Kashmir University of Agricultural Sciences and Technology, Jammu and Kashmir, India

K. A. Martin Xavier
ICAR-Central Institute of Fisheries Technology, Cochin, Kerala, India

Seemab Zehra
Beacon Development, King Abdullah University of Science and Technology, Thuwal, Jeddah, Saudi Arabia

Golam Ziauddin
Department of Fisheries Resource Management, Faculty of Fishery Science, West Bengal University of Animal and Fishery Sciences, Kolkata, India

Foreword

Fisheries is a rising sector that provides employment to millions of people across the globe and the improvement in the socioeconomic status of the people associated with it. Today, the fisheries sector envisages the involvement of trained professionals who can contribute to the research and development of the different areas of fisheries. This also demands new books and articles addressing the demanding issues of the fisheries sector and ways to improve it. With meticulous research and insightful analysis, this book can serve as an indispensable guide for scholars, practitioners, and enthusiasts alike, navigating the currents of innovation in the vital realms of Fisheries and Aquaculture.

The publication, which not only encapsulates the vast domain of aquatic sciences but also mirrors our institution's commitment to knowledge dissemination, has been compiled by dedicated and budding scientists. This compilation reflects the collaborative efforts of experts and enthusiasts alike, fostering a deeper understanding of sustainable practices in fisheries and aquaculture.

I am happy to express my views that this book is surely going to help students, researchers, and teachers across the globe in general, and particularly in India, and address their needs in connection with aquatic science. I wish the authors of this book a success in their future endeavors.

—Prof. (Dr.) Farooz Ahmad Bhat

Preface

It is with great enthusiasm and a sense of responsibility that I present this book, *Recent Advances in Sustainable Aquaculture and Fisheries*. The world of fisheries and aquaculture is undergoing a transformative journey, marked by ground-breaking developments that shape the future of our aquatic resources.

Within these pages, you will find a comprehensive exploration of the latest innovations, technologies, and sustainable practices that define the forefront of the fisheries and aquaculture landscape. From leveraging cutting-edge technologies for efficient resource management to fostering biodiversity conservation, the chapters within this book delve deep into the multifaceted aspects of this dynamic field

Our contributors, esteemed experts and researchers, have shared their insights, analyses, and experiences, offering a rich tapestry of knowledge that reflects the diverse challenges and opportunities in fisheries and aquaculture. This compilation aims to be a valuable resource for academics, industry professionals, policymakers, and anyone passionate about the responsible stewardship of aquatic ecosystems.

Recent Advances in Sustainable Aquaculture and Fisheries is a comprehensive book that delves into the latest developments and innovations in the field of aquaculture and fisheries management. Authored by experts in the domain, the book presents a thorough exploration of sustainable practices, technological advancements, and ecological considerations shaping the future of aquatic resource management. The book discusses cutting-edge techniques and methodologies for sustainable aquaculture, including recirculating aquaculture systems (RAS), integrated multitrophic aquaculture (IMTA), and biofloc technology. It emphasizes environmental friendly approaches that minimize resource depletion and pollution.

Readers are introduced to modern fisheries management strategies aimed at ensuring a long-term viability of fish stocks. Topics include stock assessment methods, ecosystem-based management, and the role of marine protected areas (MPAs) in conservation efforts. The book explores advancements in aquafeed formulations, including the use of alternative protein sources, microalgae, and probiotics. Sustainable feed practices play a crucial role in reducing dependency on wild fish stocks and minimizing environmental impacts.

A significant focus is placed on disease prevention and management in aquaculture systems. The book covers topics such as biosecurity measures, vaccination strategies, and emerging technologies for disease diagnosis and treatment. Recognizing the challenges posed by climate change, the book discusses adaptive strategies for aquaculture and fisheries resilience and addresses issues such as water quality management, species diversification, and climate-smart aquaculture practices.

The contributors to this book bring together a wealth of research-based insights, case studies, and practical recommendations for industry professionals, researchers, policymakers, and students. The book bridges the gap between theory and practice, highlighting successful sustainable aquaculture and fisheries initiatives from around the globe.

As we find the way to an era of environmental concerns, economic dynamics, and technological evolution, understanding and embracing these recent advances is paramount. I trust that this book will serve as a catalyst for informed discussions, inspire further research, and contribute to the ongoing dialogue on ensuring the sustainable future of fisheries and aquaculture.

I extend my gratitude to all the contributors whose dedication and expertise have made this endeavor possible. May this book foster a deeper understanding of the complexities inherent in our quest for a balanced and thriving aquatic world.

—Shabir Ahmad Dar

CHAPTER 1

Technologies for Sustainable Fishing

LATHA SHENOY

ICAR-Central Institute of Fisheries Education, Mumbai, Maharashtra, India

ABSTRACT

Globally, the rising threat of declining marine fish stocks due to overfishing and unsustainable fishing practices highlights the importance of adopting sustainable fishing methods to achieve social, economic, and environmental benefits. Technological innovations and developments in fishing gear and craft that are socially, environmentally, and economically sustainable have gained great relevance in achieving sustainable development goals connected with life below water; affordable and clean energy; and climate action. New-generation fishing gear materials with greater strength and durability have facilitated optimizing fishing gear design and operation. The efficiency of fishing operations is enhanced by mechanization in fishing systems and the adoption of advanced technologies. Operational costs and environmental impact of fishing can be reduced by efficient use of energy. Conventional trawl designs are modified to achieve selectivity in fishing gear. Developments in electronics and satellite navigation systems have been instrumental in enhancing safety at sea and efficient navigation and communication systems. The chapter outlines the technologies to conserve marine fish resources; improve fish catch; enhance the efficiency of fishing operations; enhance fish stocks; conserve energy in fishing; maintain safety at sea; fish in deeper waters; and reduce ecological and environmental impacts.

Recent Advances in Sustainable Aquaculture and Fisheries. Shabir Ahmad Dar, Imtiyaz Qayoom, Gohar Bilal Wani, Adnan Abubakr, and Seemab Zehra (Eds.)

DOI: 10.1201/9781998511372-1

1.1 INTRODUCTION

Fishing technology covers direct means of capture such as fishing gears, vessels, methods employed to use them, and the indirect means applied to facilitate capture such as fish-locating technology among others. Fisheries management in the past focused primarily on target-fish resources and fishing activity. Over the years, simple gears were replaced by more efficient ones capable of catching fish in bulk because of the demand from market as well as industry for more fish. The need to venture into deeper waters for exploring the possibilities of catching more fish led to the development of efficient fishing craft as well as improved fishing gears. Overfishing coupled with little regard for the sustainability of resources and its impact on the environment resulted in the depletion of many of the world's fisheries and the degradation of marine ecosystems over the years.

The focus of fishing gear development shifted from increasing production to selective fishing using gears that caused minimal damage to the aquatic environment. The FAO Code of Conduct for Responsible Fisheries (CCRF) was formulated on the concept of "Sustainable Fisheries" (FAO, 1995). Several technological innovations and developments in fishing gear and craft that are socially, environmentally, and economically sustainable have gained great relevance in achieving sustainable development goals particularly connected with life below water; affordable and clean energy; and climate action. These technologies can help to enhance the efficiency of fishing operations, conserve marine fish resources, improve fish catch, enhance fish stocks, conserve energy in fishing, maintain safety at sea, fish in deeper waters, and reduce ecological and environmental impacts while reducing costs.

1.1.1 SUSTAINABLE FISHING

Globally, there is a rising threat of declining marine fish stocks due to overfishing and unsustainable fishing practices which highlights the importance of sustainable fishing to protect ocean biodiversity and ensure a healthy and resilient ocean, cause minimum damage to the environment and ecosystem, contribute to global food security, and protect livelihoods and communities that depend on fishing and seafood. As per the latest report (Oceana, 2019), 31% of fish populations in the world are overexploited; 26 million tons are illegally fished; 58.1% of fish stocks in the world have reached their maximum exploitation; 30% of world fisheries are not reported to the United Nations (UN); and fishing stocks categorized as overexploited increased 20% in the last 15 years. Overexploitation encourages illegal fishing, which generates 36 billion dollars annually, besides degrading marine ecosystems and endangering food security. The present world fisheries situation calls for a sustainable fishing model to fish in a responsible way that prevents overfishing; minimizes bycatch; and avoids waste while contributing to food security; generating jobs and maintaining marine biodiversity.

1.2 TECHNOLOGIES FOR ENERGY CONSERVATION IN FISHING

Fishing, dependent on limited fossil fuels is considered one of the most energy-intensive food production methods (EFTP, 2012). Trawling particularly is an active energy-intensive fishing method dependent on non-renewable fossil fuel (diesel), the burning of which results in CO_2 emission that impacts climate change. Generally, fuel cost accounts for more than half of the operating costs of mechanized boats like trawlers (Boopendranath, 2012). Hence, increases in fuel prices affect the economic viability of fishing operations. Fuel efficiency is an indicator of environmental effects of fishing and fuel use efficiency of a vessel. Operational costs and environmental impact can be reduced by efficient use of energy.

1.2.1 OPTIMIZING FISHING GEAR DESIGN AND OPERATION

Within the trawl system, netting contributes more than 50% of the total drag (Saly Thomas and Sandhya, 2019). As the trawl system is the major component of the resistance responsible for fuel use, improving design, construction, and operation of trawls to reduce fuel use has great potential. Changes in the design of the trawl like larger meshes in the herding parts of the trawl; thinner and stronger netting materials; better rigging of fishing gear, use of hydrodynamically efficient trawl doors, off-bottom fishing, accessories, and optimal warp length can reduce fuel consumption. The use of strong synthetic warps, use of cambered otter boards, smart trawl design and operation, use of advanced materials that reduce twine surface area are other measures to reduce trawl resistance. Knotless netting can further reduce resistance. Other fuel saving options in trawling include the adoption of alternative fishing gears (Semi-pelagic trawl); operational changes such as speed optimization; improved navigation and fish finding; and increasing herding by artificial stimuli while trawling.

1.2.2 NEW GENERATION FISHING GEAR MATERIALS

Newer fishing gear materials with greater strength and durability have facilitated changes in the design and size of fishing nets. Synthetic fishing gear materials with high tensile strength and less drag result in fuel efficiency. Some of the new-generation materials include Aramid fibers, Ultrahigh molecular weight polyethylene (UHMWPE), and biodegradable polymers. Aramid fibers are available under the trade names Kevlar and Nomex. UHMWPE also known as high-performance polyethylene (HPPE) is available under the trade names Dyneema and Spectra. The reduced drag of Dyneema helps in better sinking speed for purse seines and fuel saving in the case of trawl nets. Biodegradable polymers are made from polybutylene succinate (PBS) which is degraded by microorganisms in due course of time. Their use for fishing gear is an eco-friendly option to counter the threat of fishing-induced plastic pollution in the aquatic environment.

1.2.3 OPTIMIZING FISHING VESSEL DESIGN AND TECHNOLOGIES FOR IMPROVING POWERING OF VESSELS

In capture fisheries, vessels and gear are the two main sources of energy consumption. The choice of vessel design, size of engine, and type of propeller determines fuel efficiency. Proper vessel length-to-width ratio, smooth hull painting and fairing, bulbous bows, high-efficiency internal combustion engines, and larger diameter propellers with nozzles are some of the important features of a highly fuel-efficient vessel. The propeller nozzle such as the Kort nozzle (a cylindrical ring of aerofoil section fitted around a ship's propeller that produces a lift force and hence a propulsive effect on the vessel) offers advantages over bare propellers. Reduction of fuel consumption can be achieved by taking advantage of technology developments such as supercharging. The use of this technology helps to increase thermal efficiency; reduce fuel consumption and result in more power output for the same cylinder size.

1.2.4 ALTERNATE ENERGY SOURCES TO POWER FISHING BOATS

The fisheries sector is highly dependent on the use of external energy, particularly in the form of fossil fuels. Fuel use varies widely with fishing type and effort level. Solar, wind, tidal, and hydropower comprise renewable energy sources, non-renewable energy sources are those that are sourced from biomass crops or derived from fossil fuels.. Solar-powered boats used for fishing in calm inland waters, backwaters, and rivers run using sun energy. They have electric motors and storage batteries charged by solar panels and photovoltaic cells. Solar power can also be used for powering lights, fans, and navigational equipment onboard mechanized fishing vessels. There is no requirement for fuel for these boats and hence no pollution from the burning of fuel. Liquefied natural gas (LNG) powered fishing vessels have been gaining momentum in the recent past in Europe, the USA, and parts of Asia (Baiju, 2019). Natural gas (stored in liquid form) and bunker fuel are used for converting chemical energy to mechanical energy. It helps in the reduction of emissions of sulfur oxide and carbon dioxide in addition to being less expensive.

1.3 TECHNOLOGIES FOR IMPROVING EFFICIENCY OF FISHING OPERATIONS

1.3.1 ADVANCES IN TESTING/MODELING FISHING GEAR

The computer-aided designs, testing in flume tanks, direct observation using underwater cameras, acoustics, and trawl-mounted sensors are some of the notable advancements in fishing gear designing. The time taken and prohibitive cost of evaluating new gear designs at sea compelled most institutes/agencies to proceed with computer simulation and testing of physical scale models (Paul et al., 2006).

Model testing is used for the evaluation of the existing gear design or the development of a new gear design. The objective of model testing is to optimize the design parameters of existing gears or develop new designs. Fishing gear development cycle includes defining objectives; modifying existing gear design/developing new gear design; computer-aided simulation; model scaling, construction, and flume tank testing; building full-scale prototype; initial sea trials to determine gear geometry, resistance, etc.; data analysis; comparative fishing experiments; commercial fishing trials followed by introduction into the fleet if found satisfactory. During field trials, the operational parameters are observed through gear monitoring instrumentation.

1.3.2 FISHING WITH LIGHT

Light fishing assists in huge aggregations of fish within a short time. Each fish has its own selective range (wave length) of light and hence species selectivity is possible. Also, the duration of fishing can be reduced by fixing a proper fishing area. Efforts to find more energy-efficient light sources led to the emergence and use of LED lights as a means of reducing energy usage. Commercial applications of light in purse seines, lift nets, and squid jigging are widespread in the Asia-Pacific region. In the context of energy conservation and climate change, fishing using LED light helps to reduce fuel consumption and greenhouse gas emissions.

1.3.3 MECHANIZATION IN FISHING SYSTEMS

The shooting and hauling of fishing gears done manually in the past has been gradually replaced by winches thereby saving both time and labor. With the introduction of net drums, it was possible to operate larger nets from smaller fishing vessels with limited space while facilitating ease of net handling and storage. Stern ramps are useful where trawls have to be changed quickly. The introduction of power block and Triplex was a spectacular development in the mechanization of purse seining to speed up the hauling process with lesser manual labor. Fish pumps which replaced the hand-operated brails, made it possible to empty the gear faster with lesser manpower in a purse seiner (Gabriel et al., 2005).

Gill netters operate longer fleets of nets with reduced manpower with the help of net haulers. Simple reels used for operating gill nets in deeper waters have been replaced by hydraulically powered ones. Modern net shakers (Shaking machines) help overcome the practical difficulty of removing each fish from the mesh.

Hauling and storing of lines has become faster requiring less manpower with the use of winding devices. Hand-operated reels were replaced by those using small electric motors. Gurdies are small line haulers that reduce the amount of labor required. Mechanization of squid jigging started with the introduction of hand-operated roller systems followed by those with a mechanical drive and finally an automatic jigging machine system. The automatic jigging machine continues jigging till the fish and hook are at the surface and squids are then removed.

Mechanization in trap fishing facilitates easy setting, hauling, and harvesting of fish. Hand-operated devices for setting large traps held by stakes have been replaced by special pole-drivers. Modern methods of using suction pumps for emptying trap nets have been found more effective. They are powered either mechanically or hydraulically. Larger and heavier pots used for king crabs are hauled using hydraulically powered pot haulers. Advanced high-speed hydraulic winches are found useful in operating a greater number of pots in a much shorter time in deeper waters.

1.3.4 ADOPTION OF ADVANCED TECHNOLOGIES

Adoption of advanced technologies such as global positioning system (GPS), use of acoustic instruments (echo sounder, SONAR, net sonde, gear monitoring system), use of remote sensing (RS) technology and geographical information system (GIS), and deployment of fish aggregating devices/artificial reefs (FADs/ARs) can improve fishing and fuel efficiency. Opportunities for reducing dependence on fossil fuel exist in the usage of renewable energy systems such as wind and solar-powered generation of electricity.

1.3.4.1 GLOBAL POSITIONING SYSTEM (GPS)

The GPS is a global navigation satellite system that provides information about location and time. In addition to latitude and longitude, GPS provides altitude information. In marine fisheries, it can be used by a survey vessel to continuously monitor water quality and record the exact location. The real-time mapping of data can be done. It is possible for fishing or survey vessels to accurately determine the location of trawl hauls and water sampling areas (Biradar and Pikle, 2005). It is possible to record the vessel's location and catch data at frequent intervals of time and thus have a complete record of catch against the location. The use of GPS is encouraged in order to monitor fishing activities.

1.3.4.2 USE OF ACOUSTIC INSTRUMENTS

Finding fish with accuracy is very important for commercial fishing operations. Traditional fishermen had their own innovative methods for fish finding, based on their observations and experience. With the introduction of newer and more sophisticated fishing vessels and methods, increasing the range of fishing operations, a strong need was felt for accurate fish-detection equipment. The use of electro-acoustic equipment like echo sounder, sonar, net sonde, and gear monitoring system has become almost indispensable in industrial fishing not only to find fish but also to collect information such as depth, nature of the bed, the position of the net, fish behavior, etc.

Modern day echo sounders generally, have electro-strictive ceramic transducers due to their advantages of lower cost and higher efficiency, and the receiver units have signal processing controls. As the strength of echoes at close range is more than those at a distance, the sensitivity of the receiver is automatically adjusted (suppressed at the time of transmission and increases with time) for the removal or reduction of echoes from close range. Sound propagation losses are compensated by incorporating the time varied gain (TVG) function in the receiver units of modern echo sounders. Earlier types of recorders used mechanically operated pens for marking on either wet or dry paper. More advanced echosounders have features that allow the detection of fish lying close to the seabed through white line and gray line effect. Modern echo sounders use CRT and liquid crystal display (LCD) monitors to enable differentiation of bottom fish and seabed echoes that are displayed in distinct colors. Red color represents strong echoes whereas weak echoes are seen in blue. Strong signals are depicted as dark lines against a white background in a monochrome display. White and gray line display is provided to differentiate between the weakfish echoes from strong bottom echoes (Joshy, 2019).

The sonar can produce pulses of different duration. The observation of echoes is carried out in three different methods, the paper recording, the cathode ray scope, and the aural system (listening). Search-light sonars and omni sonars comprise two distinct types of sonars available now for use in fishing operations. Search-light sonars transmit a sound pulse on a selected sector bearing, listen to the echo, and on completion, move to the next bearing whereas omni sonars transmit a single pulse radiating from the transducer through a complete 360 degree and electronically scan the entire area while listening to the echoes. This will give a complete picture of the vessel for each transmission.

Net sonde is a type of echo sounder developed mainly to find out the precise position of a pelagic trawl. It also helps to find out the opening of the trawl and records the fish which enter the trawl. It is also useful in bottom trawling for providing a better estimate of catching efficiency.

Gear monitoring system helps in precise measurement of operational parameters of fishing gears. In the past, they were measured using different instruments and methods. Technological developments in the field of marine electronics led to the development of efficient electronic instruments for precise measurement of operational parameters. Performance of an underwater trawl is measured by parameters such as warp load at the ship; warp angle at towing block; warp declination at towing block; boat speed; load in warp due to different components of the net; otter board angle of attack; otter board angle of pitch; distance between otter boards; otter board angle of heel; net mouth spread; vertical opening; trawl speed; and trawl depth.

Mechanical types of underwater recording instruments are attached to the gear and towed with it. Hence, the data can be obtained only after hauling the net besides being comparatively heavy and less accurate. Acoustic and non-acoustic types of wire telemetering instruments are accurate and give instantaneous data. Cable-less net monitoring systems have been developed to overcome the problem of cable-breakage during operation. Indigenous electronic instruments for measuring the operational parameters of fishing gear and craft have been developed by CIFT, Kochi which are wire telemetering instruments with comparatively lightweight transducers and the cost being low compared with imported instruments (Ramakrishnan, 2000).

1.3.4.3 APPLICATION OF REMOTE SENSING IN FISHERIES

Satellite RS has been used to support marine fisheries by locating potential fishing zones (PFZ) using temperature and productivity data to identify areas where fish tend to aggregate for feeding. Vessel monitoring systems (VMSs) using satellite RS have been applied to locate and monitor vessels for fisheries management.

Potential fishing zone (PFZ) is a technique of identifying fish shoals depending on certain oceanic features/processes, such as oceanic front, upwelling, eddies, etc., and is a short-term forecast of fish aggregation zones in the open sea. The rationale of fish finding through integration of sea surface temperature (SST) and chlorophyll images is based on the phenomena of fish accumulation in the areas of abundant food (phytoplankton/zooplankton) and suitable environment. Ocean color monitor (OCM) sensor-derived chlorophyll concentration and NOAA AVHRR-derived SST (National Oceanic and Atmospheric Administration Advanced Very High-Resolution Radiometer) are used in the forecast of PFZ along the Indian coastline. This advanced technology is beneficial to the fisher community. A reliable and timely forecast of the PFZs reduces search time and fuel cost involved in locating fishing grounds thereby enhancing the profitability of fishing operations (Kamei et al., 2014). Dissemination of weather forecasts facilitates greater safety during navigation and fishing operations. Remote sensing, combined with conventional data collection techniques, provides a powerful tool for designing harvesting strategies for living marine resources.

1.3.4.4 GEOGRAPHICAL INFORMATION SYSTEM (GIS)

GIS, GPS, and RS comprise the three important spatial technologies in fisheries. Geographic information system works with spatially referenced data which is a promising tool to solve spatially related fisheries and aquaculture management problems. It provides an electronic representation of information. It is possible to map marine habitats; assess the proportion of juveniles, bycatch and discards; fish distribution and abundance in relation to environmental parameters; prepare thematic maps; to estimate catch and effort and their variation in space and time; to obtain exact geospatial information on activities of fishing vessels etc. besides marine spatial planning. GIS technology shows a promising future for fisheries science and management. Spatial dimension forms a necessary component while designing a fisheries management plan (Meaden and Aguilar-Manjarrez, 2013). The database created in GIS platform with illustrations in the form of maps can be used as a tool by the policymakers to tackle problems in conserving and managing the fisheries resources.

1.3.4.5 ENHANCE FISH STOCKS THROUGH FADS

As per FAO (FAO, 2019), fish aggregating devices (FADs) refer to a permanent, semi-permanent, or temporary object, structure, or device of any material, man-made or natural, which is deployed, and/or tracked, and used to aggregate fish for subsequent capture. They can be either anchored or drifting objects placed in the ocean that attract and aggregate pelagic fish makes harvesting them easier. This method of attracting and aggregating fishes like tunas is an age-old practice adopted to increase fish stock populations and enhance fishing opportunities through the use of FADs. They are deployed to enhance biological productivity in coastal waters; rehabilitate and conserve depleting stocks; reduce scouting time in harvesting operations; increase catch rate with lesser cost of production; and improve the income and living standards of fishers. A typical FAD consists of three components: surface, underwater, and a marker to indicate its position. Modern markers have electronic transmitters. Industrial FADs improve the catch rate of particularly purse seine and pole and line vessels that target large shoals of tuna. Anchored buoys can also be used. Technological advances in FAD designs have led to increased fishing efficiency.

1.4 TECHNOLOGIES FOR RESPONSIBLE FISHING

1.4.1 SELECTIVITY OF FISHING GEARS

The use of selective fishing gears and methods facilitates the protection of spawners and juveniles, minimizing waste and ensuring the survival of escaping fish by incorporating bycatch reduction devices (BRDs). Species selectivity in mobile gears like trawls is achieved by using grids or separator panels while the size selectivity can be accomplished by controlling mesh size and shape. Turtle excluder device (TED) is one such type of grid incorporated in shrimp trawl that allows the escape of endangered turtles. The use of environmentally friendly fishing gears such as semi-pelagic trawls that are operated slightly off the bottom minimizes impact on the bottom fauna and prevents habitat degradation. There are possibilities of significant bycatch in fishing gears which are not modified to exclude the entanglement of fish, turtles, or seabirds. There is an increased risk of depletion of vulnerable or endangered species.

1.4.1.1 BYCATCH REDUCTION DEVICES (BRDS)

Any device that excludes or reduces non-targeted catch when fitted in a fishing gear for improving its selectivity is called a BRD. ICAR-CIFT has developed several indigenous BRDs to reduce bycatch in trawl nets.

A square mesh window panel consists of a piece of webbing of square mesh attached as a window in a trawl. Square meshes remain open during fishing facilitating escapement of fish to be eliminated through the opening. It operates on the differential behavior of shrimps and fishes. The size of the mesh, location of netting, and shape of the fish determine the escapement rates. It is simple, easy to use, and is cheap. The Big eye consists of a simple slit provided on the upper side of the cod end or belly and operates based on the differential behavior of shrimps and fishes. The slits can be made of different shapes. Radial type escapement device consists of two rings of 80 cm diameter and made of 10 mm steel rods. The rings are connected by square mesh webbing of 1.5 m in length and a guiding funnel (made of small mesh webbing fitted inside the device) is attached between the net and the cod end. The size of the fish to be eliminated determines the mesh size. Fishes swim backward reach the space between the two rings and try to escape through the mesh opening while the shrimps remain in the cod end.

TEDs are BRDs that exclude larger-sized organisms. Fish eye and square mesh cod end are BRDs that facilitate the escape of small-sized fishes. Fish eye is an eye-shaped structure made of 6 mm steel rod of 30 cm width and 10 cm height fixed in a triangular frame. It provides an escape opening at the top for small fishes. The square mesh cod end allows small fish to escape. Selection of mesh size is important. Advantages include less sorting time, an increase in shrimp catch on account of the possibility of an increase in towing time, and less dependency on fish behavior. Shape of fish affects escape rates.

Rigid grid sorting devices are the most advanced type of selective device used in many countries for eliminating turtles before entering the cod end of a trawl net (TEDs). The spherical, rectangular, or elliptical grid made of 10 mm stainless steel or aluminum rods is vertically divided. During fishing, shrimp and fish enter through the funnel and reach in front of the grid. The turtles and bigger fishes escape when encountered at the grid through an escape opening provided at the top.

Juvenile fish excluder cum shrimp sorting device (JFESSD) designed by CIFT, is a device that reduces bycatch of juveniles and small-sized non-targeted fishes but retains large sizes (Boopendranath, 2010). The net is provided with an accelerator funnel and JFESSD grid. The upper cod end has 60 mm^2 mesh and the lower cod end 20 mm^2 mesh. Mature shrimps are retained in the bottom and mature finfish in the top portion of the net. Advantages are good catch quality, less sorting time and longer tow duration.

1.4.2 LOW-ENERGY ENVIRONMENTALLY RESPONSIBLE FISHING PRACTICES

Low impact fuel-efficient (LIFE) fishing gears are examples of low-energy environmentally responsible fishing practices. Fishing with gill nets, lines, pots, and traps is beneficial from the standpoint of energy optimization and conservation (Suuronen et al., 2012). Fishing with gill nets is a highly size-selective, environmentally responsible, and low-energy method. Longlines are size-selective and energy-efficient gears. Trolling is an environmentally responsible method due to the possibility of quick release of unwanted catch back to sea. Traps and pots have lower unintended catch and less seafloor impact than mobile gear like trawls.

1.4.3 TRAWLING TECHNOLOGIES FOR SUSTAINABLE FISHING

Conventional trawl designs are modified to achieve certain special design objectives such as reducing the drag of a net, improving the herding effect, sweeping a wider area along the seabed, increasing the vertical opening of the net, facilitating harvesting of fast-swimming fishes, increasing the horizontal spread, separate shrimps from other fishes or simply make it an eco-friendly trawl. Some of the specialized trawl designs developed by ICAR-CIFT are Bulged belly trawl; Low-drag trawls, such as Rope trawl, and Large-mesh high opening trawl; Long-wing trawl; High-speed demersal trawl; and Eco-friendly semi-pelagic trawl (Ravishankar, 2019). Separator trawl is designed to separate shrimps from other fishes. Horizontal spread is increased in the Duplex and Triplex trawl by incorporating 2 and 3 cod ends in the trawl.

1.4.3.1 SMART TRAWLING TECHNOLOGY

The smart trawling technology uses special instrumentation for constant and automatic measurement of the distance of trawl doors and ground gear from the seabed. The use of dropper chains suspended from the footrope helps to reduce sea bed contact, but not at the cost of catching efficiency. It reduces seabed damage from bottom trawling (Valdemarsen and Suuronen, 2003).

1.4.3.2 CRISP TRAWL TECHNOLOGY

The key features of CRISP trawl technology developed by the Center for Research-based Innovation in Sustainable Fish Capture and Processing technology are fewer contact points between trawl components and the seabed, lighter contact of trawl components with the ground and rigging of trawl doors for off-bottom towing. These features make this bottom-habitat-friendly technology (Valdemarsen et al., 2011). The net camera gives real-time information about the fish species and sizes entering the cod end and allows the skipper to make informed decisions regarding how to continue the fishing process. The trawl may also be fitted with an active mechanism to release unwanted catch (based on image analysis). The trawl concept also includes a cable connection from the vessel to the trawl headline. The cable carries the video signal from the net camera and acoustic sensors and also increases the vertical opening of the trawl.

1.4.3.3 ECO-FRIENDLY SEMI-PELAGIC TRAWLING

Semi-pelagic trawling has been a later introduction to reduce the adverse impact of trawling on the benthic fauna. It was developed for catching fish off the seabed (He et al., 2021). In a semi-pelagic trawl, pelagic otter boards with a high aspect ratio are used to keep them off the seabed even when the net touches the bottom.

1.5 DEEP-SEA FISHING TECHNOLOGIES

1.5.1 FISHING GEARS FOR DEEP-SEA FISHING

Monofilament tuna longlining was introduced by FSI, Mumbai as an efficient technique for catching tunas. The feasibility of multifilament and monofilament tuna longline fishing has been established by locating productive tuna grounds in the oceanic sector of the Indian EEZ. Experiments were carried out on mid-water trawling, bottom set long lining, vertical long lining, squid jigging, and trap fishing as diversified eco-friendly fishing techniques for harvesting deep-sea fisheries resources (FSI, 2008).

Multifilament tuna long lining/monofilament tuna long lining; squid jigging; bottom set gillnet; trap fishing; bottom set vertical long lining; myctophid trawl; high-speed demersal trawl; large-mesh demersal trawl; and eco-friendly trawl (semi-pelagic) are some of the suitable fishing gears for deep-sea fishing.

1.6 ADVANCES IN FISHING CRAFT TECHNOLOGIES

1.6.1 COMPUTER-AIDED DESIGNS (CAD)

Visualization of the new boat design on the computer is possible using the latest technology in computer boat design software before starting the actual construction. It is possible to refine the craft design with the help of software simulation and model testing.

1.6.2 BEACH LANDING CRAFT

For surf-beaten beaches, with very few natural or artificial harbors, a mechanically propelled beach landing craft (BLC) was developed suitable to use on the beaches of the east coast of India (Jacob and Shenoy, 2014). BLC operating from the beaches are subject to a harsher environment than a craft operating from the harbor, as it has to encounter great impact forces while crossing breakers and landing on the beach. These craft have a greater carrying capacity and better protection for crew and catch than the traditional Catamaran and other craft operating from the beach. Outboard motors are capable of crossing the surf satisfactorily, but they had economic limitations due to the high operating cost and short service of the motor. In order to overcome this, the BOBP developed a liftable propulsion system connected to air or water-cooled diesel engines available locally; called BOB drive for BLC. A variety of fishing craft, FRP BLC and plywood canoes to narrow outrigger canoes have been fitted with the BOB drive and found to be acceptable to fishermen operating from open surf beaches and shallow water outlets.

1.6.3 ALTERNATE MATERIALS

1.6.3.1 RUBBER WOOD CANOES

To reduce dependency on conventional boat-building material wood which has become scarce and expensive, CIFT, Kochi has developed a technology for the construction of canoes using rubber wood, a low-cost substitute for conventional fishing craft materials (Ravishankar, 2019). In spite of its good mechanical properties and working quality, rubber wood is highly perishable under marine conditions and hence needs to be treated with chemical preservatives prior to construction. The FRP sheathing provided to the canoes enhances the dimensional stability of the canoes. It also makes the canoe waterproof, resistant to attack by marine borers, abrasion, etc., and reduces the maintenance cost. Rubber wood-treated canoes are operated by fishermen in backwaters, inland, and marine waters.

1.6.3.2 PLYWOOD BOATS

Plywood is well established as a boat-building material. It comprises as much as 80% of the material on some boats and is rarely completely missing on any vessel. Plywood is suitable for hull sheathing, decking, bulkheads, superstructure, boat patterns, and molds due to its panel or sheet form and availability in a wide range of sizes and thicknesses. Marine plywood is a type of plywood that is used for the construction of fishing boats as this plywood is waterproof and can be glued to any surface.

The advanced manufacturing process of marine plywood is pre-pressing; a preliminary process of applying pressure to glued adjacent layers at room temperature, for developing a temporary bond between different layers of veneers. Such assembly is subsequently hot-pressed to get higher shear strength and unparalleled bonding. This advanced technology improves the bonding as well as makes it warp-free, and eliminates the possibilities of defective plywood production.

Irrespective of hull materials, plywood is used in all small fishing vessels for joiners' work to decks, bulkheads and superstructure.

Considering its strength and lightweight, plywood is equal to any other material that can be used for these applications. Other considerations for the choice of plywood are maritime regulations, available labor, durability, costs, and technical knowledge. With suitable labor and adequate technical knowledge of the characteristics of the plywood, it is less costly and more effective than other materials for these applications. Marine plywood has a high strength-to-weight ratio. Also, it is relatively cheap and rot-resistant and hence does not harm the aquatic environment. Plywood is treated with preservatives like CCA (Manju Lekshmi, and Leela Edwin, 2019). The treated panels are then redried in a kiln till the moisture content is brought down to 18%. These boats are protected from abrasion by providing a coating of fiberglass.

1.6.4 DECK MACHINERY

1.6.4.1 POWER TAKE-OFF (PTO) AND WINCH DRIVE

Type of fishing and size of fishing vessels determine the deck machinery and equipment required for gear handling onboard. Modern vessels require auxiliary power for driving onboard machinery such as winches, pumps, or compressors. The present-day marine transmissions are compact, low cost, and efficient; clutched or direct-drive power take-off units, which are bolted directly to the rear of the forward gearbox.

1.6.4.2 CAPSTAN

A Capstan is a vertical axle rotating machine developed for use onboard vessels to apply force to ropes, cables, and hawsers. Modern Capstans are powered electrically, hydraulically, pneumatically, or via an engine.

1.6.4.3 GANTRIES

Conventional gallows, mast, derrick, and stays are replaced by gantry in present-day trawlers for positioning warp leads, hitching of otter boards, lifting heavy fish catch, etc. It is the most reliable and efficient means of handling heavy fish catch. Gantry also offers direct leads to the warping heads of winches, thereby reducing the wear and tear of wires.

1.6.4.4 STERN RAMP

Stern ramp is a flat or curved surface extending from the waterline to the working deck at the stern of large trawlers for hauling the fish catch directly onto the deck. It reduces the amount of catch handling and also does away with the splitting of the cod end.

1.6.4.5 WINCHES

Availability of space, fuel supply, ease of repairs and maintenance, cost tec determine the type of winch drive for fishing vessels. Generally, small vessels use mechanical drive due to its simplicity, reliability, ease of repairs and maintenance, and cost-effectiveness. The need for a large capacity generating plant and auxiliary diesel are the driving forces for the use of electric drive in larger vessels. Flexibility of operation and control are the main advantages of electric winches. Hydraulic drive is used only in large vessels. Though winches can be driven directly by a separate engine, the direct diesel drive is not popular due to requirements of more space, fuel supply, and exhaust arrangements.

1.6.4.6 NET DRUMS

Reduction of labor for operation of gill nets, trawls and seines is facilitated by installation of net drums mostly at the stern in fishing vessels. The entire net is wound on the drum. Net drums also prevent the sagging or fouling of the gear during the operation.

1.6.4.7 POWER BLOCKS

Power blocks are labor-saving devices used on purse seiners, trawlers, and drift netters.

1.7 TECHNOLOGIES FOR SAFE AND EFFICIENT NAVIGATION AND COMMUNICATION SYSTEMS

1.7.1 APPLICATION OF ELECTRONICS IN NAVIGATION

Commercial fishing like other maritime activities has benefited from the application of electronics. This is particularly apparent in the development of a wide range of fish-finding devices and navigational aids. Three methods of electronic navigation are satellite navigation, radio navigation, and radar navigation. There have been several developments in the electronic navigational systems in the last few decades with the advancements in microelectronics and the launching of specific satellites.

1.7.1.1 ELECTRONIC CHART DISPLAY AND INFORMATION SYSTEM (ECDIS)

The Electronic Chart Display and Information System (ECDIS), is one of the notable developments in the navigational chart systems used in vessels. It is primarily used to obtain directions. It complies with the IMO regulations. The ECDIS equipment complying with SOLAS requirements can be used as an alternative to traditional paper charts. Chart correction is easier in ECDIS than the paper charts. The navigational information displayed is in real-time. The ECDIS enhances navigational safety. Provision of alarms highlights dangers.

1.7.1.2 AUTOPILOT

Autopilot is an advanced navigational equipment used for steering the vessel without constant handling of the steering wheel manually. This reduces the wear and tear of the steering gear. It controls the rudder of a vessel using an automatic system. The use of autopilot can reduce fuel consumption by smoothing out the large-angle rudder movements. It can be either a mechanical, electrical or hydraulic system which maintains a vessel on a predetermined (set) course without the need for human intervention.

1.7.2 SATELLITE NAVIGATION SYSTEMS

Satellite navigation is the latest navigation that works on the satellite-based hyperbolic system, whereas the earlier Loran, Decca, and Omega are land-based hyperbolic systems. Each satellite transmits a signal which permits its position to be calculated accurately as a function of time. The calculated satellite positions, range difference measurements between these positions, and the information about the motion of the vessel are used to fix the position of a vessel. Calculations are done by a digital computer which is integral to the equipment. The satellite receiver is designed to accept signals from all commonly used log and gyrocompass inputs.

The GPS is used in finding the position, time, speed and course, waypoints, and for navigation. Several types of alarms are incorporated into GPS to assist in navigation. These include destination alarm, cross track error alarm, anchor watch alarm, hazard alarm, etc. for alerting while nearing/reaching the destination, vessel going off course, drifting, and nearing any danger zone, respectively. The VMS is essentially a global position system carried on commercial fishing boats that transmits the boat's geographic position to various state enforcement agencies. It is a tool in the monitoring, control and surveillance (MCS) programs. VMS is a key part of MCS programs at the national and international levels. It allows fisheries regulatory organizations to monitor the position, time at a position, course, and speed of fishing vessels. It is a reliable and direct means of communication between vessels and the monitoring agency. With VMS, patrol boats reach directly without wasting time on searching the vessel. VMS technology also finds application in enhancing the safety of fishing vessels.

1.7.3 SATELLITE COMMUNICATION

The development of satellite technology has made global communication possible through telephone, telegraph, telex, and fax. Special safety systems like global maritime distress and safety system (GMDSS) have been introduced for distress alerting, search, and rescue coordination, and transmission of maritime safety information including navigational and meteorological warnings.

The GMDSS includes equipment, such as EPIRB, SART, INMARSAT, NAVTEX, and DSC-MF, HF, and VHF. Emergency position indicating radio beacon (EPIRB) that works on a frequency of 406 MHz sends alerts and detects a vessel in emergency. It can be activated and deactivated automatically or manually. A search and rescue transponder (SART) operates on a frequency of 9 GHz. It is used to locate vessels in distress and rescue survivors in a life raft. The detection range is normally about 8 nautical miles. International Maritime Satellite (INMARSAT) is a satellite communication network operated by the INMARSAT. It is useful for sending emergency messages, rescue operations, weather forecasting, and data communication to the vessel in the sea. Navigational telex (NAVTEX) is an international, automated system for instantly distributing maritime navigational warnings, weather forecasts, and warnings. Digital selective calling (DSC), another component of GMDSS helps enhance vessel navigation safety with its automated digital communications capabilities. It is primarily intended to initiate ship/ship, ship/shore, and shore/ship radiotelephone (Shenoy, 2022).

1.8 TECHNOLOGIES FOR SAFETY OF FISHING VESSELS

1.8.1 DISTRESS ALERT TRANSMITTER (DAT)

Distress alert transmitter (DAT) is primarily designed for the transmission of distress messages. It also provides positional information when activated. INSAT-based DAT developed by ISRO finds wide commercial application (Shenoy, 2012). It just requires to switch on by selecting a relevant message (fire in boat/boat sinking/medical help/any other emergency) in times of distress.

1.8.2 AUTOMATIC IDENTIFICATION SYSTEM (AIS)

Vessels at sea are identified and located using an automatic tracking system termed automatic identification system (AIS). The vessels are located by electronically exchanging data with ships in the vicinity and AIS base stations. As this information supplements marine radar, AIS also helps to avoid collision in addition to the identification and location of fishing vessels.

KEYWORDS

- **advanced technologies**
- **mechanization**
- **satellite navigation and communication**
- **selectivity in fishing gear**
- **sustainable fishing**

REFERENCES

Baiju, M. V. (2019). Energy-efficient fishing vessels and use of alternate energy for fishing vessels. In L. Edwin, N. Saly Thomas, M. P. Remesan, P. Ashraf Muhamed, M. V. Baiju, N. Manju Lekshmi, V. R. Madhu (Eds.), *ICAR winter school on responsible fishing: Recent advances in resource and energy conservation;* (Chapter 11, pp. 143–150). ICAR-CIFT.

Biradar, R. S., & Pikle, M. (2005). Global positioning system. In *ICAR winter school on remote sensing and GIS applications in fisheries research and Management* (pp. 190–193). ICAR-CIFE.

Boopendranath, M. R. (2010). Bycatch reduction technologies. In B. Meenakumari, M. R. Boopendranath, L. Edwin, T. V. Sankar, N. Gopal, and G. Ninan (Eds.), *Coastal fishery resources of India: Conservation and sustainable utilisation* (pp. 4–30). Society of Fisheries Technologists.

Boopendranath, M. R. (2012, March 7–8). Modern trends in fishing technology. In UGC sponsored seminar on emerging trends in aquaculture, harvesting and post harvesting technology (pp. 1–6). Department of Zoology, Marthoma College.

EFTP. (2012, Sept 12). Key challenges to promote the energy efficiency in fisheries from a research and innovation perspective. Report from the Workshop about energy efficiency organized by the European Fisheries Technology Platform (p. 20). Secretaría general de pesca.

FAO. (1995). *Code of conduct for responsible fisheries* (p. 41). FAO.

FAO. (2019). *Voluntary guidelines on the marking of fishing gear* (p. 88). FAO.

FSI. (2008). Introduction of Monofilament Longline technology for harvesting oceanic tuna and allied resources in the Indian EEZ. Bulletin No. 30.

Gabriel, O., Lange, K., Dahm, E., & Wendt, T. (Eds.). (2005). *Von Brandt's fish catching methods of the world*, 4th edn (p. 523). Wiley-Blackwell.

He, P., Chopin, F., Suuronen, P., Ferro, R. S. T., & Lansley, J. (2021). *Classification and illustrated definition of fishing gears.* FAO; FAO Fisheries and Aquaculture Technical Paper No. 672.

Jacob, J., & Shenoy, L. (2014). *Fisheries engineering: Fishing vessels*, Vol. 2 (p. 202) CIFE.

Joshy, C. D. (2019). Marine electronic equipments used in fishing vessels. In L. Edwin, N. Saly Thomas, M. P. Remesan, P. Ashraf Muhamed, M. V. Baiju, N. Manju Lekshmi, and V. R. Madhu (Eds.), *ICAR winter school on responsible fishing: Recent advances in resource and energy conservation* (Chapter 36, pp. 411–432). ICAR-CIFT.

Kamei, G., Felix, J., Shenoy, L., Shukla, S. P., & Devi, H. (2014). Application of remote sensing in fisheries: Role of potential fishing zone advisories. In J. Sundaresan et al. (Eds.), *Geospatial technologies and climate change* (pp. 175–186). Springer Publications.

Manju Lekshmi, N., & Edwin, L. (2019). Boatbuilding materials with special reference to environmental impacts. In *ICAR winter school on responsible fishing: Recent advances in resource and energy conservation* (pp. 87–94), ICAR-CIFT.

Meaden, G. J., & Aguilar-Manjarrez, J. (Eds.). (2013). *Advances in geographic information systems and remote sensing for fisheries and aquaculture* (p. 425). FAO Fisheries and Aquaculture. Technical Paper No. 552.

Oceana. (2019). FAO. WWF y global fishing watch. https://www.iberdrola.com

Paul, D. W., Harold, D., & George, L. (2006). Designing and testing new fishing gears: The value of a flume tank. *Marine Technology Society Journal, 40*(3), 44–49.

Ramakrishnan, K. (2000). Electronic instruments for fishing gear research. In *Training course manual on fishing technology* (pp. 165–173). CIFT.

Ravishankar, C. N. (2019). Recent advancement in fish harvest sector by ICAR-CIFT. In L. Edwin, N. Saly Thomas, M. P. Remesan, P. Ashraf Muhamed, M. V. Baiju, N. Manju Lekshmi, and V. R. Madhu (Eds.), *ICAR winter school on responsible fishing: Recent advances in resource and energy conservation* (Chapter 2, pp. 9–18). ICAR-CIFT.

Saly Thomas, N., & Sandhya, K. M. (2019). Netting materials for fishing gear with special reference to resource conservation and energy saving. In L. Edwin, N. Saly Thomas, M. P. Remesan, P. Ashraf Muhamed, M. V. Baiju, N. Manju Lekshmi, & V. R. Madhu (Eds.), *ICAR Winter School on Responsible fishing: Recent Advances in resource and energy conservation* (Chapter 6, pp. 55–70). ICAR-CIFT.

Shenoy, L. (2012). Recent advances in fishing craft technology. In D. D. Nambudiri, and K. U. Peter (Eds.), *Advances in harvest and post-harvest technology of fish* (Chapter 1, pp. 5–20). New India Publishing.

Shenoy, L. (2022). *Fishing gear and craft technologies for sustainable fishing* (457 pp.). Narendra Publishing House.

Suuronen, P., Chopin, F., Glass, C., Lokkeborg, S., Matsushita, Y., Queirolo, D., & Rihan, D. (2012). Low impact and fuel-efficient fishing—Looking beyond the horizon. *Fisheries Research, 119–120*, 135–146.

Valdemarsen, J. W., & Suuronen, P. (2003). Modifying fishing gear to achieve ecosystem objectives. In M. Sinclair, M., and G. Valdermarsen (Eds.), *Responsible fisheries in the marine ecosystem* (p. 341). FAO and CABI International Publishing.

Valdemarsen, J. W., Ovredal, J. T., & Asen, A. (2011). *New semi-pelagic trawl construction (CRIPS-trawl)* (p. 17). Report from Havforskningen No. 18. Institute of Marine Research.

CHAPTER 2

Depuration of Bivalves: Fundamental Aspects and Complementary Methods to Improve Depuration Process

SANDHIYA VENKATESH[1], SANJU NEHRA[1], D. M. AFFARIN TINKU[1], SOIBAM NGASOTTER[1,*], K. A. MARTIN XAVIER[2], and SOIBAM KHOGEN SINGH[3]

[1]*ICAR-Central Institute of Fisheries Education, Mumbai, India*

[2]*ICAR-Central Institute of Fisheries Technology, Cochin, Kerala, India*

[3]*College of Fisheries, Central Agricultural University (Imphal), Tripura, India*

*Corresponding author

ABSTRACT

Bivalve aquaculture production has been increasing globally for the past 20 years with a traditional market value. World aquaculture production comprises 87.5 million metric tons, of which molluscans contributed around 17.7 million metric tons, with bivalves as the major component. Bivalve farming can be made more profitable if the produce meets the basic quality and sanitation standards. Bivalve molluscan shellfish are filter feeders; they feed by filtering large volumes of seawater, accumulating the food particles from their surrounding environment. Seawater, conversely, contains loads of water pollutants, such as bacteria, viruses, heavy metals, microplastics, and biotoxins which are considered unsafe for human health. Hence, bivalves accumulate human pathogenic bacteria, viruses, and biotoxins during filtering action when the environment is

Recent Advances in Sustainable Aquaculture and Fisheries. Shabir Ahmad Dar, Imtiyaz Qayoom, Gohar Bilal Wani, Adnan Abubakr, and Seemab Zehra (Eds.)

DOI: 10.1201/9781998511372-2

contaminated. This poses a potential health risk for bivalve consumers when the bivalves are consumed raw or lightly cooked. Three commercial techniques are used to render the bivalves fit for consumption: cooking, relaying, and depuration or controlled purification. However, depuration helps remove microbes but often fails to eliminate heavy metals, biotoxins, and other hazardous contaminants. Therefore, a proper depuration facility will aid in grabbing a good position in the international bivalves market. This chapter aims to cover the fundamental aspects of depuration and other lab-tested complementary methods and strategies to depuration, improving the process efficiency.

2.1 INTRODUCTION

Bivalves are crucial for fisheries, aquaculture, and the environment as biological monitors/ sentinels for analyzing the ecosystem's health. Since most bivalves are consumed whole, including the viscera, either raw or barely cooked; one of the main concerns of the seafood business is the health risks connected with the traditional consumption of oysters. Oysters are a high-risk food group due to their filter-feeding behavior, which causes naturally prevalent diseases in growing areas contaminated with polluted water to accumulate and are highly perishable (Pardío-Sedas, 2015). When bivalves are eaten raw/undercooked, a common practice in Western and Southeast Asian countries poses several health risks. Bivalves being unselective filter feeders, accumulate the contaminants present in the environment, such as bacteria, viruses, toxins, and pollutants in multiple folds than in the water; hence, more hazards are associated with its consumption.

Marine bivalves act as potential carriers of several groups of foodborne pathogens, such as *Salmonella, Vibrio*, pathogenic forms of *Escherichia coli, Campylobacter, Norovirus*, Hepatitis A virus, parasites such as *Giardia, cryptosporidium*, etc. As the environment is already contaminated with several types of hazards, subjecting the bivalves to a depuration process that safeguards public health becomes necessary. Harvesting them from clean and safe water is the most effective option for ensuring the safety of bivalve consumption. However, that is practically impossible as most coastal areas are polluted due to anthropogenic activities. Fundamentally, there are three processes to eliminate these hazards, each to a degree of extent: (i) Cooking/thermal treatment, which destroys the pathogens, (ii) Relaying, which is a process of placing the harvested bivalves in a clean, microbiologically safe environment that allows them to cleanse themselves by their filter-feeding habit, (iii) Depuration/controlled purification, which is placing the bivalves in a tank of disinfected seawater where the purification depends on the filter-feeding habit (Lees et al., 2010). The significant difference between relaying and depuration is that relaying is done in natural unpolluted waters, and depuration is done in artificial tanks with disinfected water. However, both processes allow the marketing of the bivalves in live conditions. Relaying and depuration remove only microbial agents, not other environmental contaminants, such as heavy metals, shellfish toxins, and pesticides. As bivalve consumption and export have increased, depuration became a comprehensive approach to meet microbiological specifications. Depuration is widely practiced in North America, Europe, Asia, and Australia. Depuration has its shortcomings, as it is ineffective in eliminating viruses, resulting in some *Norovirus* outbreaks related to bivalve consumption. Moreover, the use of chlorine as a disinfecting agent for seawater outturns in the production of metabolites such as trihalomethanes that have carcinogenic effects and similar problems with the use of ozone, a disinfectant agent where it forms metabolites such as bromates. On the other hand, the disinfecting power of UV directly depends on the water turbidity and volume. With all these drawbacks, to ensure the safety of the bivalve consumers, some complementary methods, such as gamma irradiation, UV, ozone, bacteriocins, phages, and chemicals such as chitosan can be used to eliminate the hazards associated with the bivalves (Martinez-Albores et al., 2020).

2.2 HISTORY OF DEPURATION

The history of purification of bivalves before consumption dates back to the 20th century when an outbreak of typhoid fever along the Atlantic coast existed. Phelps, in 1911, demonstrated that fecal coliforms could be removed entirely from sewage-contaminated oysters by simply relaying (Phelps, 1911). Following this, Johnstone, in 1914, tried relaying mussels, where he saw a reduction in coliform numbers (Johnstone, 1914). Further development of relaying to depuration was done by Dodgson in 1915 with the development of a depuration system for mussels consisting of large outdoor tanks using seawater treated with sodium hypochlorite followed by sodium thiosulfate neutralization. Further developments came in the depuration system by using different water treatment approaches such as ozone and UV, and design approaches such as a recirculating system, which enhances the bivalves' filter-feeding ability and decreases the accumulation of waste by-products..

2.3 DEPURATION

Depuration is a process to free light to moderately contaminated bivalve molluscan shellfish from microbial contaminants. This is achieved by placing the shellfish in tanks of clean seawater so that they can perform their regular pumping activity for a certain amount of time, which can last for several hours to days (Lee et al., 2008). Depuration is a purification technique that helps clean bivalves by utilizing their natural filtering habit. In this process, bivalves are held in clean seawater tanks; microbial contaminants are usually expelled from their body by natural filtering activity. Depuration effectively removes many fecal contaminants such as *E. coli, Staphylococcus,* and *Vibrio*. An effective depuration system requires proper design, operation, and pre-depuration conditions during harvest, transport, and storage. The principal factors that affect depuration are the design/type of the depuration system and seawater quality. The purity/quality of seawater defines the effectiveness of the depuration process, which is why various disinfection methods have been followed for disinfecting the natural/artificial seawater used for depuration.

2.4 TYPES OF DEPURATION SYSTEMS

2.4.1 RECIRCULATING DEPURATION SYSTEM

As the name indicates, seawater recirculation means that there will be an inflow and outflow of seawater in the depuration tank. Water will constantly circulate between the filtration unit, disinfection unit, and depuration tank containing living bivalves. Properties like salt concentration, temperature, dissolved oxygen, and water flow rate will be monitored throughout the process.

2.4.2 FLOW-THROUGH DEPURATION SYSTEM

Flow-through systems, unlike recirculating systems, are employed in depuration facilities situated near freely flowing seawater. These systems channel naturally flowing water into holding tanks, which are subsequently connected to filters designed to extract any particles present in the water. The filtered water is then disinfected and directed toward a depuration tank housing a live bivalve. There will be no control over water properties in this system and after processing, water will be discarded from the filtering system.

2.4.3 STATIC DEPURATION SYSTEM

In this system, there will be no inflow or outflow of water. Disinfected seawater used for depuration in a tank will be replaced with fresh water daily or at regular intervals. It is the simplest method among all three, utilizing a limited quantity of water. This system is also referred to as the Fill and Draw depuration system (FDDS).

2.5 METHODS FOR DISINFECTION OF SEAWATER

2.5.1 ULTRAVIOLET LIGHT

Since the 1970s, UV has been used to treat water. It effectively disinfects water by killing microbes and has the advantage of not producing any chemical by-products that could harm human health. UV radiation breaks the genetic material in DNA and RNA inside the microbe's cell wall, thus preventing it from reproducing. UV near 254 nm is effective in destroying pathogenic viruses and bacteria. UV dose depends on the turbidity of seawater and the presence of dissolved organic matter, as it needs optimum depth to penetrate and inactivate microorganisms sufficiently. A minimum dose of 10 mW/cm^2/sec is adequate for use in recirculating systems, which means a 30 W lamp for a system containing 2200 L of seawater.

2.5.2 CHLORINE AND CHLORINE-CONTAINING COMPOUNDS

It is an effective bactericide in seawater with low to moderate sediment and organic load. It works by diffusion, actively passing through the cell membrane and disrupting the cell elements. The gaseous form of chlorine and the hypochlorite compounds it forms with sodium and calcium are the two readily available forms of the chemical. Hypochlorous acid is formed when they instantly hydrolyze in water, and depending on the pH of the water, this acid subsequently dissociates to produce hydrogen and hypochlorite ions. Microbiological contamination can be efficiently disinfected using hypochlorite ions and hypochlorous acid, which has a concentration range of 0.2–1 mg/L. It is the least expensive, easy, and readily available option. For depuration, 2–3 mg/L free chlorine is normally used for a contact time of up to an hour. The amount of chlorine solution required may be determined using the following formula (Equation 2.1):

$$\text{Volume to add (L)} = \frac{Final\ concentration\ required\,(mg\,/\,L) \times Tank\ volume\,(L)}{concentration\ of\ stock\ solution\,(mg\,/\,L)} \qquad (2.1)$$

2.5.3 OZONE

It can effectively inactivate both bacteria and viruses. Ozone is purchased as gas in the cylinder and when it diffuses in seawater, it deactivates microorganisms. During the treatment, ozone gas is bubbled through the seawater, and ozone being an unstable gas, will break down into oxygen gas and an oxygen atom (O_3; O_2+O). This monoatomic oxygen radical is highly reactive, which removes organic matter and microorganisms from water. Ozone at a concentration not exceeding 0.5 mg/L can be used to treat seawater in batches for up to 10 min.

2.5.4 IODOPHORS

Iodophor-based systems have been practiced in Italy, and efforts have been made to market them elsewhere. Low residual iodophor levels in the shellfish's intestinal tract will directly kill microorganisms, including viruses, and purify the depuration water.

2.5.5 ELECTROLYZED WATER

It is a new method that can help reduce bacterial contaminants in water. The use of electrolyzed water is an eco-friendly approach to disinfect seawater in comparison to chemical additives. When seawater having sodium chloride is passed through electrodes, transfer of electrons occurs into the water, which results in the production of gases (oxygen, chlorine, and hydrogen), free hypochlorite ions, hypochlorous and hydrochloric acids, and sodium hydroxide. Hypochlorite, an oxidizer, has been identified as the main active bactericidal product of electrolyzed water.

2.5.6 OTHER METHODS FOR DISINFECTION

Water treatment with antimicrobials has also been found effective in depuration systems, with advantages being natural, economically feasible to produce or extract, and maybe the products of recycled waste that benefit the environment. For instance, bacteriophage serves as an antimicrobial agent that effectively targets and lyses specific bacteria in food while preserving the food's natural microflora (Moye et al., 2018). Likewise, lactic acid bacteria (LAB) are naturally present and generate substances such as bacteriocins and organic acids, which hinder the growth of spoilage and potentially harmful bacteria in food (Abee et al., 1995; Rusmana et al., 2020).

2.6 HAZARDS ASSOCIATED WITH BIVALVES

Hazards in bivalve consumption can be grouped into bacterial and viral, chemical, and biotoxins. The different hazards linked with bivalve consumption are given in Table 2.1.

TABLE 2.1 Hazards Associated with Bivalve Consumption.

Bacterial and Viral	Chemical	Biotoxins
Salmonella	Heavy metals	Paralytic shellfish poisoning
Shigella	Hydrocarbons	Amnesic shellfish poisoning
Vibrio	Pesticides	Neurotoxic shellfish poisoning
Listeria		Diarrhetic shellfish poisoning
Campylobacter		
Norovirus		
Hepatitis A		

2.6.1 BACTERIAL DISEASES

Vibrios are the major group associated with bivalves, as they are naturally present in the marine environment. They are found in large numbers during the summer months, usually year-around in tropical countries, and the summer season in temperate countries (Turner et al., 2014). *Vibrio parahaemolyticus* is widely present in the marine environment and is the major cause of bacterial gastroenteritis (strains having *trh* and *tdh* genes) resulting from consuming raw/undercooked bivalves. Most outbreaks associated with the raw consumption of bivalves are due to *V. parahaemolyticus*. *Vibrio cholerae* serotypes O1 and O139 reach humans through contaminated seawater and the consumption of raw seafood. Another important marine bacterium, *Vibrio vulnificus,* causes wound infections that are fatal and associated with the inhalation of bacteria through wave action and consumption of raw/undercooked bivalves. *V. vulnificus* is most dangerous for immunocompromised persons or people with underlying health issues.

Salmonella, a Gram-negative Enterobacter, a natural inhabitant of animal intestines, reaches marine environments due to fecal contamination. *Salmonella* causes typhoid and paratyphoid. *Salmonella*, when ingested, crosses the gastrointestinal barrier and reaches the small and large intestines, thereby proliferating and producing toxins that ultimately cause severe gastroenteritis.

Escherichia coli has been regarded as the sole indicator of fecal contamination, which implies that its presence indicates the presence of other public health significant enteric pathogens, such as *Salmonella*, *Shigella*, *Campylobacter*, *Norovirus*, hepatitis A virus, protozoan parasites, such as *Giardia, cryptosporidium* (Potasman et al., 2002). Hence, depending on the levels of *E. coli*, the fecal indicator organism in the bivalves harvesting area has been grouped into Class A, Class B, and Class C areas by the European Union (Table 2.2).

TABLE 2.2 Classification of Shellfish Growing Areas Based on *E. coli* Number.

Class	Microbiological Standards ***E. coli* in Live Bivalve Mollusks Must Not Exceed**	Post-harvest Treatment
A	230 MPN *E. coli*/100g in 80% of the sample	No treatment required
	700 MPN *E. coli*/100g in 20% of the sample	
B	4600 MPN *E. coli*/100g in 90% of the sample	Depuration/cooking/relaying
	46,000 MPN *E. coli*/100g in 10% of the sample	
C	46,000 MPN *E. coli*/100g	Relaying/cooking

Although various fecal virus types may be transmitted by shellfish, human *Norovirus* (HuNoV) and hepatitis A virus (HAV) are currently recognized as the two main risks. *Norovirus*, an unenveloped virus belonging to the family *Calciviridae*, is found to be the most common cause of viral gastroenteritis caused by the consumption of raw/under-cooked bivalves. Hepatitis A virus belonging to the *Picornaviridae* family is generally found in areas of deplorable hygiene where it reaches the host and causes fatal diseases later in life.

2.6.2 HEAVY METALS AND BIOTOXINS

Marine bivalves, such as oysters, mussels, clams, and scallops tend to accumulate heavy metals, such as Cd, Cu, Al, Co, Zn, Pb, Cr, Ni, and Hg (Wang and Lu, 2017; El Nemr et al., 2012). Heavy metals are present in the surrounding water or sediment of the bivalves. Heavy metals originate from both natural and anthropogenic sources. Natural sources encompass geological weathering, atmospheric deposition, airborne dust, and soil erosion. On the other hand, anthropogenic sources involve the use of pesticides, fertilizers, leachates from mining sites, waste generated during ore smelting, vehicular traffic emissions, brake fluid leaks, industrial waste, sewage sludge, as well as partially treated or untreated industrial wastewater (Jahromi et al., 2021). Essential metals, namely iron, copper, zinc, and manganese, hold significant significance in biological systems due to their crucial functions. In contrast, non-essential metals, such as lead, cadmium, arsenic, and mercury tend to be highly toxic, and their accumulation in tissues can result in intoxication, reduced fertility, cellular and tissue harm, cell death, and dysfunction of various organs (El Nemr et al., 2012). Heavy metals disrupt the human system in two ways: (i) bioaccumulation in any specific organ and causing long-term lethal effects, and (ii) displacement of the essential vital minerals hinders the system's functioning.

Ingesting toxic elements in low concentrations over a prolonged period can have harmful effects. Even essential metals can become toxic when consumed excessively (Suprapti et al., 2016). Lead, ranked second on the list of the most poisonous heavy metals (following arsenic), can adversely affect various organs, such as bones, brain, blood, kidneys, reproductive and cardiovascular systems, as well as the thyroid gland.

Ingestion of Cd above 350 mg causes acute renal failure and is fatal (El Nemr et al., 2012). Cd^{2+} produces problems in the immune and reproductive systems due to its solubility in acidic media facilitated by ocean acidification. Mercury is dangerous to human health, the heaviest heavy metal, with greater importance in pregnant women and children (Martinez-Albores et al., 2020). Cobalt, an indispensable nutrient, has been linked to the regulation of blood pressure and thyroid function. However, excessive intake of cobalt has been associated with the development of congestive heart failure, polycythemia, and anemia (El Nemr et al., 2012). Both Cu and Zn are essential trace elements for the bivalves. Interestingly, the accumulation of Zn in the oysters facilitates the Cd uptake in the oysters (Wang and Lu, 2017). Copper is an essential part of several enzymes and is necessary for synthesizing hemoglobin. High levels of zinc cause pancreatitis, anemia, muscle pain, and acute renal failure. A human requires 1 μg of chromium daily, and its deficiency results in impaired growth and disturbances in glucose, lipid, and protein metabolism. Nickel is essential for the enzyme urease and hydrogenase. Still, nickel is a carcinogenic metal, and overexposure to it can cause decreased body weight, heart and liver damage, and skin irritation (El Nemr et al., 2012).

Marine biotoxins encompass a diverse collection of exceedingly powerful neurotoxins and gonyautoxins generated by harmful phytoplankton. As these toxins accumulate within marine organisms, such as mussels, clams, and fish, their transfer along the food chain can lead to a severe condition capable of causing fatality in both marine mammals and humans. Specifically, paralytic shellfish poisoning toxins, which are saxitoxin derivatives, induce paralysis upon ingestion (Leal et al., 2023).

Neurological symptoms brought on by toxin-mediated shellfish poisoning can be severe and even fatal. Phytoplankton such as dinoflagellates and diatoms produce shellfish toxins. When bivalves filter the suspended particles from water, the toxins get concentrated in the bivalve; eventually, when humans ingest the bivalves in the area of harmful algal blooms, shellfish poisoning occurs. Shellfish toxins, such as paralytic shellfish poisoning, diarrhetic shellfish poisoning, amnesic shellfish poisoning, and neurotoxic shellfish poisoning consist of different toxins which have specific mechanisms causing disruption of homeostasis in humans, causing deaths (Table 2.3).

TABLE 2.3 Major Shellfish Poisoning and Their Symptoms.

Shellfish Poisoning	Name of the Toxin	Causative Agent	Symptoms
Paralytic shellfish poisoning	Saxitoxin	*Gymnodium and Alexandrium* spp.	Tingling sensation, numbness, gastrointestinal illness, unstable motor control, and paralysis
Diarrhetic shellfish poisoning	Okadoic acid, pectenotoxins, yessotoxins	*Dinophysis* spp., *and Prorocentrum* spp.	Nausea, vomiting, abdominal pain, and diarrhea
Neurotoxic shellfish poisoning	Brevetoxin	*Gymnodium breve*	Pins and needles sensation, disturbed motor control, nausea, and severe muscle pain
Amnesic shellfish poisoning	Domoic acid	*Nitzschia* spp.	Nausea, vomiting, abdominal cramps, diarrhea, and amnesia
Azaspiracid shellfish poisoning	Azaspiracid	*Protoperidinium crassipes*	Nausea, vomiting, and diarrhea

2.7 LIMITATIONS OF DEPURATION

The depuration process is more effective in removing fecal bacterial contamination and less effective in eliminating viral contaminants like *Norovirus*, hepatitis A and other naturally occurring marine Vibrio (*V. parahaemolyticus and V. vulnificus*). In the case of *Vibrio*, a major concern is that if the salinity is in the right range (e.g., 10–30 ppt) and the temperature is high enough (e.g., above 20°C), an increase in the concentration of *Vibrio* may occur during a depuration cycle (Lee et al., 2008).

Research with Pacific oysters (*Crassostrea gigas*) conducted in northern Europe suggested that viruses are eliminated more slowly than *E. coli*. In a properly constructed and managed system, about one-third of the starting concentration of the virus may still be present after 2 days at 8°C; at higher temperatures (18–21°C), they can be eliminated more quickly. Maximizing the setup and operation of systems to eliminate pathogens is crucial rather than only focusing on eliminating indicator bacteria like *E. coli*. Other limitations of depuration include the inability to eradicate biotoxins from seawater produced by dinoflagellates and diatoms entirely. Additionally, heavy metals and other organic compounds take several weeks to reduce to significant levels, which can be a primary concern.

2.8 DEPURATION METHODS

2.8.1 DEPURATION BY PHYSICAL METHODS

Physical methods, such as gamma irradiation, refrigeration, freezing, heat treatment, high-pressure processing, and ozonation can be employed in the depuration process, which will aid in eliminating pathogens of public health significance. The major disadvantage of these techniques is that the bivalves die during the process, and hence they can be effectively used for the bivalve products. Moreover, these techniques will destroy the essential nutrients and change the sensory properties of the bivalve.

High-pressure processing (HPP) is commonly used in processing raw meat and shellfish as it preserves heat-sensitive nutrients. HPP is popularly used for the shucking of bivalves. HPP does not completely lead to microbial inactivation as some injured microbial cells may revive after the HPP, resulting in a threat under ambient storage conditions. In this case, HPP can be combined with other hurdles, resulting in an efficient process for eliminating microbes of public health significance. Ahmadi et al. (2015) studied the biocontrol of *V. cholerae* in mussels using HPP and virulent bacteriophages (VP). The study revealed that the combination of HPP and VP increased the removal of *V. cholerae* more than applying the individual methods. HPP alone resulted in the complete removal of *V. cholerae* at 550 MPa for 5 min, which is an energy-efficient process. In contrast, when the bacteriophage of *V. cholerae* was used, it eliminated *V. cholerae* at 350 Mpa. Hence, this suggests that this combination of hurdles, that is, high pressure and bacteriophages, can be effectively utilized for the biocontrol of the microorganisms in foods rendering safety. There is no evidence of HPP being used to break the chemical bonds of shellfish toxins; therefore, HPP cannot be used to destroy shellfish toxins. HPP is most effective against Gram-negative bacteria compared with Gram-positive bacteria. HPP treatment at a dose of 275–300 Mpa and >300 Mpa destroys *V. vulnificus* and *V. parahaemolyticus* to undetectable levels. The mechanism of microbial inactivation is by the distortion of the lipid cell membrane, thereby causing loss of membrane integrity, and by altering the protein structure, causing protein denaturation.

Liu et al. (2009) highlighted the possible utilization of freezing technology to eliminate *V. parahaemolyticus* from the mussel sample. Flash freezing followed by storage at −21°C for five months results in a 3.52 log reduction (MPN/g) in *V. parahaemolyticus* and *V. cholerae* populations.

Rey et al. (2012) evaluated the significance of the slurry packaging system in reducing microbial load in bivalves. The study revealed that the slurry packaging system has significantly reduced the microbial load and increased the organoleptic and sensory properties, such as odor, taste, juiciness, appearance, and shelf life compared with the control group.

UV sterilized seawater combined with a low temperature of 7–15°C reduced the *V. parahaemolyticus* load by >3 log MPN/g after 5 days of depuration. Still, mere storage in that temperature without depuration does not significantly decrease the bacterial population (Phuvasate et al., 2012).

Irradiation is another possible method that can be employed to reduce the pathogen load in the bivalves. It is operational in EU countries with a limit of 3 KGy. European Food Safety Authority has recommended that 5 KGy has an upper threshold for irradiation of seafood. Ionizing radiation is employed in food irradiation to kill bacteria, mold, and pests; it causes the breakage of chemical bonds, thereby killing/stopping the multiplication of the microbes. Irradiation can also be used to destruct the shellfish toxins where a different dose–time combination is needed for various shellfish toxins. Small doses of irradiation are enough for the destruction of the domoic acid and yessotoxins (McCarron et al., 2007).

Ozone treatment of PSP-contaminated bivalves showed a positive response as it detoxifies the saxitoxin to a greater extent (Gacutan et al., 1984). Schneider and Rodrick (1992) proposed the effectiveness of ozone on the destruction of brevetoxin and ozone's effect on the dinoflagellate *Gymnodium breve*. Ozone has been approved for direct application on food products and is considered safe by USFDA. The antimicrobial effect of ozone is based on the oxidation of the glycoprotein membrane and DNA destruction of enzymes needed for homeostasis (O'Donnell et al., 2012). Super chilled storage and direct ozone depuration pretreatment work together synergistically to improve processing. Oyster depuration takes place over a shorter period, oyster survival is unaffected, and oyster microbiological safety has improved with a safe level for human consumption up to 9 days (Hernández et al., 2018).

It should be taken into account that the initial investment for these techniques to be employed as a complementary method for depuration is high, and some techniques lead to killing and changes in organoleptic and sensory characteristics of the bivalve. The mode of action of different physical methods is presented in Table 2.4.

TABLE 2.4 Mode of Action of Different Physical Methods on the Microbes.

Physical Method	Mode of Action
HPP	Modification of the structure of the cell membrane Ionization and precipitation of protein complexes Disruption of biochemical reaction
Irradiation	Direct breakage of chemical bonds in the nucleic acid Generation of reactive oxygen species
Freezing	Damage in the cell wall integrity Slows rate of molecular and chemical reaction
UV	Formation of pyrimidine dimers in the nucleic acids Disruption of cellular functions Inactivation of enzymes Alteration of proteins
Ozone	Oxidation of cell membrane components Leakage of cellular components Cell lysis

2.8.2 DEPURATION BY BIOLOGICAL METHODS

The biological method is one in which live organisms/metabolites of living organisms are used in the bioremediation process where substance degradation occurs. A biological method as a complementary method to depuration can potentially involve: (i) Live beneficial bacteria called probiotics, (ii) Bivalve's antimicrobial peptides, (iii) Bacteriocins, and (iv) Bacteriophages. The biological methods and their mode of action are given in Table 2.5.

The term probiotic was coined by Lilly and Stillwell, which refers to "for life." According to Parker, probiotics are live microorganisms that contribute to intestinal balance (Fioramonti et al., 2003). The most commonly used probiotic strains are *Lactobacillus* spp., *Bifidobacterium* spp., *Bacillus* spp., *Pediococcus* spp., and some yeast strains (Soccol et al., 2010). The primary mechanism of action behind probiotics is competitive exclusion. It colonizes the intestine and competes for space and food with pathogens, thus contributing to the well-being of the host organism (Elo et al., 1991). Therefore, probiotics can be used in bivalve aquaculture where they colonize the bivalve's digestive gland and exert their mechanism of action; hence, it is a viable aftereffect-less option to be used in bivalve aquaculture for safety (Polo et al., 2015). Abasolo-Pacheco et al. (2017) proved the efficacy of probiotics against *V. alginolyticus* infection in Catarina scallops.

Antimicrobial peptides (AMPs) play a significant role in the non-specific immune systems of various animals, inhibiting bacteria, viruses, fungi, and parasites. The mechanism of action of antimicrobial peptides includes membrane targeting and non-membrane targeting mechanisms such as inhibition of the following, such as protein synthesis, genetic material synthesis, cell division, and protease activity (Huan et al., 2020). A description of antimicrobial peptides from *Mytilus* spp. has been reported by Defer et al. (2013).

Bacteriocins are ribosome-produced polypeptides with bactericidal activity—for example, Nisin, a bacteriocin with commercial food applications. In the seafood industry, bacteriocin is mainly applied for biocontrol against *Listeria monocytogenes* (Valenzuela et al., 2010). The application of bacteriocins in the depuration system can be in a direct addition to bacteriocin-producing bacteria in the water in which UV/chlorine/ozone may interfere with the viability of the bacteria or direct application of encapsulated bacteriocin into the system.

Bacteriophages are the viruses that infect bacteria. Phage therapy is the use of lytic phages to kill bacteria. Phages have been used to treat infections in bivalve aquaculture, such as vibriosis (Wang et al., 2017). Pereira et al. (2017) studied the effect of bacteriophages in the depuration system for decontaminating *E. coli,* where phages showed 5 log CFU/g reductions in *E. coli* number.

The advantage of these biological methods is the specificity and the environmentally friendly nature as they act as an effective alternative to antibiotics. However, the use of biological agents in the depuration system has not yet been extensively studied, and there remains the scope for future studies.

TABLE 2.5 Different Biological Methods and Their Mode of Action.

Biological Method	Mode of Action
Probiotics	Competitive exclusion Modulation of immune system Stimulation of host defense system Production of organic acids and antimicrobials
Antimicrobial peptides	Disruption of cell membrane Interaction and modification of intracellular components
Bacteriophages	Adhesion and invasion Dodging of the immune system Exotoxin production Disruption of cell wall
Bacteriocins	Pore formation on the cytoplasmic membrane Degradation of cell wall membrane Inhibition of protein synthesis Inhibition of nucleic acid synthesis

2.8.3 DEPURATION BY CHEMICAL METHODS

As filter feeders, bivalves accumulate heavy metals and biotoxins slowly throughout their lives (Martinez-Albores et al., 2020). Their consumption causes food safety concerns threatening human health and reducing the availability of protein-rich food (Leal et al., 2023). Heavy metals and biotoxins are eliminated when contaminated bivalves are exposed to purified seawater, and their tissues are cleansed by the passage of water through the filter-feeding mechanism on exposure, called depuration (Blogoslawski, 2018). However, elimination is tremendously slow when treated only with clean seawater (Zhang et al., 2017). Hence, chemical methods such as treatment with chelating agents are done to enhance depuration.

Chelation refers to the process of creating multiple coordination bonds between organic molecules and a transition metal ion, resulting in the metal being sequestered (Van Lith and Ameer, 2016). When metals interact with chelating agents, they can form intricate ring-shaped structures (Hamin and Bakar, 2017). These chelating agents generate ring-like formations known as "chelates," which attach to metal ions and form complexes that are subsequently eliminated from the system. Chelating agents possess significant potential for application in the food and health domains due to their beneficial effects, including immunomodulatory, antimicrobial, antioxidant, and antitumor properties. Moreover, they have the ability to chelate and decrease the availability of heavy metals, making them highly valuable (Martinez-Albores et al., 2020). The use of catalysts enhances the chelation technique, further optimizing its effectiveness (Hamin and Bakar, 2017). As a result, it has been suggested that chelating agents could be employed to extract heavy metals or other toxins, thereby reducing their presence or eliminating them from seafood, such as bivalves (Ali et al., 2022; Martinez-Albores et al., 2020).

Among the chelating substances, metallothioneins (MTs), naturally produced by bivalves, are widely used. MTs are low molecular weight proteins rich in cysteine. They are found in various organisms and exhibit a remarkable affinity for both essential and non-essential metals possessing the d10 electronic configuration, including Zn^{2+}, Cd^{2+}, and Cu^{+} (Zhang et al., 2017). Within bivalves, such as mussels (M. edulis), MTs play a crucial role in the transportation and storage of heavy metals.

Additionally, they offer a protective function by detoxifying excessive amounts of non-essential metals like Cd^{2+}, Ag^{2+}, or Hg^{2+} through the formation of metallated MT-metal complexes (Martinez-Albores et al., 2020). The in vitro affinity of MT decreases in the hierarchical sequence $Hg^{2+} > Cu^{+}$, $Ag^{+} > Cd^{2+} > Pb^{2+} > Zn^{2+} > Co^{2+}$ (Zhang et al., 2017). In *Mytilus* sp., MTs have been discovered and characterized in two forms: monomeric (MT-10, 10 kDa, with 21 cysteines) and dimeric (MT-20, 20 kDa, with 23 cysteines) isoforms. The tertiary structure of MTs is primarily determined by the arrangement of metal-thiolate clusters. The metallated MT-Cd^{2+} form adopts a monomeric dumbbell shape, consisting of α- and β-domains, connected by a flexible linker containing a Lys-Lys segment. The α-domain (located at the C-terminal) is described as "adamantane-like," with 4 Cd^{2+} ions coordinated through 11 metal-thiolate bonds. The β-domain, on the other hand, is referred to as "distorted chair-like," binding 3 Cd^{2+} ions through 9 metal-thiolate bonds (Zhang et al., 2017).

Other chelating agents include chitosan, organic acids (acetic acid), sodium acetate, trisodium citrate and disodium oxalate, edible oxidant (peracetic acid), and Chelex (polymer ligands) for chelating the heavy metals (Suprapti et al., 2016; Martinez-Albores et al., 2020; Ali et al., 2022). Chitosan is a long-chain polysaccharide obtained by the deacetylation of chitin from crustacean shells (Martinez-Albores et al., 2020). Chitosan is a long linear polymeric molecule of β(1,4)-linked glycans (Onsøyen and Skaugrud, 1990). The chelating organic acids can dislodge the exchangeable, carbonate, and reducible fractions of heavy metals via washing procedures. The reduction of heavy metals in bivalves treated with acetic acid could be due to the formation of insoluble acetate salts of these metals (Suprapti et al., 2016). The catalysts, such as Al_2O_3, CuO, Fe_2O_3, and ZnO are used along with chelating agents to remove heavy metals and toxins (Ali et al., 2022).

Depuration using a chelating agent involves collecting bivalves from the marine environment or aquaculture ponds. Determine heavy metal content using inductively coupled plasma mass spectrometry (ICP-MS) and atomic absorption spectroscopy (AAS). Maintain the bivalves at neutral pH and the source temperature and salinity. Finally, add the chelating agent to the bivalves at the optimized concentration for optimized duration. Determine the heavy metal content for reduction to the permissible limit (Azelee et al., 2014; Ali et al., 2022).

2.8.3.1 EFFECT OF CHELATING AGENTS ON HEAVY METALS AND TOXINS

The different chelating agents and their efficacy in removing chemical hazards are given in Table 2.6. The utilization of MTs extracted from fishing by-products resulted in a significant reduction in Cd^{2+} concentrations in different tissues of mussels (*M. edulis*) when complexed with the MTs. After a 15-day exposure, the gills, mantle, and viscera exhibited reductions of ~ 30, 40, and 25% respectively. Moreover, the complexation of MTs protein hydrolysate with Fe^{2+} led to the elimination of 50% of Cd^{2+}, highlighting the effectiveness of the catalyst in conjunction with MTs (Zhang et al., 2015). In a study by Baudrimont et al. (2003), positive depuration outcomes were observed for Cd^{2+}, Al^{2+}, and Hg^{2+} using MTs in Asian clams (*Corbicula fluminea*).

Chitosan exhibits solubility solely under acidic conditions, necessitating its usage in a manner that promotes its solubility (Huang et al., 2012). Consequently, chitosan can be dissolved in 5% acetic acid to form a gel. This gel, containing chitosan, is administered by combining it with ozonized seawater (with concentrations of 1.5 mg/L ozone and 0.5 mg/L chitosan) and pumping it at a rate of 1.3 m/s for a duration of 1 day. This treatment led to a reduction of over 90% in Hg^{2+} content in mussels (*Perna viridis*) and ~ 85% in cockles (*A. granosa*) (Ningrum et al., 2016). Suprapti et al. (2016) reported that soaking green mussels in acetic acid solution at a concentration of 25% for 90 min reduces the heavy metals Pb, Cr, and Cd above 38%. In a study conducted by Azelee et al. (2014), the effectiveness of three chelating agents, namely, trisodium citrate, disodium oxalate, and sodium acetate, in the removal of heavy metals, such as As, Pb, Cd, and Ni from *P. viridis* were examined. The results revealed that sodium acetate exhibited the ability to effectively chelate and eliminate all the investigated heavy metals, reducing their levels below the permissible limit.

The study conducted by Xie et al. (2013) employed chitosan for purifying oysters (*Ostrea rivularis*) in an experimental setting that spanned 7 days. Throughout this duration, the researchers successfully achieved a reduction of over 60% in the quantity of paralytic shellfish poisoning (PSP) toxins within the oysters using chitosan alone. Moreover, when chitosan was combined with *Chlorella* microalgae, the reduction surpassed 85%. Table 2.6 presents the impact of chelating agents on the depuration process.

TABLE 2.6 Efficacy of Chelating Agents in Depuration.

Chelating Agent	Inhibition Against	Dosage and Time	Bivalve Species	Efficacy	References
Metallothioneins (MTs)	Cd^{2+}	ND	Asiatic clams	Sequestered 40% of the total Cd^{2+} bioaccumulated	Baudrimont et al. (2003)
Chitosan oligosaccharide + Ca^{2+} (COS-Ca)	Cd^{2+}	1.75–8.75 mg/L for 6 days	Scallops	Depuration rate of 47%	Huang et al. (2012)
Chitosan, *Chlorella* and Chitosan + *Chlorella*	Paralytic shellfish poisoning toxin	8 × 103 cells/mL *Chlorella*, 0.05 g/L chitosan, and a combination of both	Oysters	Toxicity decreased from 9.07 mouse units (MUs) to 1.41 MUs using chitosan and 0.12 mouse units using chitosan plus *Chlorella*	Xie et al. (2013)
MTs (protein hydrolysate-Fe^{2+})	Cd^{2+}	40 mg/L protein hydrolysate-Fe^{2+} for 15 days	Blue mussels	Cd^{2+} concentration decreased from 46.1 to 23.3 µg/g	Zhang et al. (2017)
MTs (hydrolysis peptide–metal element complexes (Fe^{2+}, Zn^{2+}, Ca^{2+}, or Hg^{2+})	Cd^{2+}	5, 10, 15, and 20 mg/L for 8 days	Blue mussels	Cd^{2+} decreased in the range 25–40% after exposure to 20 mg/L of hydrolysis peptide–metal element complexed to Fe^{2+}, Zn^{2+}, and Ca^{2+} No significant decrease was found for hydrolysis peptide–metal element complexed to Hg^{2+}	Zhang et al. (2017)
Combinations between chitosan, ozone, and hydrodynamic treatment	Hg^{2+}	1.5 mg/L ozone, 0.5 mg/L chitosan and 1.3 m/s hydrodynamic treatment for 60 min	Green mussels and blood cockles	chitosan-ozone combination achieved a Hg^{2+} depuration of 96.5% in green mussels and 87% in blood cockles	Ningrum et al. (2016)
Sodium acetate	As, Pb, Cd, Ni		*P. viridis*	Removed As 59.50%, Pb 88.57%, Cd 68.01%, and Ni 79.67%	Azelee et al. (2014)

TABLE 2.6 *(Continued)*

Chelating Agent	Inhibition Against	Dosage and Time	Bivalve Species	Efficacy	References
Trisodium citrate,	As, Pb, Cd, Ni		*P. viridis*	Removed As 38.13%, Pb 68.90%, Cd 70.49% and Ni 36.92%	Azelee et al. (2014)
Disodium oxalate	As, Pb, Cd, Ni		*P. viridis*	Removed As 46.89%, Pb 85.46%, Cd 60.41%, and Ni 47.80%	Azelee et al. (2014)
Acetic acid	Pb, Cr, Cd	10–25% for 0–90 min	*P. viridis*	25% acetic acid at 90 min removed 50% of metals	Suprapti et al. (2016)
Cation-exchange resin	Paralytic shellfish toxins	1 g/L for 24 to 70 h	*Mytilus edulis*	80% decrease in overall toxicity upon exposure to the resin after 48 h	Leal et al. (2023)

2.9 CONCLUSION

There is not a single technique that can be successfully used to defend against all sanitary concerns and be applied to all bivalve species. Even for removing extremely persistent contaminants, a few of the approaches listed demonstrated excellent outcomes in laboratory settings. They have not, however, been examined at the industrial level. Using complementary methods, such as physical, chemical, and biological methods for depuration along with the traditional depuration system aids in efficiently removing contaminants and thereby ensures food safety.

KEYWORDS

- **bivalves**
- **depuration**
- **harmful microbes**
- **hazardous contaminants**
- **methods**
- **strategies**

REFERENCES

Abasolo-Pacheco, F., Campa-Córdova, Á. I., Mazón-Suástegui, J. M., Tovar-Ramírez, D., Araya, R., & Saucedo, P. E. (2017). Enhancing growth and resistance to *Vibrio alginolyticus* disease in catarina scallop (*Argopecten ventricosus*) with *Bacillus* and *Lactobacillus* probiotic strains during early development. *Aquaculture Research, 48*(9), 4597–4607. https://doi.org/10.1111/are.13283

Abee, T., Krockel, L., & Hill, C. (1995). Bacteriocins: Modes of action and potentials in food preservation and control of food poisoning. *International Journal of Food Microbiology, 28*(2), 169–185. https://doi.org/10.1016/0168-1605(95)00055-0

Ahmadi, H., Anany, H., Walkling-Ribeiro, M., & Griffiths, M. W. (2015). Biocontrol of *Shigella flexneri* in ground beef and *Vibrio cholerae* in seafood with bacteriophage-assisted high hydrostatic pressure (HHP) treatment. *Food and Bioprocess Technology, 8*, 1160–1167. https://doi.org/10.1007/s11947-015-1471-6

Ali, T. G., Abdul Keyon, A. S., & Mahat, N. A. (2022). Occurrence of heavy metals and their removal in *Perna viridis* mussels using chemical methods: A review. *Environmental Science and Pollution Research, 29*, 4803–4821. https://doi.org/10.1007/s11356-021-17343-3

Azelee, I. W., Ismail, R., Ali, R., & Bakar, W. A. W. A. (2014). Chelation technique for the removal of heavy metals (As, Pb, Cd and Ni) from green mussel, *Perna viridis*. *Indian Journal of Geo-Marine Sciences 43*(03), 372–376.

Baudrimont, M., Andres, S., Durrieu, G., & Boudou, A. (2003). The key role of metallothioneins in the bivalve *Corbicula fluminea* during the depuration phase, after in situ exposure to Cd and Zn. *Aquatic Toxicology, 63*(2), 89–102. https://doi.org/10.1016/S0166-445X(02)00134-0

Blogoslawski, W. J. (2018). Enhancing shellfish depuration. In *Molluscan shellfish depuration* (pp. 145–149). CRC Press. https://doi.org/10.1201/9781351074810-13

Defer, D., Desriac, F., Henry, J., Bourgougnon, N., Baudy-Floc'H, M., Brillet, B., Le Chevalier, P., & Fleury, Y. (2013). Antimicrobial peptides in oyster hemolymph: The bacterial connection. *Fish & Shellfish Immunology, 34*(6), 1439–1447. https://doi.org/10.1016/j.fsi.2013.03.357

El Nemr, A., Khaled, A., Moneer, A. A., & El Sikaily, A. (2012). Risk probability due to heavy metals in bivalve from Egyptian Mediterranean coast. *The Egyptian Journal of Aquatic Research, 38*(2), 67–75. https://doi.org/10.1016/j.ejar.2012.11.001

Elo, S., Saxelin, M., & Salminen, S. (1991). Attachment of *Lactobacillus casei* strain GG to human colon carcinoma cell line Caco-2: Comparison with other dairy strains. *Letters in Applied Microbiology 13*(3), 154–156. https://doi.org/10.1111/j.1472-765X.1991.tb00595.x

Fioramonti, J., Theodorou, V., & Bueno, L. (2003). Probiotics: What are they? What are their effects on gut physiology? *Best Practice & Research Clinical Gastroenterology, 17*(5), 711–724. https://doi.org/10.1016/S1521-6918(03)00075-1

Gacutan, R. Q., Tabbu, M. Y., de Castro, T., Gallego, A. B., Bulalacao, M., Arafiles, L., & Icatlo Jr., F. (1984). Detoxification of pyrodinium-generated paralytic shellfish poisoning toxin in *Perna viridis* from western Samar, Philippines. Toxic Red Tides and Shellfish Toxicity in Southeast Asia: Proceedings of a Consultative Meeting held in Singapore 11–14 September 1984 (pp. 80–85). Singapore: Marine Fisheries Research Department, Southeast Asian Fisheries Development Center; Ottawa, Canada: International Development Research Centre. http://hdl.handle.net/20.500.12066/4869

Hamin, S., & Bakar, W. A. W. A. (2017). Demetallization of heavy metals in clam, *Corbicula fluminea* utilizing catalytic chelation technique. *eProceedings Chemistry, 2*(1).

Hernández, K. L., Sedas, V. P., Dehaibes, S. R., Valencia, V. S., Mozo, I. R., Herrera, D. M., Primo, A. F., & Serrano, R. U. (2018). Improved microbial safety of direct ozone-depurated shellstock Eastern oysters (*Crassostrea virginica*) by superchilled storage. *Frontiers in Microbiology, 9*, 2802. https://doi.org/10.3389/fmicb.2018.02802

Huan, Y., Kong, Q., Mou, H., & Yi, H. (2020). Antimicrobial peptides: Classification, design, application and research progress in multiple fields. *Frontiers in Microbiology, 11*, 2559. https://doi.org/10.3389/fmicb.2020.582779

Huang, G., Sun, J., Wang, D., Xu, Y., & Xu, W. (2012). Chitosan oligosaccharide-Ca complex accelerates the depuration of cadmium from *Chlamys ferrari*. *Journal of Ocean University of China, 11*, 219–226. https://doi.org/10.1007/s11802-012-1893-y

Jahromi, F. A., Keshavarzi, B., Moore, F., Abbasi, S., Busquets, R., Hooda, P. S., & Jaafarzadeh, N. (2021). Source and risk assessment of heavy metals and microplastics in bivalves and coastal sediments of the Northern Persian Gulf, Hormogzan Province. *Environmental Research, 196*, 110963. https://doi.org/10.1016/j.envres.2021.110963

Johnstone, J. (1914). The methods of cleansing living mussels from ingested sewage bacteria. *Report for 1914 on the Lancashire Sea Fisheries Laboratory No. 23*, 57–108.

Leal, J. F., Bombo, G., Amado, P. S., Pereira, H., & Cristiano, M. L. (2023). Cation-exchange resin applied to paralytic shellfish toxins depuration from bivalves exposed to *Gymnodinium catenatum*. *Foods, 12*(4), 768. https://doi.org/10.3390/foods12040768

Lee, R., Lovatelli, A., & Ababouch, L. (2008). Bivalve depuration: Fundamental and practical aspects. FAO Fisheries Technical Paper No. 511. Food and Agriculture Organization of the United Nations, Rome.

Lees, D., Younger, A., & Dore, B. (2010). Depuration and relaying. *Safe Management of Shellfish and Harvest Waters, 9*, 145–181.

Liu, C., Lu, J., & Su, Y. C. (2009). Effects of flash freezing, followed by frozen storage, on reducing *Vibrio parahaemolyticus* in Pacific raw oysters (*Crassostrea gigas*). *Journal of Food Protection, 72*(1), 174–177. https://doi.org/10.4315/0362-028x-72.1.174

Martinez-Albores, A., Lopez-Santamarina, A., Rodriguez, J. A., Ibarra, I. S., Mondragón, A. D. C., Miranda, J. M., Lamas, A., & Cepeda, A. (2020). Complementary methods to improve the depuration of bivalves: A review. *Foods, 9*(2), 129. https://doi.org/10.3390/foods9020129

McCarron, P., Kotterman, M., de Boer, J., Rehmann, N., & Hess, P. (2007). Feasibility of gamma irradiation as a stabilisation technique in the preparation of tissue reference materials for a range of shellfish toxins. *Analytical and Bioanalytical Chemistry, 387*, 2487–2493.

Moye, Z. D., Woolston, J., & Sulakvelidze, A. (2018). Bacteriophage applications for food production and processing. *Viruses, 10*(4), 205. https://doi.org/10.3390/v10040205

Ningrum, E. W., Solihin, D. D., & Butet, N. A. (2016). Mercury depuration effectiveness on green mussel (*Perna viridis* L.) and blood cockle (*Anadara granosa* L.) from Jakarta Bay using ozone, chitosan and hydrodynamic technique. *IOP Conference Series: Earth and Environmental Science, 31*(1), 012041. https://doi.org/10.1088/1755-1315/31/1/012041

O'Donnell, C., Tiwari, B., Cullen, P., & Rice, R. (2012). *Ozone in food processing*. John Wiley & Sons, Ltd. https://doi.org/10.1016/B978-0-12-404699-3.00055

Onsøyen, E., & Skaugrud, O. (1990). Metal recovery using chitosan. *Journal of Chemical Technology and Biotechnology 49*(4), 395–404. https://doi.org/10.1002/jctb.280490410

Pardío-Sedas, V. (2015). Influence of ozone depuration on the physical properties of fresh American oysters (*Crassostrea virginica*). In *Processing and impact on active components in food* (pp. 455–461). Academic Press.

Pereira, C., Moreirinha, C., Teles, L., Rocha, R. J., Calado, R., Romalde, J. L., Nunes, M. L., & Almeida, A. (2017). Application of phage therapy during bivalve depuration improves *Escherichia coli* decontamination. *Food Microbiology 61*, 102–112. https://doi.org/10.1016/j.fm.2016.09.003

Phelps, E. B. (1911). Some experiments upon the removal of oysters from polluted to unpolluted waters. *Journal of American Public Health Association, 1*(5), 305.

Phuvasate, S., Chen, M. H., & Su, Y. C. (2012). Reductions of *Vibrio parahaemolyticus* in Pacific oysters (*Crassostrea gigas*) by depuration at various temperatures. *Food Microbiology, 31*(1), 51–56. https://doi.org/10.1016/j.fm.2012.02.004

Polo, D., Feal, X., & Romalde, J. L. (2015). Mathematical model for viral depuration kinetics in shellfish: An useful tool to estimate the risk for the consumers. *Food Microbiology, 49*, 220–225. https://doi.org/10.1016/j.fm.2015.02.015

Potasman, I., Paz, A., & Odeh, M. (2002). Infectious outbreaks associated with bivalve shellfish consumption: A worldwide perspective. *Clinical Infectious Diseases, 35*(8), 921–928. https://doi.org/10.1086/342330

Rey, M. S., Miranda, J. M., Aubourg, S., & Barros-Velázquez, J. (2012). Improved microbial and sensory quality of clams (*Venerupis rhomboideus*), oysters (*Ostrea edulis*) and mussels (*Mytilus galloprovincialis*) by refrigeration in a slurry ice packaging system. *International Journal of Food Science & Technology, 47*(4), 861–869. https://doi.org/10.1111/j.1365-2621.2011.02919.x

Rusmana, I., Suwanto, A., & Mubarik, N. R. (2020). Organic acid produced by lactic acid bacteria from bekasam as food biopreservatives. *IOP Conference Series: Earth and Environmental Science, 414*(1), 012003. https://doi.org/10.1088/1755-1315/414/1/012003

Schneider, K., & Rodrick, G. (1992). The use of ozone to degrade *Gymnodinium breve* toxins. In *2: Conference Internationale sur la Purification des Coquillages, Rennes (France), 6–8 Apr.*

Soccol, C. R., de Souza Vandenberghe, L. P., Spier, M. R., Medeiros, A. P., Yamaguishi, C. T., De Dea Lindner, J., Pandey, A., & Thomaz-Soccol, V. (2010). The potential of probiotics: A review. *Food Technology and Biotechnology, 48*(4), 413–434.

Suprapti, N. H., Bambang, A. N., Swastawati, F., & Kurniasih, R. A. (2016). Removal of heavy metals from a contaminated green mussel [*Perna viridis* (Linneaus, 1758)] using acetic acid as chelating agents. *Aquatic Procedia, 7*, 154–159. https://doi.org/10.1016/j.aqpro.2016.07.021

Turner, J. W., Malayil, L., Guadagnoli, D., Cole, D., & Lipp, E. K. (2014). Detection of *Vibrio parahaemolyticus*, *Vibrio vulnificus* and *Vibrio cholerae* with respect to seasonal fluctuations in temperature and plankton abundance. *Environmental Microbiology, 16*(4), 1019–1028. https://doi.org/10.1111/1462-2920.12246

Valenzuela, A. S., Benomar, N., Abriouel, H., Cañamero, M. M., & Gálvez, A. (2010). Isolation and identification of *Enterococcus faecium* from seafoods: Antimicrobial resistance and production of bacteriocin-like substances. *Food Microbiology, 27*(7), 955–961. https://doi.org/10.1016/j.fm.2010.05.033

Van Lith, R., & Ameer, G. A. (2016). Antioxidant polymers as biomaterial. In *Oxidative stress and biomaterials* (pp. 251–296). Academic Press. https://doi.org/10.1016/B978-0-12-803269-5.00010-3

Wang, W. X., & Lu, G. (2017). Heavy metals in bivalve mollusks. In *Chemical contaminants and residues in food* (pp. 553–594). Woodhead Publishing. https://doi.org/10.1016/B978-0-08-100674-0.00021-7

Wang, Y., Barton, M., Elliott, L., Li, X., Abraham, S., O'Dea, M., & Munro, J. (2017). Bacteriophage therapy for the control of *Vibrio harveyi* in greenlip abalone (*Haliotis laevigata*). *Aquaculture, 473*, 251–258. https://doi.org/10.1016/j.aquaculture.2017.01.003

Xie, W., Liu, X., Yang, X., Zhang, C., & Bian, Z. (2013). Accumulation and depuration of paralytic shellfish poisoning toxins in the oyster *Ostrea rivularis* Gould–Chitosan facilitates the toxin depuration. *Food Control, 30*(2), 446–452. https://doi.org/10.1016/j.foodcont.2012.07.035

Zhang, B., Fang, C. D., Xu, J. H., & Cao, H. J. (2017). Depuration of cadmium from blue mussel (*Mytilus edulis*) by protein hydrolysate–Fe^{2+} complex: The role of metallothionein. *Journal of Food Science, 82*(11), 2767–2773. https://doi.org/10.1111/1750-3841.13939

Zhang, B., Shi, Z. R., Wang, X. L., Deng, S. G., & Lin, H. M. (2015). Depuration of cadmium from blue mussel (*Mytilus edulis*) by hydrolysis peptides and chelating metal elements. *Food Research International, 73*, 162–168. https://doi.org/10.1016/j.foodres.2014.12.043

CHAPTER 3

Antibiotic Resistance in Aquaculture and Its Potential Solution

INAIN JAIES[1,*], FEROZ AHMAD SHAH[2], IMTIYAZ QAYOOM[3], and SHABIR A. DAR[4]

[1]*Division of Aquatic Animal Health Management, Faculty of Fisheries, SKUAST-K, Rangil, Ganderbal, Jammu and Kashmir, India*

[2]*Division of Aquatic Animal Health Management, Faculty of Fisheries, Sher-e-Kashmir University of Agricultural Sciences and Technology of Kashmir (SKUAST-K), Rangil, Ganderbal, India*

[3]*Division of Aquatic Environmental Management, Faculty of Fisheries, SKUAST-K, Rangil, Ganderbal, Jammu and Kashmir, India*

[4]*Division of Fishery Engineering, Faculty of Fisheries-Rangil, Sher-e-Kashmir University of Agricultural Sciences and Technology, Rangil, Jammu and Kashmir, India*

*Corresponding author

ABSTRACT

Controlling dangerous diseases is one of the major concerns in aquaculture. Numerous disease-causing bacteria in fish have developed resistance to frequently used antibiotics as a result of the indiscriminate use of antibiotics in fish farming. Some microorganisms are now immune to even the strongest drugs currently available. Due to biomagnification, the improper use of antibiotics in aquaculture has an impact on human health in addition to aquatic life. Antibiotic alternatives are therefore needed now more than ever in aquaculture. In this chapter, the subject of resistance in aquaculture will be briefly discussed, along with some potential options.

Recent Advances in Sustainable Aquaculture and Fisheries. Shabir Ahmad Dar, Imtiyaz Qayoom, Gohar Bilal Wani, Adnan Abubakr, and Seemab Zehra (Eds.)

DOI: 10.1201/9781998511372-3

3.1 INTRODUCTION

Antibiotics are medications with either a natural or synthetic origin designed to inhibit the growth and to kill pathogenic bacteria. An antibiotic is a chemical substance created by microorganisms that has the power to stop the growth of bacteria and other microbes and even to kill them (Waksman, 1947). They generally act either by damaging the cell membranes or by disrupting protein/DNA synthesis or by inhibiting the enzyme activity (Vignesh et al., 2011). In aquaculture, antibiotics are employed as preventative or therapeutic interventions. Antimicrobial chemotherapy is still essential for treating and, in some cases, preventing some bacterial diseases, despite the fact that bacterial diseases can be controlled in some situations by eradication, the upkeep of animals in a certain state of health, immunization, and excellent hygiene (Smith et al., 2003). The antibiotics prescribed by Serrano (2005) that have been recommended for use with aquatic animals, are given in Table 3.1.

TABLE 3.1 Antibiotics Approved for Aquaculture Usage (Serrano, 2005).

Disease	Drugs
Furunculosis	Sulphonamides, Sarafloxacin, Floefenicol, Oxytetracycline
Bacterial kidney disease	Erythromycin
Hemorrhagic septicaemia	Oxytetracycline
Columnaris disease	Oxytetracycline
Enteric red mouth	Sulphonamides, Sarafloxacin, Oxytetracycline
Vibriosis	Sulphonamides, Sarafloxacin
Cold water disease	Oxytetracycline

Antibiotics often function through one of two mechanisms:

1. A bactericidal action, which means that the antibiotic typically interferes with the creation of the cell wall, killing the bacteria or the contents of the cell. Penicillin, fluoroquinolones, and metronidazole are a few examples.
2. A bacteriostatic action, in which the antibiotic prevents the growth of bacteria by interfering with their ability to produce proteins, replicate DNA, or engage in other metabolic processes. Tetracyclines, sulfonamides, chloramphenicol, and macrolides are a few examples (Loree and Lappin, 2019).

Antibiotic Resistance:

Antimicrobials are frequently added to the fish feed in aquaculture, which is then poured into the aquatic ecosystem. Antimicrobials may in some situations be added straight to the water. The exposed environments experience a selective pressure as a result of these procedures. Many times, the antibiotics employed are non-biodegradable and last a very long time in the aquaculture environment. As a result, bacteria flourish and develop a resistance to these antibiotics that is passed down to the next generations (Vignesh et al., 2011). The intestinal tract of fish absorbs oxytetracycline, one of the most frequently used antibiotics in fish farms and hatcheries, rather poorly. For 10–15 days, it must be given at the rate of 100–150 mg/kg of fish. Due to the delayed excretion of high concentrations of this antibiotic as a result of this treatment, there may be a greater push for the development of bacteria in the gut that are resistant to oxytetracycline (Navarrete et al., 2008). About 80% of the antimicrobials used in aquaculture make their way into the environment, where they choose microorganisms whose resistance is the result of mutations or, more critically, from mobile genetic elements with numerous resistance determinants that can be transmitted to other bacteria. Such selection modifies the typical flora of fish and shellfish as well as the biodiversity in aquatic habitats. Existence of terrestrial microorganisms in an aquatic environment, along with lingering antimicrobials, biofilms, and highly concentrated bacteriophages carrying viruses with human and animal origins, can facilitate (Aly and Albutti, 2014). In addition, mobile genetic components like plasmids and transposable elements are utilized by bacteria. These components enable bacteria to have access to a sizable pool of itinerant genes, which is transferable between bacterial cells and propagate throughout bacterial populations (Romero et al., 2012). The different forms of antibiotic resistance as given by Romero et al. (2012) are:

1. **Inherent Resistance:** This might be due to the consequence of inability of antibacterial agent to penetrate the bacteria and reach its intended target location, or as a result of a lack of affinity between the antibacterial agent and its intended target, or even because the target is absent in the cell.

2. **Acquired Resistance:** In this instance, the bacterial species is often sensitive to a specific medication, but some strains exhibit treatment resistance.

Sites used for aquaculture are "hotspots for antibiotic-resistant genes" (Anastasiou et al., 2019). Antibiotic-resistant bacteria can adapt to their environment and grow by distributing their genes to new locations (Guo et al., 2017). Resistant bacteria that can infect and transmit to humans can be found in aquatic environments. These bacteria are tough to treat due to their resistant characteristics. Human pathogens like *Vibrio* spp., *Shigella* spp., *Salmonella* spp., or opportunistic pathogens like *Aeromonas hydrophila*, *Edwardsiella tarda*, *Streptococcus iniae*, and *Escherichia coli* may be involved in such direct transmission of resistance from such resistant bacteria and can spread to people directly through aquatic species, water consumption, or handling or eating seafood (Aly and Albutti, 2014).

Bacteria can exchange antibiotic resistance genes by mechanisms including transformation, transduction, or conjugation that include lateral DNA transfer. In transformation, bacteria take up (foreign) DNA from the outside environment. Transduction, in which bacteria take up genes through viral DNA infection. Due to the large viral (bacteriophage) concentrations in aquatic ecosystems, seawater, and marine sediment, this alternative may be crucial in the transfer of resistance and Conjugation, which is when bacteria mate and exchange genes. A plasmid is transferred from one organism to another during this process via a pilus (Hobot, 2015).

Due to the emergence of acquired resistance in pathogens attacking fish, these resilient microbes have the potential to serve as a source of resistance genes that could eventually be transferred to human infections. Because resistance can spread among bacteria and for antibiotic-resistant diseases, antibacterial therapies might not be effective, the presence of antibiotic-resistant microorganisms in foods of animal origin poses a possible health risk (Economou and Gousia, 2015). Duran and Marshall (2005) investigated six brands of ready-to-eat shrimp purchased from supermarkets as part of a microbiological investigation of market items. *E. coli*, *Salmonella*, *Shigella*, and *Vibrio* spp. were found among the resistant isolates, along with other human diseases.

Antibiotic residues, antibiotic-resistant bacteria, and resistance genes are all regarded as environmental contaminants that are to blame for a persistent public health concern around the world (Polianciuc et al., 2020). Antibiotic use in food-producing animals not only promotes antibiotic resistance but also raises the possibility that goods obtained from animals, such as meat, milk, eggs, kidney, liver, and fat, that are intended for human consumption, may include antibiotic residues. Aquatic environments are thought to be a fundamental setting for the environmental release, transformation, mixing, and persistence of antibiotic residues, antibiotic-resistant bacteria, and antibiotic resistance genes (Taylor et al., 2011). According to report/s, antibiotic-resistant bacteria and resistance genes originate from and are stored in the aquatic environment (Baquero et al., 2008). Despite the fact that there is confirmation that some medications, antibiotics might cause nephrotoxicity (Hentschel et al., 2005), immunomodulation is the most well-known adverse effect (Tafalla et al., 1999).

3.1.1 ANTIBIOTIC USAGE IN AQUACULTURE IS RELATED WITH THE FOLLOWING CONCERNS TO HUMAN HEALTH

Two risks must be taken into account separately in order to determine the likelihood of the risk of resistance transfer: the risk associated with the introduction of non-pathogenic bacteria containing antimicrobial resistance into the human environment and the risk associated with the transmission of resistant bacteria from aquaculture environments to humans, which is the increase in resistance in human bacterial pathogens as a direct result of the use of antibiotics in aquaculture (Serrano, 2005). Knowing about residual bio-accumulation, chronic toxicity, and related issues is essential because animals are fed antibiotics for extended periods of time. These compounds could be poisonous or allergic-provoking. The target animal species, non-target species ingested accidentally, humans, or both, may exhibit toxic effects. With the exception of chloramphenicol, which causes toxic aplastic anaemia and can occasionally result in fatal aplastic anaemia, the chance of direct poisoning from antibiotics or their metabolites in animals is incredibly low. Numerous antibiotics are powerful antigens, including macrolides, quinoxalines, and bacitracin, and daily occupational contact to these substances can cause sensitization. The majority of allergic reactions that have been recorded have been linked to ß-lactam antibiotic residues in milk or meat, and the allergic reaction has also been linked to exposure of food to antibiotic residues (Serrano, 2005). The majority of allergic responses are linked to intramuscular medication delivery followed by oral treatment, and the protein-bound antibiotic's epitope distribution is so negligibly small that it is unlikely to be the cause of an allergic reaction when consumed.

3.1.2 ANTIMICROBIAL RESISTANT BACTERIA IN AQUACULTURE AS A RISK TO HUMAN HEALTH

1. Due to the emergence of acquired resistance in fish pathogens, these resistant microbes have the potential to serve as a source of resistance genes that could eventually spread to human infections. In other instances, the human cholera-causing bacteria *Vibrio cholera* has received plasmid-borne drug resistance genes from the fish pathogen *Vibrio anguillarum* (Nakajima et al., 1983).
2. When people eat fish that has been treated with antibiotics, antibiotic-resistant bacteria enter the human gut and likely pass their resistant genes to the gut flora. The development of antibiotic resistance in human intestinal bacteria may result from the dissemination of antibiotic-resistant genes (Ventola, 2015).

3.2 ALTERNATIVES TO ANTIBIOTICS IN AQUACULTURE

3.2.1 BACTERIOPHAGES

Bacteriophages were first identified as bacterial viral infections in the early 1920s, and their significance for antibacterial therapy was quickly realized. Bacteriophages are viruses with the capacity to infect, replicate, and kill vulnerable bacteria. Because phages are exceedingly selective in the bacteria they choose to infect, the collateral damage to unaffected bacteria is negligible. Recent reports provide evidence supporting the use of a bacteriophage mixture to successfully shield brook trout from *Aeromonas salmonicida* disease (Imbeault et al., 2006). *Pseudomonas plecoglossicida* infection was prevented in Ayu (*Plecoglossus altivelis*) when fed with phage impregnated feed and it immediately vanished from the kidneys of the phage-treated fish (Park et al., 2000). In a study, Wu et al. (1981) identified and partially described eight bacteriophages that attack *A. hydrophila*. Coldwater fish are frequently affected by the bacterial infection *Flavobacterium psychrophilum*, which results in cold water disease in salmonids and rainbow trout fry syndrome in trout. Numerous phages have been discovered and studied in relation to *F. Psychrophilum* (Stenholm et al., 2008; Kim et al., 2010).

Fish whose infection is brought on by bacterial colonization on their skin and gills will benefit from bath delivery of phages (Nakai and Park, 2002). Infections where oral transmission is the primary mode of pathogen transmission, such as those caused by *Lactococcus garvieae* infection of yellowtail and *Psedomonas plecoglossicida* infection of ayu, will benefit from phage administration through feed (Nakai and Park, 2002). Two lytic bacteriophages with a focus on the enteric septicemia-causing *Edwardsiella ictaluri* in catfish were discovered. Each strain of *E. ictaluri* examined was susceptible to phage infection, however, the effectiveness of these infections varied, but the phages displayed no signs of plaques in any other bacterial species, indicating its potential application in biotherapy and diagnostic tools for catfish enteric septicaemia (Walakira et al., 2008).

3.2.1.1 MAJOR ADVANTAGES OF BACTERIOPHAGES

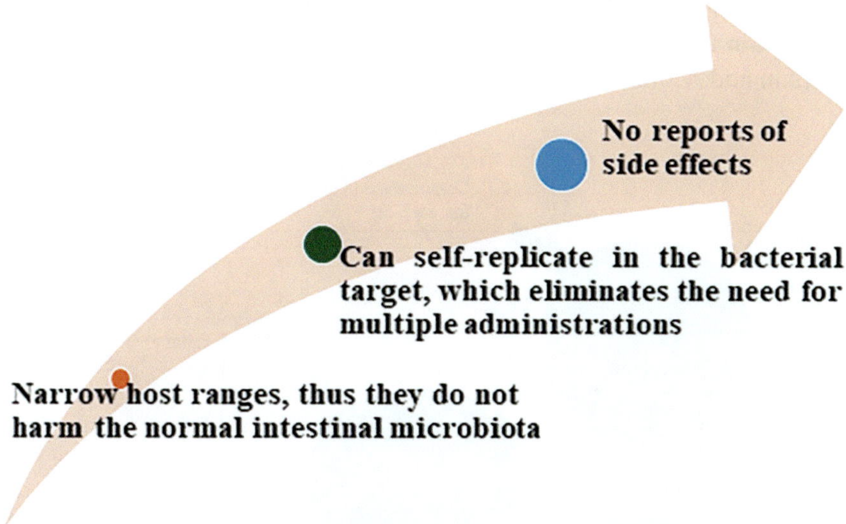

Bacteriophages as alternative to antibiotics

3.2.2 PROBIOTICS

Alternative to antibiotics, the use of probiotics to support health maintenance and illness prevention has lately grown in popularity. Probiotics are living bacteria that give the host health benefits when given in sufficient doses (Reid et al., 2003). In recent years, interest in using probiotics as an antibiotic alternative to maintain health and prevent disease has grown. Lactic acid bacteria, *Bacillus*, *Vibrio*, *Pseudomonas*, and *Aeromonas* were the microorganisms that had been assessed as probiotics in aquaculture.

Due to their strong metabolic capacity, yeasts (*Saccharomyces* and *Debaryomyces*), bacterial spore formers, and more recently *Actinobacteria*, have also attracted attention. A probiotic with a variety of microorganisms may be more effective at reducing bacterial pathogens, according to some research (Verschuere et al., 2000; Chapman et al., 2011). The assumption behind this is that it would be challenging for a single bacterium to maintain dominance in a constantly shifting environment.

The effectiveness of these probiotics has been demonstrated in a variety of aquatic organisms, including teleost fish (Merrifield et al., 2010), prawns (Van Hai and Fotedar, 2009), shrimp (Farzanfar, 2006; Ninawe and Selvin, 2009), and bivalve molluscs (Kesarcodi-Watson et al., 2008; Prado et al., 2010). Rainbow trout were shown to have a higher survival rate than control when challenged with *Yersinia ruckeri* after being fed on a multistrain feed containing *Enterobacter cloacae* and *Bacillus mojavensis* (Capkin and Altinok, 2009).

FIGURE 3.1 Different means by which probiotics provide protection against pathogens.

Source: Adapted from Romero et al., 2012

3.2.3 ESSENTIAL OILS

The utilization of vegetable extracts is related to another potential remedy for chemotherapies in aquaculture. Natural substances such as phenolic compounds, polysaccharides, proteoglycans, and flavonoids that are able to boost the fish immune system may be important in the prevention or management of pathogenic microorganisms. These substances are known to be present in herbal extracts from plants or algae. These products, which combine sustainable production methods with high-quality seafood, are acknowledged as environmentally beneficial options for therapeutic and preventative purposes in the health management of aquatic animals.

There are a few but promising examples of essential oils (EOs) being used in aquaculture systems *in vivo*. In the scant study that has been conducted, the impact of EOs on prawns and different fish has been demonstrated. *Vibrio alginolyticus* susceptibility in *Litopenaeus vannamei* subjected to hot water extracts from *Cinnamomum kanehirae* twigs was significantly reduced, according to a recent study by Yeh et al. (2009). The EOs of stout camphor tree also exhibited antibacterial effects against various pathogens of aquatic animals (Yeh et al., 2009). Similar to this, EOs of *C. fragrans* decreased *Vibrio* concentrations and total heterotrophic aerobic bacteria in the larval culture of *Penaeus monodon* (Sarter et al., 2011). The development of *Vibrio* spp. and *Photobacterium damselae* was most strongly inhibited by the acetone extract from *Withanias omnifera* (Immanuel et al., 2009). It is interesting to note that channel catfish (*Ictalurus punctatus*) fed on oregano EO obtained from *Origanum heracleoticum* had low mortality rate following an *A. hydrophila* infection when compared to fish fed a mixture of carvacrol and thymol, the primary active ingredients in oregano EO (Zheng et al., 2009).

3.3 CONCLUSION

Since antibiotics only provide selection pressure to pathogens, they have a smaller possibility of leading to the emergence of resistance than alternative methods of controlling bacterial infections in aquaculture. If immediate action is not done, we will soon return to the so-called "post-antibiotic period," where quite little bacterial infections may cause significant economic losses in aquaculture.

KEYWORDS

- **alternative**
- **antibiotics**
- **aquaculture**
- **bacteria**
- **biomagnification**
- **resistance**

REFERENCES

Aly, S. M., & Albutti, A. (2014). Antimicrobials use in aquaculture and their public health impact. *Journal of Aquaculture Research & Development, 5*(4), 1.

Anastasiou, T. I., Mandalakis, M., Krigas, N., Vézignol, T., Lazari, D., Katharios, P. et al. (2019). Comparative evaluation of essential oils from medicinal-aromatic plants of Greece: Chemical composition, antioxidant capacity and antimicrobial activity against bacterial fish pathogens. *Molecules, 25*(1), 148.

Baquero, F., Martínez, J. L., & Cantón, R. (2008). Antibiotics and antibiotic resistance in water environments. *Current Opinion in Biotechnology, 19*(3), 260–265.

Capkin, E., & Altinok, I. (2009). Effects of dietary probiotic supplementations on prevention/treatment of yersiniosis disease. *Journal of Applied Microbiology, 106*(4), 1147–1153.

Chapman, C. M. C., Gibson, G. R., & Rowland, I. (2011). Health benefits of probiotics: Are mixtures more effective than single strains? *European Journal of Nutrition, 50*(1), 1–17.

Duran, G. M., & Marshall, D. L. (2005). Ready-to-eat shrimp as an international vehicle of antibiotic-resistant bacteria. *Journal of Food Protection, 68*(11), 2395–2401.

Economou, V., & Gousia, P. (2015). Agriculture and food animals as a source of antimicrobial-resistant bacteria. *Infection and Drug Resistance,* 49–61.

Farzanfar, A. (2006). The use of probiotics in shrimp aquaculture. *FEMS Immunology & Medical Microbiology, 48*(2), 149–158.

Guo, J., Li, J., Chen, H., Bond, P. L., & Yuan, Z. (2017). Metagenomic analysis reveals wastewater treatment plants as hotspots of antibiotic resistance genes and mobile genetic elements. *Water Research, 123*, 468–478.

Hentschel, D. M., Park, K. M., Cilenti, L., Zervos, A. S., Drummond, I., & Bonventre, J. V. (2005). Acute renal failure in zebrafish: A novel system to study a complex disease. *American Journal of Physiology-Renal Physiology, 288*(5), F923–F929.

Hobot, J. A. (2015). Bacterial ultrastructure. In *Molecular medical microbiology* (pp. 7–32). Academic Press.

Imbeault, S., Parent, S., Lagacé, M., Uhland, C. F., &Blais, J. F. (2006). Using bacteriophages to prevent furunculosis caused by *Aeromonas salmonicida* in farmed brook trout. *Journal of Aquatic Animal Health, 18*(3), 203–214.

Immanuel, G., Uma, R. P., Iyapparaj, P., Citarasu, T., Punitha Peter, S. M., Michael Babu, M., &Palavesam, A. (2009). Dietary medicinal plant extracts improve growth, immune activity and survival of tilapia *Oreochromis mossambicus*. *Journal of Fish Biology, 74*(7), 1462–1475.

Kesarcodi-Watson, A., Kaspar, H., Lategan, M. J., & Gibson, L. (2008). Probiotics in aquaculture: The need, principles and mechanisms of action and screening processes. *Aquaculture, 274*(1), 1–14.

Kim, J. H., Gomez, D. K., Nakai, T., & Park, S. C. (2010). Isolation and identification of bacteriophages infecting ayu *Plecoglossus altivelisaltivelis* specific *Flavobacterium psychrophilum*. *Veterinary Microbiology, 140*(1–2), 109–115.

Loree, J., & Lappin, S. L. (2019). Bacteriostatic antibiotics.

Merrifield, D. L., Bradley, G., Baker, R. T. M., & Davies, S. J. (2010). Probiotic applications for rainbow trout (*Oncorhynchusmy kiss*Walbaum) II. Effects on growth performance, feed utilization, intestinal microbiota and related health criteria postantibiotic treatment. *Aquaculture Nutrition, 16*(5), 496–503.

Nakai, T., & Park, S. C. (2002). Bacteriophage therapy of infectious diseases in aquaculture. *Research in Microbiology, 153*(1), 13–18.

Nakajima, T., Suzuki, M., Harada, K., Inoue, M., & Mitsuhashi, S. (1983). Transmission of R plasmids in *Vibrio anguillarum* to *Vibrio cholerae*. *Microbiology and Immunology, 27*(2), 195–198.

Navarrete, P., Mardones, P., Opazo, R., Espejo, R., & Romero, J. (2008). Oxytetracycline treatment reduces bacterial diversity of intestinal microbiota of Atlantic salmon. *Journal of Aquatic Animal Health, 20*(3), 177–183.

Ninawe, A. S., & Selvin, J. (2009). Probiotics in shrimp aquaculture: Avenues and challenges. *Critical Reviews in Microbiology, 35*(1), 43–66.

Park, S. C., Shimamura, I., Fukunaga, M., Mori, K. I., & Nakai, T. (2000). Isolation of bacteriophages specific to a fish pathogen, *Pseudomonas plecoglossicida*, as a candidate for disease control. *Applied and Environmental Microbiology, 66*(4), 1416–1422.

Polianciuc, S. I., Gurzău, A. E., Kiss, B., Ştefan, M. G., & Loghin, F. (2020). Antibiotics in the environment: Causes and consequences. *Medicine and Pharmacy Reports, 93*(3), 231.

Prado, S., Romalde, J. L., & Barja, J. L. (2010). Review of probiotics for use in bivalve hatcheries. *Veterinary Microbiology, 145*(3–4), 187–197.

Reid, G., Jass, J., Sebulsky, M. T., & McCormick, J. K. (2003). Potential uses of probiotics in clinical practice. *Clinical Microbiology Reviews, 16*(4), 658–672.

Romero, J., Feijoó, C. G., & Navarrete, P. (2012). Antibiotics in aquaculture—Use, abuse and alternatives. *Health and Environment in Aquaculture, 159*, 159–198.

Sarter, S., Randrianarivelo, R., Ruez, P., Raherimandimby, M., & Danthu, P. (2011). Antimicrobial effects of essential oils of *Cinnamosmafragrans* on the bacterial communities in the rearing water of *Penaeusmonodon* larvae. *Vector-Borne and Zoonotic Diseases, 11*(4), 433–437.

Serrano, P. H. (2005). *Responsible use of antibiotics in aquaculture*, Vol. 469. Food & Agriculture Org.

Smith, D. L., Johnson, J. A., Harris, A. D., Furuno, J. P., Perencevich, E. N., & Morris Jr, J. G. (2003). Assessing risks for a pre-emergent pathogen: Virginiamycin use and the emergence of streptogramin resistance in *Enterococcus faecium*. *The Lancet Infectious Diseases, 3*(4), 241–249.

Stenholm, A. R., Dalsgaard, I., & Middelboe, M. (2008). Isolation and characterization of bacteriophages infecting the fish pathogen F*lavobacterium psychrophilum*. *Applied and Environmental Microbiology, 74*(13), 4070–4078.

Tafalla, C., Novoa, B., Alvarez, J. M., & Figueras, A. (1999). *In vivo* and *in vitro* effect of oxytetracycline treatment on the immune response of turbot, *Scophthalmus maximus* (L.). *Journal of Fish Diseases, 22*(4), 271–276.

Taylor, N. G., Verner-Jeffreys, D. W., & Baker-Austin, C. (2011). Aquatic systems: Maintaining, mixing and mobilising antimicrobial resistance? *Trends in Ecology & Evolution, 26*(6), 278–284.

Van Hai, N., & Fotedar, R. (2009). Comparison of the effects of the prebiotics (Bio-Mos® and β-1, 3-D-glucan) and the customised probiotics (*Pseudomonas synxantha* and *P. aeruginosa*) on the culture of juvenile western king prawns (PenaeuslatisulcatusKishinouye, 1896). *Aquaculture, 289*(3–4), 310–316.

Ventola, C. L. (2015). The antibiotic resistance crisis: Part 1: Causes and threats. *Pharmacy and Therapeutics, 40*(4), 277.

Verschuere, L., Rombaut, G., Sorgeloos, P., & Verstraete, W. (2000). Probiotic bacteria as biological control agents in aquaculture. *Microbiology and Molecular Biology Reviews, 64*(4), 655–671.

Vignesh, R., Karthikeyan, B. S., Periyasamy, N., & Devanathan, K. (2011). Antibiotics in aquaculture: An overview. *South Asian Journal of Experimental Biology, 1*(3), 114–120.

Waksman, S. A. (1947). Antibiotics and tuberculosis: A microbiologic approach. *Journal of the American Medical Association, 135*(8), 478–485.

Walakira, J. K., Carrias, A. A., Hossain, M. J., Jones, E., Terhune, J. S., & Liles, M. R. (2008). Identification and characterization of bacteriophages specific to the catfish pathogen, *Edwardsiella ictaluri*. *Journal of Applied Microbiology, 105*(6), 2133–2142.

Wu, J. L., Lin, H. M., Jan, L., Hsu, Y. L., & Chang, L. H. (1981). Biological control of fish bacterial pathogen, *Aeromonashydrophila*, by bacteriophage AH 1. *Fish Pathology, 15*(3–4), 271–276.

Yeh, R. Y., Shiu, Y. L., Shei, S. C., Cheng, S. C., Huang, S. Y., Lin, J. C., & Liu, C. H. (2009). Evaluation of the antibacterial activity of leaf and twig extracts of stout camphor tree, *Cinnamomum kanehirae*, and the effects on immunity and disease resistance of white shrimp, *Litopenaeus vannamei*. *Fish & Shellfish Immunology, 27*(1), 26–32.

Zheng, Z. L., Tan, J. Y., Liu, H. Y., Zhou, X. H., Xiang, X., & Wang, K. Y. (2009). Evaluation of oregano essential oil (*Origanumheracleoticum* L.) on growth, antioxidant effect and resistance against *Aeromonas hydrophila* in channel catfish (*Ictalurus punctatus*). *Aquaculture, 292*(3–4), 214–218.

CHAPTER 4

Use of Microalgae as a Sustainable Source for Aquaculture

SEEMAB ZEHRA[1,*], ASAAD H. MOHAMED[1], EDOARDO PANTANELLA[1], and A. AL SHAIKHI[2]

[1]*Beacon Development, King Abdullah University of Science and Technology, Thuwal, Jeddah, Saudi Arabia*

[2]*Ministry of Environment, Water and Agriculture, Riyadh, Saudi Arabia*

*Corresponding author

ABSTRACT

One of the food production industries with the highest worldwide growth is aquaculture, which is crucial for bridging the gap between the supply and demand for food. Feed has a key role in increasing aquaculture productivity. For sustainable aquaculture, it is essential to find other sources for fish meal and fish oil since the current ones are not sustainable. A reliable, organic source of bioactive chemicals is microalgae. Microalgae are a fantastic source of polyunsaturated fatty acids (PUFA), especially the n-3 series, and have a roughly 50% crude protein content and an amino acid profile comparable to fishmeal. As a result, microalgae may substitute fish oil and fish meal in diets and raise the n-3 PUFA deposition quality of meat. The literature on the protein, amino acid, lipid, and fatty acid content of microalgae is examined in this chapter. Additionally, this chapter compiled the literature on using microalgae in lieu of fish meal and fish oil. The current state of knowledge has been presented in this chapter, along with important areas for the future growth of the microalgae-based aquafeed business and the general sustainability of the aquaculture sector.

Recent Advances in Sustainable Aquaculture and Fisheries. Shabir Ahmad Dar, Imtiyaz Qayoom, Gohar Bilal Wani, Adnan Abubakr, and Seemab Zehra (Eds.)

DOI: 10.1201/9781998511372-4

4.1 INTRODUCTION

Systems of intensive aquaculture meet the rising demand brought on by the expanding population (FAO, 2021). However, the use of fishmeal in the production methods renders intensive aquaculture unsustainable (NRC, 2011). Additionally, there are recurring changes in both the price and availability of fishmeal (Gasco et al., 2018). Several researchers are working on substituting fishmeal with other high-quality protein sources, such as microalgae (Christaki et al., 2011). Microalgae are ideal for aquaculture feed due to a variety of factors, including their high nutritional content and growth rate, antioxidant capabilities, and potential to strengthen the animals' defenses against infections (Roy and Pal, 2015). The microorganism class known as microalgae is incredibly varied, with an estimated 200,000 species (Norton et al., 1996). Cyanophyceae (blue-green algae), Chlorophyceae (green algae), Bacillariophyceae, including the diatoms, and Chrysophyceae containing the golden algae are the most well-known and possibly more ecologically significant microalgal classes/divisions (Chacón-Lee and González-Mariño, 2010). They are utilized either directly for eating or indirectly as prepared feed and are necessary for the sustenance of larvae during a short time of the life cycle. The majority of the time, entire algae are utilized as feed or a supplement to feed. Additionally, live algae enhance water quality. Microalgae are becoming more and more popular as aquaculture feed owing to their ideal size, high nutritional content, rapid growth rate, antioxidant property, disease resistance power, and so forth. Microalgae's size, shape, digestion, and biochemical makeup, all affect how nutritious they are (Brown et al., 1997). A reliable, organic source of bioactive chemicals is microalgae. Microalgae are a fantastic source of polyunsaturated fatty acids (PUFA), especially the n-3 series, and have a roughly 50% crude protein content and an amino acid profile comparable to fishmeal. Thus, via deposition of n-3 PUFA, microalgae may substitute fish oil and fish meal in diets and improve the quality of meat (Roy and Pal, 2015, Sarker et al., 2016a,b, Alagawany et al., 2021). Microalgae have the potential to develop into a substantial renewable energy source since they are a large source of oils and other compounds that may be utilized as feedstock to make biofuels and high-value goods (Moreno-Garcia et al., 2017). On a variety of wastewaters, including those produced in aquaculture by the operation of hatcheries and farming systems, microalgae may grow successfully (Malibari et al., 2018, Dourou et al., 2020). By 2020, the market for microalgae is anticipated to reach US $3.4 billion and to increase by 4.3% annually (Kumar et al., 2021).

The use of microalgae as a substitute for fishmeal and fish oil in the aquaculture industry has some disadvantages and difficulties, such as the high cost of production of microalgae (Becker, 2007, Sarker et al., 2016a), the possibility that some microalgae have difficult-to-digest cell walls (Skrede et al., 2011), the possibility that a significant amount of biomass may be contaminated (Hannon et al., 2010), and so forth. Until recently, scientific advancements and commercial uses of microalgae as a micro feed component have mostly concentrated on their specialized beneficial features rather than their overall nutritional value to the receiving animal. Research on the utilization of microalgae as a potential component of aquafeed for various fish species has been growing rapidly. Priyadarshani and Rath (2012), Yaakob et al. (2014), and Roy and Pal (2015) have all shown that microalgae have a high potential for use as a bulk feedstock for aquaculture diets. However, there is still a dearth of current information and recent advancements on the use of microalgal biomass in lieu of fishmeal and fish oil for the growth of a sustainable aquaculture business.

4.2 CURRENT STATUS OF THE MICROALGAE PRODUCTION

The term "microalgae" refers to a wide range of unicellular algae species, including diatoms, *Nannochloropsis* species, *Schizochytrium* species, and *Crypthecodinium* species, to mention a few intimately connected to aquaculture. *Spirulina* and *Arthrospira*, two genera that include roughly 15 species of cyanobacteria (also known as blue-green algae), are often considered to be microalgae (Cai et al., 2021). Compared to seaweeds, the output of commercial microalgae is relatively low. Commercial microalgae cultivation, according to FAO figures, accounted for less than 0.2% of the world's algal culture tonnage in 2019. This suggests that microalgae cultivation is significantly less than seaweed farming (FAO, 2021). Spirulina (*Arthrospira*) production in Mexico from 1962 to 1993, with a peak of 4375 ton in 1989, was the only wild microalgae production documented by FAO data. The first significant microalgae cultivation to be included in FAO data began in 2003 with the production of 16,483 ton of spirulina (*Arthrospira*) in China. Global microalgae production peaked in 2010 at 93,756 ton, but it fell to 56,456 ton in 2019, mostly due to changes in spirulina farming in China (Cai et al., 2021) (Table 4.1). The 56,456 ton of microalgae produced globally in 2019 were mostly made up of 248 ton of green microalgae and 56,208 ton of spirulina (*Arthrospira*), both of which were grown in four different nations (Cai et al., 2021). Only *Spirulina-Athrospira* produces substantial amounts of microalgae (measured in live weight, or LW) according to FAO data (Figures 4.1 and 4.2).

TABLE 4.1 Microalgae Production.

Country/Area	Microalgae (ton)	*Spirulina-Athrospira (ton)*	Green Microalgae (ton)				
			Total	*Haematococcus pluvialis*	*Chlorella vulgaris*	*Tetraselmis spp.*	*Dunaliella salina*
World	**56,456**	**56,208**	**248**	**242**	**4.77**	**1.45**	**0.22**
1. China	54,850	54,650	200	200			
2. Chile	903	861	42	42			
3. France	207	201	6.22		4.77	1.45	
4. Greece	142	142					
5. Tunisia	140	140					
6. Burkina Faso	140	140					
7. Central African Republic	50	50					
8. Chad	20	20					
9. Bulgaria	2.65	2.65					
10. Spain	1.52	1.30	0.22				0.22

Source: Adapted from FAO., 2021.

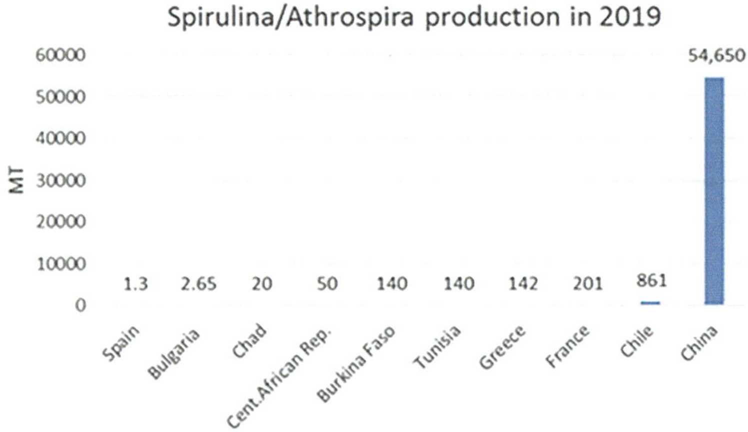

FIGURE 4.1 Spirulina production per country in 2019 (LW).

Source: Reprinted from FAO, 2021.

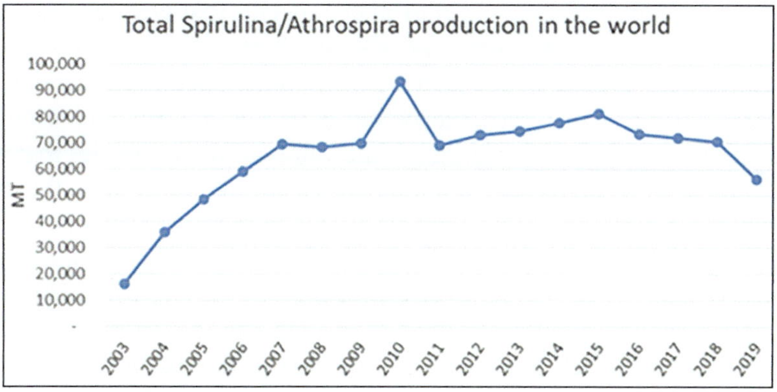

FIGURE 4.2 Spirulina production trend.

Source: Reprinted from FAO, 2021.

Other microalgae present in the FAO database are *Haematococcus pluvialis*, largely cultivated to produce astaxanthin but only listed as being cultivated in China (200 MT LW) and Chile (20 MT LW); *Chlorella* (France, 4.22 MT LW); *Tetraselmis* (France, 1.45 MT LW); and *Dunaliella* (Spain, 0.22 MT LW).

The majority of commercial microalgae cultivation is concentrated on the extraction of various bioactive or biochemical compounds, such as pigments for dietary supplements (e.g., carotenes) and dried microalgae biomass for functional foods or dietary supplements (e.g., Chlorella or spirulina powder). For example, canthaxanthin and astaxanthin obtained from *Haematococcus*, docosahexaenoic acid (DHA) derived from *Cryptodinium*, the addition of polysaccharides to cosmetic goods; and naturally occurring food dyes, such as spirulina (Spolaore et al., 2006, Barkia et al., 2019).

Other possible uses for microalgae include wastewater remediation, algae meal, carbon sequestration, biofuels, and algae oils (Spolaore et al., 2006, Barkia et al., 2019, Ragaza et al., 2020). The commercialization of these applications may not yet be complete. The markets for several of them are developing quickly and increasing. There is significant microalgae cultivation to provide food for animal species in other aquaculture operations; however, such production is seldom included in official statistics.

4.3 MICROALGAE AS THE SOURCE OF PROTEIN

The environmental settings in which microalgal species and strains are developed have a significant impact on the protein contents of such organisms. Similar to the values of many species that were examined by others under cultivation, many species may have significant amounts of protein, often 40–60% of dry matter, while other species have relatively low levels of protein, particularly those chosen for oil and biodiesel production (Becker, 2007), similar to the values of many species that were reviewed by others (Acquah et al., 2020). According to a study, the crude protein content of microalgae biomass varies from 6% to 63% on the basis of dry mass, with the majority of species having above 40% crude protein content (Chacón-Lee and González-Mariño, 2010). The investigation of the protein content in 16 microalgae showed that the protein content (% of dry cell matter) varied from 12% (in the case of the diatom *Chaetoceros gracilis*) to 35% (in the case of the eustigmatophyte *Nannochloropsis oculata*) (Brown, 1991). In 2007, Becker gave a summary of the 13 microalgae's main components (Becker, 2007). The protein concentration ranged widely, from 6% to 71% of dry matter, with above 50% in the majority of the species. In contrast to the values of 12–24% reported by Tipnee et al. (2015) and 18% by Saragih et al. (2019), *Spirogyra* sp. showed a lower but wider range of protein content (6–20% of dry matter). In 33 different microalgal species, Becker reported in 2013 that the protein concentration ranged from 10% to 71%. (Becker, 2013). There have been reports of 17 different microalgal species with protein contents ranging from 6% to 58% (Acquah et al., 2020). The protein composition of 22 microalgal species, researched in various labs, was provided in another review. A range of 28–71% of protein in dry biomass was found, and most species possessed above 50% protein, with the exception of *Spirogyra* sp. (6–20%) and *Scenedesmus dimorphus* (8–18%) (Koyande et al., 2019). It is clear that the majority of microalgal species have significant protein content, and some of them, like the cyanobacteria *Arthrospira* sp. and the chlorophyceae *Chlorella* sp., have been created and employed as food proteins or nutritional supplements (Becker, 2007). Due to their high protein concentrations (51–58% of dry matter) and advantageous essential amino acid profiles, *Chlorella vulgaris* and *Arthrospira* sp. seem to be the most often used industrial species (Becker, 2007, Ismail et al., 2020). When compared to other chlorophycean species, diatoms have more protein, according to Renaud et al. (1994). Due to their 40–70% protein levels, *Spirulina, Scenedesmus,* and *Chlorella* were regarded as sources of single cell protein (Venkataraman and Becker, 1985). In microalgae, the nutritional value of the protein component (protein plus free amino acids) is deemed to be high if it closely matches the requirements of the feeding animals for essential amino acids (Roy and Pal, 2015).

The most popular microalgal species that is produced commercially is spirulina (Farag et al., 2016). They are cyanobacteria and contain between 56% and 69% of their dry weight in protein. Spirulina contains all nine necessary amino acids and satisfies the FAO's requirements for equiponderant proteins (Velasquez et al., 2016). In addition, spirulina contains a variety of bioactive substances, such as vitamins, minerals, carotenes, vital fatty acids, and antioxidants (Vonshak, 1997). Spirulina has a phenomenal growth rate in comparison to other plants, which makes it a potent instrument for addressing the worldwide protein deficit and the global food crisis (Shay, 1993). In addition, spirulina is a potent substitute for animal-derived proteins in fish diets and has been shown to exhibit promising results in a variety of fish species, including *Oreochromis niloticus* (Lu and Takeuchi, 2004), *Paralichthys olivaceus* (Lee et al., 2015), *Huso huso* (Adel et al., 2016), African catfish (Raji et al., 2020), barramundi, and *Lates calcarifer* (Van Vo et al., 2020). These studies show that the positive effects of spirulina rely on the amount of microalgae in fish diets and are not just restricted to certain species.

4.4 MICROALGAE AS THE SOURCE OF CARBOHYDRATE

The starch in the chloroplast and the cellulose on the inner cell walls make up the majority of the carbohydrates in microalgae. Lignin and cellulose together are absent from the algal cell (Ho et al., 2012). Glucose, which accounts for 28–86% of the total carbohydrate in polysaccharides of various algae species, is the most prevalent sugar. As a source of carbohydrates, *Isochrysis galbana, P. lutheri, C. vulgaris, PseudoIsochrysis paradoxa*, and *Pyramimonas virginica* are the most often employed microalgal species in aquafeeds (Roy and Pal, 2015). Under autotrophic circumstances, *C. vulgaris* acquires 50% (DCW) carbohydrates (Wang et al., 2014). Therefore, there is great potential for using microalgal carbohydrate as a component of aquaculture feedstock. To locate acceptable microalgal species with desirable characteristics, such as rich carbohydrate and high growth rate while growing on affordable medium for large-scale culture, further study is still needed. Aquafeed that contains heterotrophically grown (capable of growing without light) algae meal might increase feed intake and growth rates (Kupchinsky et al., 2015).

4.5 MICROALGAE AS THE LIPID AND FATTY ACID SOURCES

Essential fatty acids, such as the omega-3 fatty acids EPA, ALA, and DHA, are found in microalgal lipids. Microalgae's lipid content varies from strain to strain, according to the method of growth, and depending on the stage of harvesting. The microalgal strain and culture conditions have an impact on the fatty acid content. Microalgae may store up to 70% of lipid when under stress (Stephenson et al., 2011). According to D'Alessandro and Antoniosi Filho (2016), the lipid yield in *C. vulgaris* varies from 12% to 26% and in *Bortryococcus braunii* from 14% to 75%. In comparison to freshwater species, certain marine microalgae have much more DHA. Higher DHA concentrations may be seen in heterotrophically farmed algae, such as *Thrautocytrium* (16.1%), *Amphidium caryerea* (17%), and *Schizochytrium* mangrove (33–39%). About 30–50% of the total components' DHA was generated by *Crypthecodinium cohnii*. In comparison to freshwater species, certain marine microalgae have much more DHA (Yaakob et al., 2014). The addition of ALA, DHA, and EPA in aquaculture feed must be prioritized since these nutrients are necessary for healthy growth and development. Due to its high EPA concentration, *Nannochloropsis* sp. is mostly employed as a supplement in aquaculture. In the future, there may be a sizable demand for fish feed, which might be satisfied by choosing microalgal strains as feed that have the necessary cellular composition and a reduced production cost.

Different microalgae's proximate (crude protein and crude lipid), amino acid, and fatty acid compositions are provided in Tables 4.2–4.4, respectively.

TABLE 4.2 Crude Protein and Lipid of Various Microalgae Species and Fish Meal.

Microalgae	CL	CP	Citation
Algae meal, BioProcess Algae LLC., USA	8.2	51.7	https://app.iaffd.com/ficd
Algae oil, Veramaris®	98.8	0	https://app.iaffd.com/ficd
Algae, dried (*Nannochloropsis* sp. and *Isochrysis* sp.)	18.2	42.1	https://app.iaffd.com/ficd
Algae, *Schizochytrium* sp	51.4	16.3	https://app.iaffd.com/ficd
Algae, Spirulina, 55% CP	4	55.3	https://app.iaffd.com/ficd
Algae, Spirulina, 69% CP	4.8	69.1	https://app.iaffd.com/ficd
C. pyrenoidosa	13.65	40.92	Prajapati et al. (2014)

TABLE 4.2 *(Continued)*

Microalgae	CL	CP	Citation
C. sorokiniana	26	29.46	Guldhe et al. (2017)
C. sorokiniana	31.85	28.81	Ansari et al. (2017)
C. vulgaris	14–22	51–58	Becker (2007)
C. vulgaris	21.5	52	Ren et al. (2017)
C. vulgaris	14.4	37.5	Viegas et al. (2021)
Chlorella minutissima	16.32	43.78	Prajapati et al. (2014)
Chlorella sp.	14	52	Yadav et al. (2020)
DHA Gold, algae meal, DSM	53.4	15.1	https://app.iaffd.com/ficd
Isochrysis galbana	36.6	23.2	He et al. (2018)
N. gaditana	20–25	30–40	Sepúlveda et al. (2015)
N. oculata	5.6	42.2	Gamboa-Delgado et al. (2019)
Nannochloropsis salina	28	35	Chen et al. (2013)
S. obliquus	21.43	50	Ansari et al. (2015)
S. obliquus	22.6	21	Viegas et al. (2021)
Spirulina obliquus	30.85	19.52	Ansari et al. (2017)
Spirulina plantensis	1.75	66.93	Macias-Sancho et al. (2014)
Fish meal, Alaskan pollock, processing waste	7.6	69	https://app.iaffd.com/ficd

4.6 MICROALGAE AS AN ALTERNATIVE TO FISH MEAL

Microalgae may be employed as an important, viable, and alternative feed element in aquaculture feed because they are a significant source of proteins with essential amino acids, lipids with PUFA, carbohydrates, pigments, and other bioactive substances. A small amount of microalgae in aquafeed (5% dry weight) leads to a slight improvement in survival rates compared to the commercial control diet; this is due to a better immune response and a noticeably higher antioxidant content in prawn muscle, which has been demonstrated in prior studies to be a promising alternative source of fish meal and fish oil due to their biochemical composition (Li et al., 2016). Various fish and microalgal oils have been tested for their resistance to oxidation in various research, and it has been shown that microalgal oils may be much more stable than other fish oils and can even be as stable as krill oil (Ryckebosch et al., 2013). Even though the majority of the lipid and lipid-soluble nutrients have been removed, the defatted microalgae meal (DMM) may still include many useful elements for animal feed, including proteins, carbohydrates, minerals, water-soluble vitamins, and bioactive compounds (Ju et al., 2009). Researchers (Hanel et al., 2007, Ju et al., 2009) discovered that microalgae proteins may be a good supplement to aquafeeds or a viable replacement for fishmeal protein. Several species of microalgae such as *Nannochloropsis*, *Chlorella*, *Isochrysis*, *Tetraselmis*, and *Scenedesmus*, as well as other algae such as *Spirulina* and *Schizochytrium* have shown their viability as ingredients in aquaculture feeds (Shah et al., 2018, Yarnold et al., 2019, Skalli et al., 2020).

Cyprinus carpio has previously been tested using spirulina species as an alternative protein source, and diets containing 25% algal meal in place of 80% fishmeal were shown to produce equivalent or even greater growth rates (Sandbank and Hepher, 1978). Spirulina meal significantly increased growth, survival, and feed utilization in the giant freshwater shrimp (*Macrobrachium rosenbergii*) regardless of the supplementation level in the range of 5–20% in the feed, which was another species in which the effect of spirulina supplementation in the diet was studied (Nakagawa and Gómez-Díaz, 1995). Different authors replaced fish meal partially or totally with *S. platensis* in the diet of different fish and shellfish species, including Indian carps, *Catla catla* and *Labeo rohita* (Nandeesha et al., 2001), tilapia, *Oreochromis mossambicus* (Olvera-Novoa et al., 1998), red swordtail, *Xiphophorus helleri* (James et al., 2006), Siberian sturgeon, *Acipenser baeri* (Palmegiano et al., 2005), and white sturgeon, *Acipenser transmontanus* (Palmegiano et al., 2008).

Research database on the replacement of fish meal by different microalgae is given in Table 4.5.

4.7 CONCLUSION

Nutrition plays a vital role in aquaculture development. At present, there is plenty of information available on microalgae inclusion in fish feeds. In this chapter, available information on the inclusion of microalgae in fish feed has been compiled. This chapter of the book is a valuable database that will be beneficial for sustainable aquaculture production by formulating fish feed with different inclusion of microalgae.

ACKNOWLEDGMENT

The authors are thankful to the Ministry of Environment Water and Agriculture, Saudi Arabia (Applied Research Support for Enhancing Fisheries Production, Initiative No. 368) awarded to Beacon Development Department, King Abdullah University of Science and Technology, Thuwal, Jeddah, Saudi Arabia.

KEYWORDS

- **amino acid**
- **aquaculture**
- **carbohydrate**
- **microalgae**
- **supplement**
- **sustainable**

REFERENCES

Acquah, C., Tibbetts, S. M., Pan, S., & Udenigwe, C. (2020). Nutritional quality and bioactive properties of proteins and peptides from microalgae. In E. Jacob-Lopes, M. M. Maroneze, M. I. Queiroz, & L. Q. Zepka (Eds.), *Handbook of microalgae-based processes and products* (pp. 493–531). Academic Press.

Adel, M., Yeganeh, S., Dadar, M., Sakai, M., & Dawood, M. A. (2016). Effects of dietary Spirulina platensis on growth performance, humoral and mucosal immune responses and disease resistance in juvenile great sturgeon (Huso huso Linnaeus, 1754). *Fish and Shellfish Immunology, 56*, 436–444.

Alagawany, M., Taha, A.E., Noreldin, A., El-Tarabily, K.A., & Abd El-Hack, M.E. (2021. Nutritional applications of species of Spirulina and Chlorella in farmed fish: A review. *Aquaculture (Amsterdam, Netherlands), 542*, 736841.

Ansari, F. A., Shriwastav, A., Gupta, S. K., Rawat, I., Guldhe, A., & Bux, F. (2015). Lipid extracted algae as a source for protein and reduced sugar: A step closer to the biorefinery. *Bioresource Technology, 179*, 559–564.

Ansari, F. A., Singh, P., Guldhe, A., & Bux, F. (2017). Microalgal cultivation using aquaculture wastewater: Integrated biomass generation and nutrient remediation. *Algal Research, 21*, 169–177.

Barkia, I., Saari, N., & Manning, S. R. (2019). Microalgae for high-value products towards human health and nutrition. *Marine Drugs, 17*(5), 304.

Batista, S., Pereira, R., Oliveira, B., Baião, L. F., Jessen, F., Tulli, F., Messina, M., Silva, J. L., Abreu, H., & Valente, L. M. (2020). Exploring the potential of seaweed Gracilaria gracilis and microalga Nannochloropsis oceanica, single or blended, as natural dietary ingredients for European seabass Dicentrarchus labrax. *Journal of Applied Phycology, 32*(3), 2041–2059.

Becker, E. W. (2007). Micro-algae as a source of protein. *Biotechnology Advances, 25*(2), 207–210.

Becker, E. W. (2013). Microalgae for human and animal nutrition. In A. Richmond & Q. Hu (Eds.), *Handbook of microalgal culture: Applied phycology and biotechnology* (pp. 461–503). Wiley-Blackwell.

Brown, M., Jeffrey, S., Volkman, J., & Dunstan, G. (1997). Nutritional properties of microalgae for mariculture. *Aquaculture (Amsterdam, Netherlands), 151*(1–4), 315–331.

Brown, M. R. (1991). The amino-acid and sugar composition of 16 species of microalgae used in mariculture. *Journal of Experimental Marine Biology and Ecology, 145*(1), 79–99.

Cai, J., Lovatelli, A., Aguilar-Manjarrez, J., Cornish, L., Dabbadie, L., Desrochers, A., Diffey, S., Garrido Gamarro, E., Geehan, J., & Hurtado, A. (2021). Seaweeds and microalgae: An overview for unlocking their potential in global aquaculture development. *FAO Fisheries and Aquaculture Circular*, (1229), 48.

Cardinaletti, G., Messina, M., Bruno, M., Tulli, F., Poli, B., Giorgi, G., Chini-Zittelli, G., Tredici, M., & Tibaldi, E. (2018). Effects of graded levels of a blend of Tisochrysis lutea and Tetraselmis suecica dried biomass on growth and muscle tissue composition of European sea bass (Dicentrarchus labrax) fed diets low in fish meal and oil. *Aquaculture (Amsterdam, Netherlands), 485*, 173–182.

Carrero, A., Vicente, G., Rodríguez, R., del Peso, G.L., & Santos, C. (2015). Synthesis of fatty acids methyl esters (FAMEs) from Nannochloropsis gaditana microalga using heterogeneous acid catalysts. *Biochemical Engineering Journal, 97*, 119–124.

Chacón-Lee, T., & González-Mariño, G. (2010). Microalgae for "healthy" foods—possibilities and challenges. *Comprehensive Reviews in Food Science and Food Safety, 9*(6), 655–675.

Chen, C. Y., Zhao, X. Q., Yen, H. W., Ho, S. H., Cheng, C. L., Lee, D. J., Bai, F. W., & Chang, J. S. (2013). Microalgae-based carbohydrates for biofuel production. *Biochemical Engineering Journal, 78*, 1–10.

Chen, S. M., Tseng, K. Y., & Huang, C. H. (2015). Fatty acid composition, sarcoplasmic reticular lipid oxidation, and immunity of hard clam (Meretrix lusoria) fed different dietary microalgae. *Fish and Shellfish Immunology, 45*(1), 141–145.

Christaki, E., Florou-Paneri, P., & Bonos, E. (2011). Microalgae: A novel ingredient in nutrition. *International Journal of Food Sciences and Nutrition, 62*(8), 794–799.

D'Alessandro, E. B., & Antoniosi Filho, N. R. (2016). Concepts and studies on lipid and pigments of microalgae: A review. *Renewable and Sustainable Energy Reviews, 58*, 832–841.

Daroch, M., Shao, C., Liu, Y., Geng, S., & Cheng, J. J. (2013). Induction of lipids and resultant FAME profiles of microalgae from coastal waters of Pearl River Delta. *Bioresource Technology, 146*, 192–199.

de Cruz, C. R., Lubrano, A., & Gatlin III, D. M. (2018). Evaluation of microalgae concentrates as partial fishmeal replacements for hybrid striped bass Morone sp. *Aquaculture (Amsterdam, Netherlands), 493*, 130–136.

Delaporte, M., Soudant, P., Moal, J., Lambert, C., Quéré, C., Miner, P., Choquet, G., Paillard, C., & Samain, J. F. (2003). Effect of a mono-specific algal diet on immune functions in two bivalve species-Crassostrea gigas and Ruditapes philippinarum. *Journal of Experimental Biology, 206*(17), 3053–3064.

Dourou, M., Dritsas, P., Baeshen, M.N., Elazzazy, A., Al-Farga, A., & Aggelis, G. (2020). High-added value products from microalgae and prospects of aquaculture wastewaters as microalgae growth media. *FEMS Microbiology Letters, 367*(12), fnaa081.

Fadl, S. E., Elsadany, A. Y., El-Shenawy, A. M., Sakr, O. A., El Gammal, G. A., Gad, D. M., Norag, M. A. A., & Eissa, I. (2020). Efficacy of cyanobacterium Anabaena sp. as a feed supplement on productive performance and immune status in cultured Nile tilapia. *Aquaculture Reports, 17*, 100406.

FAO. (2021). *Fisheries and aquaculture software. FishStat plus - universal software for fishery statistical time series*. Rome: FAO Fisheries Division.

Farag, M. R., Alagawany, M., El-Hack, M., & Dhama, K. (2016). Nutritional and healthical aspects of Spirulina (Arthrospira) for poultry, animals and human. *International Journal of Pharmacology, 12*(1), 36–51.

Gamboa-Delgado, J., Morales-Navarro, Y. I., Nieto-López, M. G., Villarreal-Cavazos, D. A., & Cruz-Suárez, L. E. (2019). Assimilation of dietary nitrogen supplied by fish meal and microalgal biomass from Spirulina (Arthrospira platensis) and Nannochloropsis oculata in shrimp Litopenaeus vannamei fed compound diets. *Journal of Applied Phycology, 31*(4), 2379–2389.

Gasco, L., Gai, F., Maricchiolo, G., Genovese, L., Ragonese, S., Bottari, T., & Caruso, G. (2018). Supplementation of vitamins, minerals, enzymes and antioxidants in fish feeds. In L. Gasco, F. Gai, G. Maricchiolo, L. Genovese, S. Ragonese, T. Bottari & G. Caruso (Eds.), *Feeds for the aquaculture sector: Current situation and alternative sources* (pp. 63–103). Springer International Publishing, Cham.

Gbadamosi, O. K., & Lupatsch, I. (2018). Effects of dietary Nannochloropsis salina on the nutritional performance and fatty acid profile of Nile tilapia, Oreochromis niloticus. *Algal Research, 33*, 48–54.

Gladyshev, M. I., Makhutova, O. N., Kravchuk, E. S., Anishchenko, O. V., & Sushchik, N. N. (2016). Stable isotope fractionation of fatty acids of Daphnia fed laboratory cultures of microalgae. *Limnologica, 56*, 23–29.

Gong, Y., Bandara, T., Huntley, M., Johnson, Z. I., Dias, J., Dahle, D., Sørensen, M., & Kiron, V. (2019). Microalgae Scenedesmus sp. as a potential ingredient in low fishmeal diets for Atlantic salmon (Salmo salar L.). *Aquaculture (Amsterdam, Netherlands), 501*, 455–464.

Gong, Y., Guterres, H., Huntley, M., Sørensen, M., & Kiron, V. (2018). Digestibility of the defatted microalgae Nannochloropsis sp., and Desmodesmus sp. when fed to Atlantic salmon, Salmo salar. *Aquaculture Nutrition, 24*(1), 56–64.

Gong, Y., Sørensen, S. L., Dahle, D., Nadanasabesan, N., Dias, J., Valente, L. M., Sørensen, M., & Kiron, V. (2020). Approaches to improve utilization of Nannochloropsis oceanica in plant-based feeds for Atlantic salmon. *Aquaculture (Amsterdam, Netherlands), 522*, 735122.

Guldhe, A., Ansari, F. A., Singh, P., & Bux, F. (2017). Heterotrophic cultivation of microalgae using aquaculture wastewater: A biorefinery concept for biomass production and nutrient remediation. *Ecological Engineering, 99*, 47–53.

Hanel, R., Broekman, D., De Graaf, S., & Schnack, D. (2007). Partial replacement of fishmeal by lyophylized powder of the microalgae Spirulina platensis in Pacific white shrimp diets. *The Open Marine Biology Journal, 1*(1), 1–5.

Hannon, M., Gimpel, J., Tran, M., Rasala, B., & Mayfield, S. (2010). Biofuels from algae: Challenges and potential. *Biofuels, 1*(5), 763–784.

He, Y., Lin, G., Rao, X., Chen, L., Jian, H., Wang, M., Guo, Z., & Chen, B. (2018). Microalga Isochrysis galbana in feed for Trachinotus ovatus: Effect on growth performance and fatty acid composition of fish fillet and liver. *Aquaculture International, 26*(5), 1261–1280.

Ho, S. H., Chen, C. Y., & Chang, J. S. (2012). Effect of light intensity and nitrogen starvation on CO_2 fixation and lipid/carbohydrate production of an indigenous microalga Scenedesmus obliquus CNW-N. *Bioresource Technology, 113*, 244–252.

Ismail, I., Hwang, Y. H., & Joo, S. T. (2020). Meat analog as future food: A review. *Journal of Animal Science and Technology, 62*(2), 111.

James, R., Sampath, K., Thangarathinam, R., & Vasudevan, I. (2006). Effect of dietary Spirulina level on growth, fertility, coloration and leucocyte count in red swordtail, Xiphophorus helleri. *Israeli Journal of Aquaculture-Bamidgeh, 58*(2), 97–104.

Jiang, M., Zhao, H., Zai, S., Shepherd, B., Wen, H., & Deng, D. (2019). A defatted microalgae meal (Haematococcus pluvialis) as a partial protein source to replace fishmeal for feeding juvenile yellow perch Perca flavescens. *Journal of Applied Phycology, 31*(2), 1197–1205.

Jiang, X., Han, Q., Gao, X., & Gao, G. (2016). Conditions optimising on the yield of biomass, total lipid, and valuable fatty acids in two strains of Skeletonema menzelii. *Food Chemistry, 194*, 723–732.

Ju, Z., Forster, I., & Dominy, W. (2009). Effects of supplementing two species of marine algae or their fractions to a formulated diet on growth, survival and composition of shrimp (Litopenaeus vannamei). *Aquaculture (Amsterdam, Netherlands),* 292(3–4), 237–243.

Kent, M., Welladsen, H. M., Mangott, A., & Li, Y. (2015). Nutritional evaluation of Australian microalgae as potential human health supplements. *PLoS One, 10*(2), e0118985.

Knutsen, H. R., Ottesen, O. H., Palihawadana, A. M., Sandaa, W., Sørensen, M., & Hagen, Ø. (2019). Muscle growth and changes in chemical composition of spotted wolffish juveniles (Anarhichas minor) fed diets with and without microalgae (Scenedesmus obliquus). *Aquaculture Reports, 13*, 100175.

Koyande, A. K., Chew, K. W., Rambabu, K., Tao, Y., Chu, D. T., & Show, P. L. (2019). Microalgae: A potential alternative to health supplementation for humans. *Food Science and Human Wellness, 8*(1), 16–24.

Kumar, G., Nagappan, S., Das, P., AbdulQuadir, M., Thaher, M., Khan, S., Mahata, C., Al-Jabri, H., & Vatland, A. K. (2021). Potential of microalgae as a sustainable feed ingredient for aquaculture.

Kupchinsky, Z. A., Coyle, S. D., Bright, L. A., & Tidwell, J. H. (2015). Evaluation of heterotrophic algae meal as a diet ingredient for channel catfish, Ictalurus punctatus. *Journal of the World Aquaculture Society, 46*(4), 445–452.

Lee, K. J., Kim, J. D., Han, H. S., Shin, S. J., & Kim, S. S. (2015). Effects of dietary Spirulina pacifica on innate immunity and disease resistance against Edwardsiella tarda in Olive Flounder Paralichthys olivaceus. *Israeli Journal of Aquaculture-Bamidgeh, 67*, 20716.

Li, Y., Xiao, G., Mangott, A., Kent, M., & Pirozzi, I. (2016). Nutrient efficacy of microalgae as aquafeed additives for the adult black tiger prawn, P enaeus monodon. *Aquaculture Research, 47*(11), 3625–3635.

Lu, J., & Takeuchi, T. (2004). Spawning and egg quality of the tilapia Oreochromis niloticus fed solely on raw Spirulina throughout three generations. *Aquaculture (Amsterdam, Netherlands), 234*(1–4), 625–640.

Macias-Sancho, J., Poersch, L. H., Bauer, W., Romano, L. A., Wasielesky, W., & Tesser, M. B. (2014). Fishmeal substitution with Arthrospira (Spirulina platensis) in a practical diet for Litopenaeus vannamei: Effects on growth and immunological parameters. *Aquaculture (Amsterdam, Netherlands), 426*, 120–125.

Malibari, R., Sayegh, F., Elazzazy, A. M., Baeshen, M. N., Dourou, M., & Aggelis, G. (2018). Reuse of shrimp farm wastewater as growth medium for marine microalgae isolated from Red Sea-Jeddah. *Journal of Cleaner Production, 198*, 160–169.

Maliwat, G. C., Velasquez, S., Robil, J. L., Chan, M., Traifalgar, R. F., Tayamen, M., & Ragaza, J. A. (2017). Growth and immune response of giant freshwater prawn Macrobrachium rosenbergii (De Man) postlarvae fed diets containing Chlorella vulgaris (Beijerinck). *Aquaculture Research, 48*(4), 1666–1676.

Matos, Â.P., Feller, R., Moecke, E. H. S., & Sant'Anna, E. S. (2015). Biomass, lipid productivities and fatty acids composition of marine Nannochloropsis gaditana cultured in desalination concentrate. *Bioresource Technology, 197*, 48–55.

Mišurcová, L., Buňka, F., Ambrožová, J. V., Machů, L., Samek, D., & Kráčmar, S. (2014). Amino acid composition of algal products and its contribution to RDI. *Food Chemistry, 151*, 120–125.

Moreno-Garcia, L., Adjallé, K., Barnabé, S., & Raghavan, G. (2017). Microalgae biomass production for a biorefinery system: Recent advances and the way towards sustainability. *Renewable and Sustainable Energy Reviews, 76,* 493–506.

Nakagawa, H., & Gómez-Díaz, G. (1995). Usefulness of Spirulina sp. meal as feed additive for giant freshwater prawn, Macrobrachium rosenbergii. *Aquaculture Science, 43*(4), 521–526.

Nandeesha, M., Gangadhara, B., Manissery, J., & Venkataraman, L. (2001). Growth performance of two Indian major carps, catla (Catlacatla) and rohu (Labeorohita) fed diets containing different levels of Spirulina platensis. *Bioresource Technology, 80*(2), 117–120.

Norton, T. A., Melkonian, M., & Andersen, R. A. (1996). Algal biodiversity. *Phycologia, 35*(4), 308–326.

NRC. (2011). *Nutrient requirements of fish and shrimp*. National Academies Press.

Olvera-Novoa, M., Domínguez-Cen, L., Olivera-Castillo, L., & Martínez-Palacios, C. A. (1998). Effect of the use of the microalga Spirulina maxima as fish meal replacement in diets for tilapia, Oreochromis mossambicus (Peters), fry. *Aquaculture Research, 29*(10), 709–715.

Palmegiano, G. B., Agradi, E., Forneris, G., Gai, F., Gasco, L., Rigamonti, E., Sicuro, B., & Zoccarato, I. (2005). Spirulina as a nutrient source in diets for growing sturgeon (Acipenser baeri). *Aquaculture Research, 36*(2), 188–195.

Palmegiano, G. B., Gai, F., Daprà, F., Gasco, L., Pazzaglia, M., & Peiretti, P. G. (2008). Effects of Spirulina and plant oil on the growth and lipid traits of white sturgeon (Acipenser transmontanus) fingerlings. *Aquaculture Research, 39*(6), 587–595.

Parimi, N. S., Singh, M., Kastner, J. R., Das, K. C., Forsberg, L. S., & Azadi, P. (2015). Optimization of protein extraction from Spirulina platensis to generate a potential co-product and a biofuel feedstock with reduced nitrogen content. *Frontiers in Energy Research, 3*, 30.

Pascon, G., Messina, M., Petit, L., Valente, L. M. P., Oliveira, B., Przybyla, C., Dutto, G., & Tulli, F. (2021). Potential application and beneficial effects of a marine microalgal biomass produced in a high-rate algal pond (HRAP) in diets of European sea bass, Dicentrarchus labrax. *Environmental Science and Pollution Research, 28*(44), 62185–62199.

Pereira, H., Sardinha, M., Santos, T., Gouveia, L., Barreira, L., Dias, J., & Varela, J. (2020). Incorporation of defatted microalgal biomass (Tetraselmis sp. CTP4) at the expense of soybean meal as a feed ingredient for juvenile gilthead seabream (Sparus aurata). *Algal Research, 47*, 101869.

Prajapati, S. K., Malik, A., & Vijay, V. K. (2014). Comparative evaluation of biomass production and bioenergy generation potential of Chlorella spp. through anaerobic digestion. *Applied Energy, 114*, 790–797.

Priyadarshani, I., & Rath, B. (2012). Commercial and industrial applications of micro algae–A review. *Journal of Algal Biomass Utilization, 3*(4), 89–100.

Qiao, H., Cong, C., Sun, C., Li, B., Wang, J., & Zhang, L. (2016). Effect of culture conditions on growth, fatty acid composition and DHA/EPA ratio of Phaeodactylum tricornutum. *Aquaculture (Amsterdam, Netherlands), 452*, 311–317.

Qiao, H., Hu, D., Ma, J., Wang, X., Wu, H., & Wang, J. (2019). Feeding effects of the microalga Nannochloropsis sp. on juvenile turbot (Scophthalmus maximus L.). *Algal Research, 41*, 101540.

Ragaza, J. A., Hossain, M. S., Meiler, K. A., Velasquez, S. F., & Kumar, V. (2020). A review on Spirulina: Alternative media for cultivation and nutritive value as an aquafeed. *Reviews in Aquaculture, 12*(4), 2371–2395.

Raji, A. A., Jimoh, W. A., Bakar, N., Taufek, N., Muin, H., Alias, Z., Milow, P., & Razak, S. A. (2020). Dietary use of Spirulina (Arthrospira) and Chlorella instead of fish meal on growth and digestibility of nutrients, amino acids and fatty acids by African catfish. *Journal of Applied Phycology, 32*(3), 1763–1770.

Ren, H., Tuo, J., Addy, M. M., Zhang, R., Lu, Q., Anderson, E., Chen, P., & Ruan, R. (2017). Cultivation of Chlorella vulgaris in a pilot-scale photobioreactor using real centrate wastewater with waste glycerol for improving microalgae biomass production and wastewater nutrients removal. *Bioresource Technology, 245*, 1130–1138.

Renaud, S. M., Parry, D. L., & Thinh, L. V. (1994). Microalgae for use in tropical aquaculture I: Gross chemical and fatty acid composition of twelve species of microalgae from the Northern Territory, Australia. *Journal of Applied Phycology, 6*(3), 337–345.

Roy, S. S., & Pal, R. (2015). Microalgae in aquaculture: A review with special references to nutritional value and fish dietetics. *Proceedings of the Zoological Society, 68*, 1–8.

Ryckebosch, E., Bruneel, C., Termote-Verhalle, R., Lemahieu, C., Muylaert, K., Van Durme, J., Goiris, K., & Foubert, I. (2013). Stability of omega-3 LC-PUFA-rich photoautotrophic microalgal oils compared to commercially available omega-3 LC-PUFA oils. *Journal of Agricultural and Food Chemistry, 61*(42), 10145–10155.

Sahu, A., Pancha, I., Jain, D., Paliwal, C., Ghosh, T., Patidar, S., Bhattacharya, S., & Mishra, S. (2013). Fatty acids as biomarkers of microalgae. *Phytochemistry, 89*, 53–58.

Sandbank, E., & Hepher, B. (1978). The utilization of microalgae as a feed for fish. *Ergebnisse der Limnologie, 11*, 108–120.

Saragih, H., Muhamad, A., Alfianto, A., Viniwidihastuti, F., Untari, L., Lesmana, I., Widyatmoko, H., & Rohmah, Z. (2019). Effects of Spirogyra jaoensis as a dietary supplement on growth, pectoralis muscle performance, and small intestine morphology of broiler chickens. *Veterinary World, 12*(8), 1233.

Sarker, P., Gamble, M., Kelson, S., & Kapuscinski, A. (2016a). Nile tilapia (O reochromis niloticus) show high digestibility of lipid and fatty acids from marine S chizochytrium sp., and of protein and essential amino acids from freshwater S pirulina sp. feed ingredients. *Aquaculture Nutrition, 22*(1), 109–119.

Sarker, P. K., Kapuscinski, A. R., Lanois, A. J., Livesey, E. D., Bernhard, K. P., & Coley, M. L. (2016b). Towards sustainable aquafeeds: complete substitution of fish oil with marine microalga Schizochytrium sp. improves growth and fatty acid deposition in juvenile Nile tilapia (Oreochromis niloticus). *PLoS One, 11*(6), e0156684.

Sarker, P. K., Kapuscinski, A. R., Vandenberg, G. W., Proulx, E., Sitek, A. J., & Thomsen, L. (2020). Towards sustainable and ocean-friendly aquafeeds: Evaluating a fish-free feed for rainbow trout (Oncorhynchus mykiss) using three marine microalgae species. *Elementa: Science of the Anthropocene, 8*, 5.

Schwenzfeier, A., Wierenga, P. A., & Gruppen, H. (2011). Isolation and characterization of soluble protein from the green microalgae Tetraselmis sp. *Bioresource Technology, 102*(19), 9121–9127.

Seong, T., Matsutani, H., Haga, Y., Kitagima, R., & Satoh, S. (2019). First step of non-fish meal, non-fish oil diet development for red seabream, (Pagrus major), with plant protein sources and microalgae Schizochytrium sp. *Aquaculture Research, 50*(9), 2460–2468.

Sepúlveda C., Acien, F. G., Gómez, C., Jiménez-Ruiz, N., Riquelme, C., & Molina-Grima, E. (2015). Utilization of centrate for the production of the marine microalgae Nannochloropsis gaditana. *Algal Research, 9*, 107–116.

Shah, M. R., Lutzu, G. A., Alam, A., Sarker, P., Chowdhury, K., Parsaeimehr, A., Liang, Y., & Daroch, M. (2018). Microalgae in aquafeeds for a sustainable aquaculture industry. *Journal of Applied Phycology, 30*(1), 197–213.

Shay, E. G. (1993). Diesel fuel from vegetable oils: Status and opportunities. *Biomass and Bioenergy, 4*(4), 227–242.

Shi, X., Luo, Z., Chen, F., Wei, C. C., Wu, K., Zhu, X. M., & Liu, X. (2017). Effect of fish meal replacement by Chlorella meal with dietary cellulase addition on growth performance, digestive enzymatic activities, histology and myogenic genes' expression for crucian carp Carassius auratus. *Aquaculture Research, 48*(6), 3244–3256.

Skalli, A., Firmino, J. P., Andree, K. B., Salomón, R., Estévez, A., Puig, P., Sabater-Martínez, M., Hechavarria, T., & Gisbert, E. (2020). The inclusion of the microalga Scenedesmus sp. in diets for rainbow trout, Onchorhynchus mykiss, juveniles. *Animals, 10*(9), 1656.

Skrede, A., Mydland, L., Ahlstrøm, Ø., Reitan, K., Gislerød, H., & Øverland, M. (2011). Evaluation of microalgae as sources of digestible nutrients for monogastric animals. *Journal of Animal and Feed Sciences, 20*(1), 131–142.

Solana, M., Rizza, C., & Bertucco, A. (2014). Exploiting microalgae as a source of essential fatty acids by supercritical fluid extraction of lipids: Comparison between Scenedesmus obliquus, Chlorella protothecoides and Nannochloropsis salina. *The Journal of Supercritical Fluids, 92*, 311–318.

Sørensen, M., Gong, Y., Bjarnason, F., Vasanth, G. K., Dahle, D., Huntley, M., & Kiron, V. (2017). Nannochloropsis oceania-derived defatted meal as an alternative to fishmeal in Atlantic salmon feeds. *PLoS One, 12*(7), e0179907.

Spolaore, P., Joannis-Cassan, C., Duran, E., & Isambert, A. (2006). Commercial applications of microalgae. *Journal of Bioscience and Bioengineering, 101*(2), 87–96.

Stephenson, P. G., Moore, C. M., Terry, M. J., Zubkov, M. V., & Bibby, T. S. (2011). Improving photosynthesis for algal biofuels: Toward a green revolution. *Trends in Biotechnology, 29*(12), 615–623.

Suh, S. S., Kim, S. J., Hwang, J., Park, M., Lee, T. K., Kil, E. J., & Lee, S. (2015). Fatty acid methyl ester profiles and nutritive values of 20 marine microalgae in Korea. *Asian Pacific Journal of Tropical Medicine, 8*(3), 191–196.

Sultana, R., Khatoon, H., Rahman, M. R., Haque, M. E., Nayma, Z., & Mukta, F. A. (2022). Potentiality of Nannochloropsis sp. as partial dietary replacement of fishmeal on growth, proximate composition, pigment and breeding performance in guppy (Poecilia reticulata). *Bioresource Technology Reports*, 101112.

Tibbetts, S. M., Bjornsson, W. J., & McGinn, P. J. (2015). Biochemical composition and amino acid profiles of Nannochloropsis granulata algal biomass before and after supercritical fluid CO_2 extraction at two processing temperatures. *Animal Feed Science and Technology, 204*, 62–71.

Tibbetts, S. M., Patelakis, S. J., Whitney-Lalonde, C. G., Garrison, L. L., Wall, C. L., & MacQuarrie, S. P. (2020). Nutrient composition and protein quality of microalgae meals produced from the marine prymnesiophyte Pavlova sp. 459 mass-cultivated in enclosed photobioreactors for potential use in salmonid aquafeeds. *Journal of Applied Phycology, 32*(1), 299–318.

Tipnee, S., Ramaraj, R., & Unpaprom, Y. (2015). Nutritional evaluation of edible freshwater green macroalga Spirogyra varians. *Emergent Life Sciences Research, 1*(2), 1–7.

Tomás-Almenar, C., Larrán, A. M., de Mercado, E., Sanz-Calvo, M. A., Hernández, D., Riaño, B., & García-González, M. C. (2018). Scenedesmus almeriensis from an integrated system waste-nutrient, as sustainable protein source for feed to rainbow trout (Oncorhynchus mykiss). *Aquaculture, 497*, 422–430. https://doi.org/10.1016/j.aquaculture.2018.08.011.

Tsai, H. P., Chuang, L. T., & Chen, C. N. N. (2016). Production of long chain omega-3 fatty acids and carotenoids in tropical areas by a new heat-tolerant microalga Tetraselmis sp. DS3. *Food Chemistry, 192*, 682–690.

Valente, L. M. P., Custódio, M., Batista, S., Fernandes, H., & Kiron, V. (2019). Defatted microalgae (Nannochloropsis sp.) from biorefinery as a potential feed protein source to replace fishmeal in European sea bass diets. *Fish Physiology and Biochemistry, 45*(3), 1067–1081.

Van Vo, B., Siddik, M. A., Fotedar, R., Chaklader, M. R., Hanif, M. A., Foysal, M. J., & Nguyen, H. Q. (2020). Progressive replacement of fishmeal by raw and enzyme-treated alga, Spirulina platensis influences growth, intestinal micromorphology and stress response in juvenile barramundi, Lates calcarifer. *Aquaculture (Amsterdam, Netherlands), 529*, 735741.

Velasquez, S. F., Chan, M. A., Abisado, R. G., Traifalgar, R. F. M., Tayamen, M. M., Maliwat, G. C. F., & Ragaza, J. A. (2016). Dietary Spirulina (Arthrospira platensis) replacement enhances performance of juvenile Nile tilapia (Oreochromis niloticus). *Journal of Applied Phycology, 28*(2), 1023–1030.

Venkata Subhash, G., Chugh, N., Iyer, S., Waghmare, A., Musale, A. S., Nandru, R., Dixit, R. B., Gaikwad, M. S., Menon, D., & Thorat, R. (2020). Application of *in vitro* protein solubility for selection of microalgae biomass as protein ingredient in animal and aquafeed. *Journal of Applied Phycology, 32*(6), 3955–3970.

Venkataraman, L. V., & Becker, E. W. (1985). *Biotechnology and utilization of algae: The Indian experience.* Department of Science and Technology and Central Food Technological Research.

Viegas, C., Gouveia, L., & Gonçalves, M. (2021). Aquaculture wastewater treatment through microalgal. Biomass potential applications on animal feed, agriculture, and energy. *Journal of Environmental Management, 286*, 112187.

Vizcaíno, A., Rodiles, A., López, G., Sáez, M., Herrera, M., Hachero, I., Martínez, T., Cerón-García, M., & Alarcón, F. J. (2018). Growth performance, body composition, and digestive functionality of Senegalese sole (Solea senegalensis Kaup, 1858) juveniles fed diets including microalgae freeze-dried biomass. *Fish Physiology and Biochemistry, 44*(2), 661–677.

Vonshak, A. (1997). *Spirulina platensis arthrospira: Physiology, cell-biology and biotechnology*. CRC press.

Wang, Y., Guo, W. Q., Lo, Y. C., Chang, J. S., & Ren, N. Q. (2014). Characterization and kinetics of bio-butanol production with Clostridium acetobutylicum ATCC824 using mixed sugar medium simulating microalgae-based carbohydrates. *Biochemical Engineering Journal, 91*, 220–230.

Wei, M., Parrish, C. C., Guerra, N. I., Armenta, R. E., & Colombo, S. M. (2021). Extracted microbial oil from a novel Schizochytrium sp. (T18) as a sustainable high DHA source for Atlantic salmon feed: Impacts on growth and tissue lipids. *Aquaculture, 534*, 736249.

Xi, L., Lu, Q., Liu, Y., Su, J., Chen, W., Gong, Y., Han, D., Yang, Y., Zhang, Z., & Jin, J. (2022). Effects of fish meal replacement with Chlorella meal on growth performance, pigmentation, and liver health of largemouth bass (Micropterus salmoides). *Animal Nutrition,* 10, 26–40.

Yaakob, Z., Ali, E., Zainal, A., Mohamad, M., & Takriff, M. S. (2014). An overview: Biomolecules from microalgae for animal feed and aquaculture. *Journal of Biological Research-Thessaloniki, 21*(1), 1–10.

Yadav, G., Meena, D. K., Sahoo, A. K., Das, B. K., & Sen, R. (2020). Effective valorization of microalgal biomass for the production of nutritional fish-feed supplements. *Journal of Cleaner Production,* 243, 118697.

Yarnold, J., Karan, H., Oey, M., & Hankamer, B. (2019). Microalgal aquafeeds as part of a circular bioeconomy. *Trends in Plant Science, 24*(10), 959–970.

Zhou, W., Wang, Z., Xu, J., & Ma, L. (2018). Cultivation of microalgae Chlorella zofingiensis on municipal wastewater and biogas slurry towards bioenergy. *Journal of Bioscience and Bioengineering, 126*(5), 644–648.

CHAPTER 5

Application of Artificial Intelligence in Aquaculture and Fisheries

AMAL C. THAMPI[1], TAO KARA[1], SOIBAM NGASOTTER[1], SOIBAM KHOGEN SINGH[2,*], and VIDYA SHREE BHARTI[1]

[1]ICAR - Central Institute of Fisheries Education, Mumbai, Maharashtra, India

[2]College of Fisheries, Central Agricultural University (Imphal), Tripura, India

*Corresponding author

ABSTRACT

Food consumption is rising in lockstep with global population growth. It is anticipated that the demand for the world's fisheries will continue to rise as the population grows. On the other side, issues including high mortality, disease, feed spills, poor fish welfare, and a high-water usage rate plague the aquaculture business. A new branch of science known as "smart fish farming" aims to use resources wisely and advance sustainable aquaculture by thoroughly integrating the Internet of Things (IoT), big data, cloud computing, and artificial intelligence (AI). Because AI makes aquaculture a less labor-intensive business, it enables the fisheries sector to grow and improve productivity quickly. Starting with hatcheries and finishing with packing in processing facilities, AI may be used in all facets of fisheries science. AI aids in monitoring fishing activities worldwide and promotes open sea fisheries' sustainability. AI may be used in aquaculture to limit input waste and cut expenses by up to 30%. Machine learning in the form of deep learning (DL) might be applied to several issues in AI.

Recent Advances in Sustainable Aquaculture and Fisheries. Shabir Ahmad Dar, Imtiyaz Qayoom, Gohar Bilal Wani, Adnan Abubakr, and Seemab Zehra (Eds.)

DOI: 10.1201/9781998511372-5

The ability of machine learning algorithms to learn from experience is a greater benefit. As the condition of the cultural system often alters in response to the surroundings, this will aid in the amazing expansion of the fisheries sector. Thus, it appears that the application of AI in the expansion and intensification of fisheries and aquaculture cannot be avoided.

5.1 INTRODUCTION

The United Nations estimates that the global population will exceed 8.5 billion in 2030 and expand to 11.2 billion by 2100 (https://population.un.org/wpp/). The expected colossal population explosion is demanding more food in the immediate future. Fish and fishery products, which are often viewed as inexpensive sources of protein, continue to be one of the most marketed food commodities around the globe. Worldwide fish consumption grew at a rate of 3.1% per year on average between 1961 and 2017, which is nearly twice as fast as the rate of yearly global population growth (1.6% per year) (FAO, 2020). With a growing population comes a greater demand for fisheries throughout the globe. At present, the total fish production at the global level is estimated to be 178.5 million tons (FAO, 2020), with 82.1 million tons being contributed by the aquaculture sector. The annual growth rate in the world production of aquatic animals has slowed gradually in the third millennium and is estimated to be at 4.5% in 2011–2018. Thus, aquaculture has evolved into a field with rising demand and falling productivity, involving a wide range of difficulties, such as susceptibility to disease, water quality deterioration due to feed spill, poor fish welfare, high mortality of cultured fish, and a high-water utilization rate, where much scientific innovation is still required to enhance farming practices.

In the automated fish farming concept, different sensors collect relevant information like water quality, the animal's health, appetite, and machines are used to assist and carry out major operations on the farm. We can create a more stable environment for the stock with the help of modern artificial intelligence (AI) technologies. Devices will eliminate the need for a supervisor or manual labor to maintain water quality criteria, making this job considerably simpler. In AI, machines were given the ability to analyze and interpret data and make decisions. AI has been implemented in various fields, covering everything from agriculture to the fully automated production of industrial goods. Adopting AI in fisheries can bring a new frontier in the aquaculture sector capable of achieving high production within a short time.

AI, especially computer vision applications and machine learning, is the future technology for fishery data systems.

5.2 CONCEPT OF MACHINE LEARNING

When a computer program's performance on a task increases with experience, it is said to learn. A computer cannot immediately interpret the meaning inherent in raw sensory information. Direct programming cannot readily differentiate a fish from other fish or items in an underwater picture of a fish. To tackle this difficulty, deep learning (DL) divides this complicated mapping into a series of nested basic mappings.

5.3 APPLICATIONS OF DEEP LEARNING IN SMART FISH FARMING

5.3.1 IDENTIFICATION OF LIVE FISH

Photos of the desired fish taken in an uncontrolled environment may be analyzed with a DL model to extract information about the fish's characteristics. The advantages of machine vision lie in the fact that it permits cheap, long-term, non-destructive, and remote monitoring (Zhou et al., 2018a; Hartill et al., 2020).

5.3.2 CLASSIFICATION OF SPECIES

The categorization of fish species is helpful in aquaculture for managing output, forecasting yield, and monitoring the environment (Alcaraz et al., 2015; dos Santos and Goncalves, 2019). Due to the prevalence of using size, shape, and color to identify a fish species, DL models may be used for this purpose (dos Santos and Goncalves, 2019; Hu et al., 2012).

5.3.3 BEHAVIORAL RESEARCH

Fish welfare and yield are effectively determined by behavior (Zion, 2012). DL techniques have demonstrated a significant capacity to identify picture patterns (Wang et al., 2017). Meaningful behavioral monitoring can give non-destructive information and early warning of fish health, especially for atypical actions (Rillahan et al., 2011).

5.3.4 ESTIMATION OF SIZE OR BIOMASS

Estimating fish biomass without human involvement is difficult because fish are extremely sensitive and move freely in surroundings where visibility, light, and stability are frequently out of control (Li et al., 2019). In intelligent fish farming, computer vision mixed with DL can provide a more accurate assessment of fish morphological features such as length, breadth, weight, and area (Díaz-Gil et al., 2017; Marini et al., 2018).

5.3.5 FEEDING DECISION-MAKING

Feed expenses for some fish types account for more than 60% of overall expenditures in aquaculture (de Verdal et al., 2017; Føre et al., 2016; Wu et al., 2015). As a result, poor feeding reduces production efficiency, and insufficient feeding affects fish development. Overfeeding also lowers feed conversion efficiency and pollutes the environment (Zhou et al., 2018b). Consequently, improving the feeding process can result in large economic gains (Zhou et al., 2018c). Machine vision may be used to build a better feeding plan based on fish behavior. A system like this may stop feeding at more sensible periods, minimizing wasteful labor and enhancing fish well-being (Zhou et al., 2018b).

5.3.6 PREDICTION OF WATER QUALITY

The aquatic environment in practical aquaculture is characterized by several characteristics that impact each other, resulting in considerable restrictions in the process of forecasting (Liu et al., 2014). DL-based models, such as convolutional neural network (CNN) or deep belief network (DBN), can be used to extract relationships between the quantitative features of water and its quality factors (Lin et al., 2018). Water quality parameters for extensive shrimp or fish farming have been predicted using these models.

5.4 AI—A STEP TO INTELLIGENT AQUACULTURE

An AI technology platform that incorporates digitization, industrialization, mechanization, and big data is at the heart of smart aquaculture (Figure 5.1). Information is gathered using a wide variety of sensors and other forms of digital picture data acquisition devices. Data is sent to a control centre for analysis via communication nodes. The cloud is used for data processing and making decisions. Aquaculture's future might be dominated by gigantic autonomous mobile robot cages known as aquapods, such as InnovaSea's SeaStation. The "Aquapod" is a floating fish farm that can house hundreds of thousands of fish.

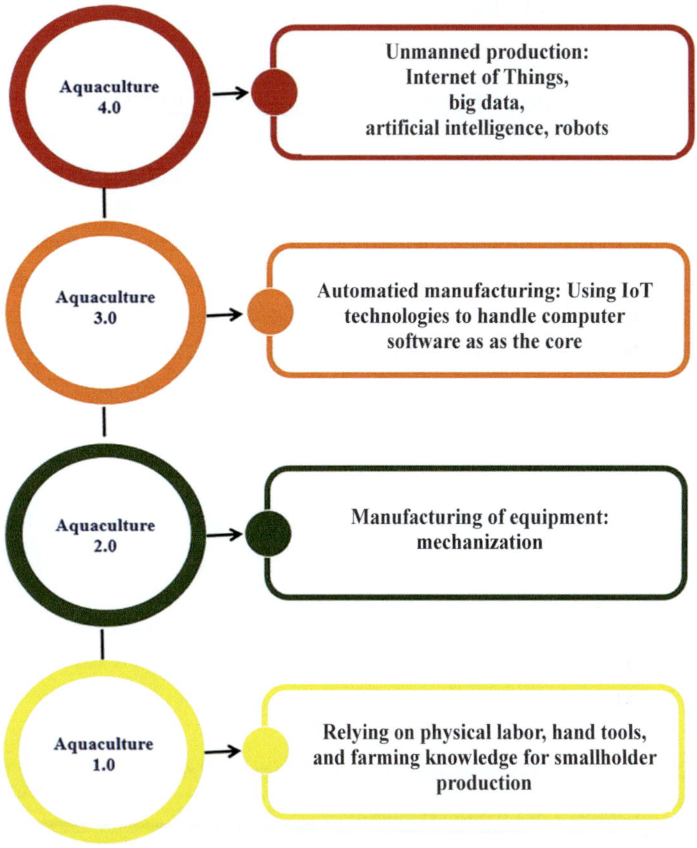

FIGURE 5.1 Evolution of aquaculture.

5.5 SENSORS FOR INTELLIGENT AQUACULTURE

Many underwater robots and drones employ sensors to direct and gather information like water oxygen levels, pH, salinity, turbidity, and contaminants. Sensorex, for example, is used by shrimp farms in India to adjust water pH and monitor dissolved oxygen levels to provide an optimal environment for enhanced shrimp efficiency and its production.

Drones connected with sensors may gather and evaluate data on water quality parameters like temperature, dissolved oxygen (DO), turbidity, and even fish heartbeat rates. Scientists created "shoal," a robotic fish that detects pollutants around the farm site. Shoal's robotic fishes use AI to locate pollution sources underwater. These data are simply accessible via a smartphone linked to the drone.

5.6 AI-BASED FISH SIZE MEASUREMENT SYSTEM

A sufficient number of fish must be sampled periodically to measure fish development, which is time-consuming for the farmer and can also stress the fish, sometimes leading to injury or death. The lens automatically measures the size of the fish in the cage using a tiny stereo camera and AI, and it connects to the internet to store the data in the cloud. This makes it considerably simpler for the farmer to monitor the fish's growth rate.

5.7 AI-BASED FEEDING DEVICES

Fish producers spend the majority of their budget on feeding; therefore, improving this area always results in increased profitability. The proper amount of food at the right time may be fed by measuring appetite. The vibration-based sensor and audio signals used to read the fish mostly depend on AI. This will aid in distinguishing between hungry and full fish.

TABLE 5.1 Global Competitors in AI Feed Dispenser Technology.

Platform	Country of Origin	Technology	Benefits
eFishery	Indonesia	AI feed dispenser	Reduce the feed cost by 21%
Observe technologies	Indonesia	Feed dispenser	Determines the amount of feed that needs to be provided by the farmers by tracking the feeding patterns of the stocks
Umitron cell	Japan and Singapore	A remote-operated, smart fish feeder	Reducing feeding costs and maintains water quality
Eruvaka	India	Appetite-based intelligent feeder	Reduce the cost of feed

5.8 VIRTUAL REALITY (VR)—ENHANCING THE VISION OF THE NEXT GENERATION

VR has various applications in the aquaculture sector, notably for training and teaching. The Norwegian University of Science and Technology (NTNU) is utilizing VR to pique the interest of the next generation in aquaculture. A new aquaculture simulator created by NTNU employs VR to let students visually tour a fish farm.

5.9 BLOCKCHAIN TO ENHANCE THE TRACEABILITY OF PRODUCTS

Blockchain is an openly available, non-enforceable digital ledger of all transactions. The ability to conduct transactions quickly and securely between suppliers and customers would greatly benefit the aquaculture sector. For tracking shrimp all the way from Andhra Pradesh, India, to the United States, Walmart Inc. has launched a blockchain technology test. Their objectives were to improve transparency and food traceability for American customers while strengthening the shrimp supply chain.

5.10 AI IN OPEN SEA FISHERIES

Overfishing and poaching have risen rapidly due to the rising population and need for food. Illegal, unreported, unregulated (IUU) fishing has significantly grown. Employing observers to keep an eye on the fishing operations taking place aboard ships is the conventional method of stopping these illicit activities. Though, in places like the Arctic, the topography and climate made it difficult for observers to trace IUU vessels. In these fields, AI is extremely important. The fishing boats may be observed through satellite technology, AI applications, and automated video inspection. Independent organization Global Fishing Watch Platform collaborated with Google, Oceana, and Sky Truth (a non-profit digital mapping company) to analyze global fishing patterns using AI and satellite data. Through this agreement, it was possible to more precisely trace IUU boats, overfishing, sea transhipments, and poaching (the transfer of cargo from one ship to another).

5.11 AI SMARTPHONE APPLICATIONS

The IoT connects all disruptive technologies. This computing and communications technology revolution enables the robot to do duties specified by a distant user or to communicate information gathered through sensors to producers for examination on smartphones, tablets, or laptops. Farm MOJO is a smartphone application developed by "Aquaconnect," an Indian aquaculture technology business that assists shrimp producers in analyzing water quality and predicting illnesses. Farmers and developers contribute photographs of shrimp illnesses, parasites, and so on a regular basis to the app. A database of disease pictures will be created for the software to study and utilize in the future.

TABLE 5.2 Overview of Global Competitors in AI.

Company/ Platform	Country of Origin	Technology	Benefits
AquaCloud	Norway	Prevention of sea lice infestation in cage culture	Successfully decreased fish mortality and reduced the need for costly interventions
Microsoft Azure Machine Learning Studio	Japan	Fish seed screening	Improved fish rearing by identifying and removing fish seeds with unusual shapes from the cage, and minimized labor
Xpertsea	Canada	Routine check-up of stocks	Uses a camera and machine learning to quickly measure and quantify (weigh, count, image, and size) shrimp in different ways
Marel	Iceland	AI-based robots	Carry out all tasks required in a manufacturing plant, including but not limited to cleaning, filleting, and packaging
SeaVax	UK	Solar-powered robotic vacuum cleaner	Pick up tons of plastic from the ocean

5.12 ADVANTAGES OF AI IN FISHERIES

1. It aids in efficiently managing aquaculture and maintains a high degree of precision in disaster forecasting (disease outbreak or depletion of water quality).

2. The use of AI in the aquaculture industry has the potential to reduce costs while increasing output, product quality, and safety.
3. Resource use is improved with intelligent aquaculture. It can reduce the amount of feed used and improve control of waste and water quality by looking at big data and making changes immediately.

5.13 LIMITATIONS OF AI IN FISHERIES

1. More resilient models must be developed to attain a fully autonomous operation system. A sensor failure or failure of one of the system's other parts can also result in crop loss and catastrophic error.
2. The expense of maintaining an AI system is also high. Additionally, a major drawback of AI is that it makes laborers unemployed.
3. Different kinds of AI dangers recognized by the World Economic Forum report are: (i) Performance risks, for applications like early warning systems for natural disasters requiring more assurance, DL could be problematic, (ii) Security risks, malicious actors may use vulnerabilities in AI to compromise critical infrastructure in energy, transportation, early warning, or other sectors, and (iii) Control risks, since AI systems operate in a mostly independent mode, they often produce unexpected outcomes. As an illustration, two systems developed a human-incomprehensible language.

5.14 CONCLUSION

AI is a rapidly growing field of technology that has the potential to revolutionize the way humanity interacts with the world. AI has given robots the capacity to evaluate, comprehend, and make decisions based on data. It has been used in many different industries, including the fisheries industry. AI is adopted in fisheries to make processes more efficient, improve safety and accuracy, and give more insight into the behavior of fish. Different sensors are utilized in an automated fish farming idea to gather pertinent data on the water quality, the health of the animals, and their appetites. Machines are also employed to help and carry out important farm tasks. Some gadgets with AI can provide a more stable environment for the stock.

The gadgets' efficient performance of both the supervisor's and the laborer's duties will simplify maintaining the water quality criteria considerably. AI-based technologies are used to identify and monitor different fish species, detect and track fish movement, map ocean and coastal environments, reduce bycatch, analyze and predict changes in the environment and the fish populations. AI and machine learning can potentially revolutionize the fisheries and aquaculture industries. These industries could see increased efficiency, economic growth, and improved fish welfare by utilizing computer vision and other AI applications. Despite potential benefits, there are still hurdles to overcome, such as the high cost of implementing AI systems and the energy consumption they require. Addressing these challenges will be crucial for successfully integrating AI into the fisheries and aquaculture industries.

KEYWORDS

- **artificial intelligence**
- **aquaculture**
- **deep learning**
- **feeding**
- **machine learning**
- **smart fish farming**
- **species**

REFERENCES

Alcaraz, C., Gholami, Z., Esmaeili, H. R., & García-Berthou, E. (2015). Herbivory and seasonal changes in diet of a highly endemic cyprinodontid fish (*Aphanius farsicus*). *Environmental Biology of Fishes, 98*, 1541–1554. https://doi.org/10.1007/s10641-015-0381-9.

De Verdal, H., Komen, H., Quillet, E., Chatain, B., Allal, F., Benzie, J. A., & Vandeputte, M. (2018). Improving feed efficiency in fish using selective breeding: A review. *Reviews in Aquaculture, 10*(4), 833–851. https://doi.org/10.1111/raq.12202

Díaz-Gil, C., Smee, S. L., Cotgrove, L., Follana-Berná, G., Hinz, H., Marti-Puig, P., Grau, A., Palmer, M., & Catalán, I. A. (2017). Using stereoscopic video cameras to evaluate seagrass meadows nursery function in the Mediterranean. *Marine Biology, 164*(6), 137. https://doi.org/10.1007/s00227-017-3169-y.

dos Santos, A. A., & Gonçalves, W. N. (2019). Improving pantanal fish species recognition through taxonomic ranks in convolutional neural networks. *Ecological Informatics, 53*, 100977. https://doi.org/10.1016/j.ecoinf.2019.100977

FAO. (2020). *Statistics-introduction*. Rome: Fisheries and Aquaculture Department. Food and Agriculture Organization of the United Nations. https://www.fao.org/publications/sofia/2020/en/ (accessed January 1, 2023)

Føre, M., Alver, M., Alfredsen, J. A., Marafioti, G., Senneset, G., Birkevold, J., Willumsen, F. V., Lange, G., Espmark, Å., & Terjesen, B. F. (2016). Modelling growth performance and feeding behaviour of Atlantic salmon (*Salmo salar* L.) in commercial-size aquaculture net pens: Model details and validation through full-scale experiments. *Aquaculture, 464*, 268–278. https://doi.org/10.1016/j.aquaculture.2016.06.045.

Hartill, B. W., Taylor, S. M., Keller, K., & Weltersbach, M. S. (2020). Digital camera monitoring of recreational fishing effort: Applications and challenges. *Fish and Fisheries, 21*(1), 204–215. https://doi.org/10.1111/faf.12413.

Hu, H., Wen, Y., Chua, T. S., & Li, X. (2014). Toward scalable systems for big data analytics: A technology tutorial. *IEEE Access, 2*, 652–687. https://doi.org/10.1109/ACCESS.2014.2332453.

Li, D., & Bao, J. (2018). Research progress on key technologies of underwater operation robot for aquaculture. *Transactions of the Chinese Society of Agricultural Engineering, 34*(16), 1–9.

Lin, Q., Yang, W., Zheng, C., Lu, K., Zheng, Z., Wang, J., & Zhu, J. (2018). Deep-learning based approach for forecast of water quality in intensive shrimp ponds. *Indian Journal of Fisheries, 65*(4), 75–80. http://dx.doi.org/10.21077/ijf.2018.65.4.72559-09.

Liu, S., Xu, L., Jiang, Y., Li, D., Chen, Y., & Li, Z. (2014). A hybrid WA–CPSO-LSSVR model for dissolved oxygen content prediction in crab culture. *Engineering Applications of Artificial Intelligence, 29*, 114–124. https://doi.org/10.1016/j.engappai.2013.09.019.

Marini, S., Fanelli, E., Sbragaglia, V., Azzurro, E., Del Rio Fernandez, J., & Aguzzi, J. (2018). Tracking fish abundance by underwater image recognition. *Scientific Reports, 8*(1), 13748. https://doi.org/10.1038/s41598-018-32089-8.

Rillahan, C., Chambers, M. D., Howell, W. H., & Watson III, W. H. (2011). The behavior of cod (*Gadus morhua*) in an offshore aquaculture net pen. *Aquaculture, 310*(3–4), 361–368. https://doi.org/10.1016/j.aquaculture.2010.10.038.

Wang, S. H., Zhao, J. W., & Chen, Y. Q. (2017). Robust tracking of fish schools using CNN for head identification. *Multimedia Tools and Applications, 76*, 23679–23697. https://doi.org/10.1007/s11042-016-4045-3.

Wu, T. H., Huang, Y. I., & Chen, J. M. (2015). Development of an adaptive neural-based fuzzy inference system for feeding decision-making assessment in silver perch (*Bidyanus bidyanus*) culture. *Aquacultural Engineering, 66*, 41–51. https://doi.org/10.1016/j.aquaeng.2015.02.001.

Zhou, C., Lin, K., Xu, D., Chen, L., Guo, Q., Sun, C., & Yang, X. (2018a). Near infrared computer vision and neuro-fuzzy model-based feeding decision system for fish in aquaculture. *Computers and Electronics in Agriculture, 146*, 114–124. https://doi.org/10.1016/j.compag.2018.02.006.

Zhou, C., Sun, C., Lin, K., Xu, D., Guo, Q., Chen, L., & Yang, X. (2018b). Handling water reflections for computer vision in aquaculture. *Transactions of the ASABE, 61*(2), 469–479.

Zhou, C., Xu, D., Lin, K., Sun, C., & Yang, X. (2018c). Intelligent feeding control methods in aquaculture with an emphasis on fish: A review. *Reviews in Aquaculture, 10*(4), 975–993. https://doi.org/10.1111/raq.12218.

Zion, B. (2012). The use of computer vision technologies in aquaculture–a review. *Computers and Electronics in Agriculture, 88*, 125–132. https://doi.org/10.1016/j.compag.2012.07.010.

CHAPTER 6

Nanobubble Technology for Sustainable Aquaculture

MADHULIKA[1], SOIBAM NGASOTTER[1], K. A. MARTIN XAVIER[2], and SOIBAM KHOGEN SINGH[3,*]

[1]*ICAR - Central Institute of Fisheries Education, Mumbai, India*

[2]*ICAR - Central Institute of Fisheries Technology, Cochin, India*

[3]*College of Fisheries, Central Agricultural University (Imphal), Tripura, India*

*Corresponding author

ABSTRACT

Nanobubbles (NBs) are nanoscopic gaseous zones that can exist in large liquids or on solid surfaces and are distinguished by their extraordinary stability over time. The exceptional stability, large surface-to-volume ratio, rapid gas dissolution rate, and ability to generate free radicals are unique properties of NBs; these properties offer a plethora of opportunities in various sectors. Due to its potential applications across several industries, NB technology has recently attracted a lot of attention. In aquaculture, NBs present a promising way to implement sustainable practices and improve aquatic organism performance by contributing to water treatment, water management, disease control, better aquaculture production, health of fish, and many other applications. Overall, the use of NB technology in sustainable aquaculture practices offers a viable path to improve performance, strengthen environmental conservation, and promote aquatic species' general well-being. This chapter explores the

Recent Advances in Sustainable Aquaculture and Fisheries. Shabir Ahmad Dar, Imtiyaz Qayoom, Gohar Bilal Wani, Adnan Abubakr, and Seemab Zehra (Eds.)

DOI: 10.1201/9781998511372-6

above-mentioned unique characteristics and potential applications of NBs for sustainable aquaculture.

6.1 INTRODUCTION

The aquaculture industry is rapidly growing day by day and it plays a vital role in meeting the growing global demand for seafood. However, it faces numerous challenges related to water quality, disease control, and environmental sustainability. In this context, nanobubble (NB) technology offers unique advantages and addresses several critical needs in aquaculture. NB technology has garnered significant attention due to their distinctive physicochemical properties, including their minute size ($< 1\ \mu m$), surface charge, and high internal pressure (Phan et al., 2020). These properties present new possibilities in various fields, such as environmental aquaculture, hydroponics, and related domains. Moreover, the utilization of reactive oxygen species (ROS) derived from NBs has shown potential in water treatment by inactivating pathogens, eliminating harmful microorganisms from food, and removing persistent organic pollutants from wastewater (with removal efficiencies exceeding 60%) (Marcelino et al., 2022). Furthermore, NB technology enables the maintenance of higher levels of dissolved oxygen (DO) in aqueous environments compared to conventional aeration methods.

NBs, which are nanoscopic gaseous domains can be observed either on solid surfaces or within liquid environments, have particularly captured interest due to their long-term stability and promising practical applications (Phan et al., 2020; Nazari et al., 2022; Haris et al., 2020). NBs, also known as ultrafine bubbles, are defined as nanoscopic gaseous domains enclosed within an interface by the International Organization for Standardization (ISO) (ISO, 2017). The production and utilization of NBs have gained significant attention in the past two decades, with a growing focus on understanding the principles and fundamentals of NBs (Nazari et al., 2022; Haris et al., 2020). While the topic of "Applications of NBs" is popular, much of the current scientific research has primarily focused on studying the basic properties and characteristics of NBs (Nazari et al., 2022; Haris et al., 2020). The advancement of NB technology has sparked significant interest and excitement in aquaculture. With their unique characteristics, NBs offer a range of potential applications in improving various aspects of aquaculture operations. From water quality management to fish health and growth enhancement, NB technology holds great promise for optimizing production systems, mitigating environmental risks, and promoting sustainable aquaculture practices. The continuous development and exploration of NB technology in aquaculture have the potential to revolutionize the industry and contribute to its long-term success.

6.2 FUNDAMENTALS AND PRODUCTION OF NANOBUBBLES

NBs refer to minuscule gaseous pockets, usually containing air, that exist within aqueous solutions. These bubbles have the ability to alter the typical properties of water. NBs can be categorized into two types: surface NBs and bulk NBs. Surface NBs and bulk NBs are extremely small gas cavities with diameters ranging from 10 to 200 nm (Demangeat, 2015). Some researchers define NBs as ultrafine bubbles, with diameters ranging from 1 to 1000 nm, while micro- and macro-bubbles have larger diameters of 10–50 μm or more (Phan et al., 2020; Ghadimkhani et al., 2016). NBs display various distinct physical and mechanical traits, including a reduced or almost absent buoyancy, an exceptionally high surface area to volume ratio, zeta potentials, improved oxygen solubility in water, the production of free radicals, and a slow rising velocity. The fact that NBs could exist for a few weeks or more is one of their unique characteristics (Demangeat, 2015). Because of their small size and versatility, NBs are excellent for various biological applications.

In order to create NBs, liquids must first become oversaturated with dissolved gas. This can be accomplished using a variety of techniques, including adjusting the temperature and pressure, mechanical agitation, solvent exchange, and gas purging (Hansen et al., 2022). NBs frequently build up when a homogeneous liquid experiences a phase shift as a result of a sharp drop in pressure below a critical level, a process known as cavitation (Zhu et al., 2018). Acoustic cavitation, hydrodynamic cavitation caused by high-pressure changes in moving liquids, and chemical reactions like electrolysis and conical-geometry methods are all examples of ways to cause cavitation. As illustrated schematically in Figure 6.1, bulk nanobubbles (BNBs) can be produced utilizing the self-designed periodic pressure change device, which is primarily composed of an electric motor, piston, and transparent U-tube. According to working theory, the periodic fluctuation of internal pressure within the U-tube is induced by the linear motion of the piston, which is converted into movement by the eccentric wheel driven by a motor. This dynamic process causes a modulation in the solubility of gas within the aqueous solution, ultimately leading to the generation of NB (Wang et al., 2019). Another intriguing technique includes creating external static electric fields based on electrostriction, which draws gas molecules to gas–liquid interfaces (English, 2022). It is possible to create "ultra-dense" NBs at the liquid–solvent interface owing to the electrostrictively produced transient negative-pressure area, which are more stable and have longer lives than BNBs (Tan et al., 2021).

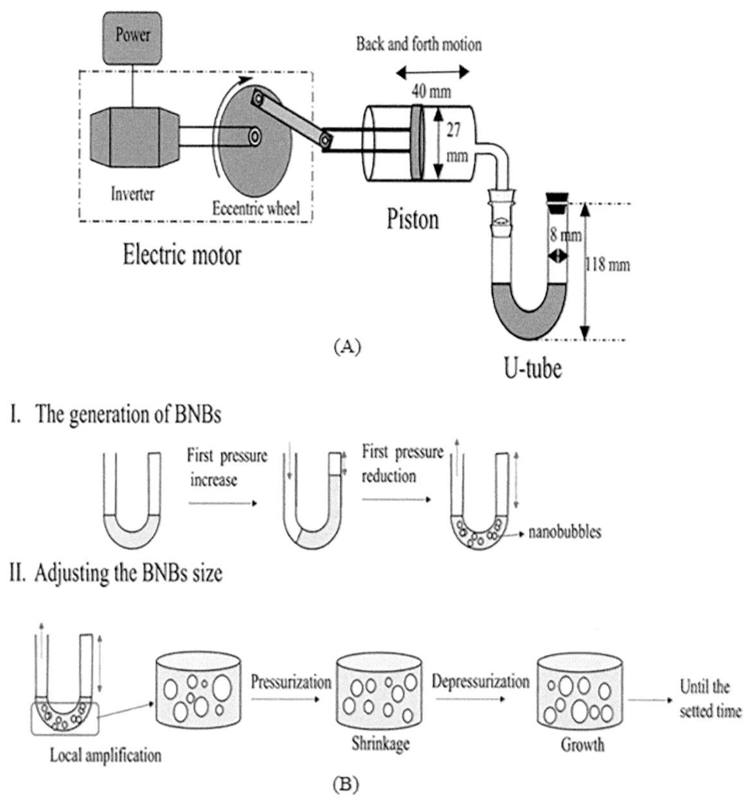

FIGURE 6.1 The schematic diagram of generating bulk nanobubbles (A). The diagram of the possible mechanism of generating bulk nanobubbles by periodic pressure change at room temperature (B).

Source: Reprinted from Wang et al. (2019). https://creativecommons.org/licenses/by/4.0/

6.3 NANOBUBBLES APPLICATIONS IN AQUACULTURE

6.3.1 WATER MANAGEMENT

NBs possess numerous attributes that make them suitable for a wide range of applications. Among these applications, water treatment is experiencing rapid growth as a technology. Various contaminants, such as heavy metals, fluoride, salts, selenium, natural organic matter, nitrate, phosphate, solvents, fuel additives, pharmaceuticals, viruses, bacteria, clays, and more, contribute to the toxicity or aesthetic issues that render rivers, lakes, seawater, groundwater, and wastewater unsuitable for beneficial use or reuse within intricate and constantly changing urban and rural water systems (Favvas et al., 2021). Additionally, NBs can aid in the processing of both organic and inorganic waste within cultivation containers. By supplying DO to the water, NBs support the activity of microbes responsible for decomposing organic waste (Foundas et al., 2022). Over the past few decades, remediation techniques for each of the aforementioned sources of pollution have been developed, including flotation, flocculation, aeration, and other treatment technologies (Qadri et al., 2020). In order to do this, floating island solar-powered NB generation technology, illustrated in Figure 6.2, can be used as a solution or a supplement to traditional technologies.

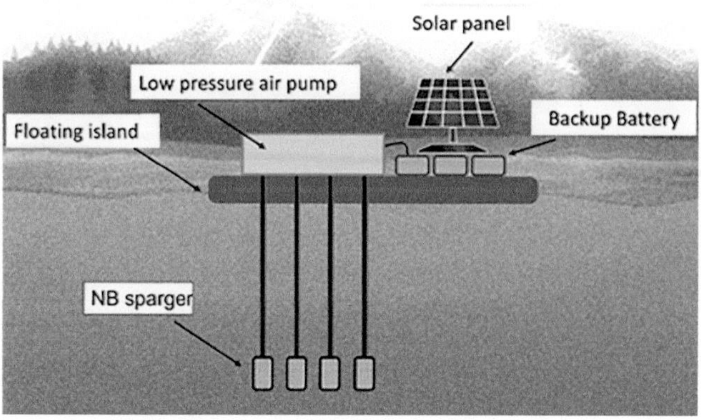

FIGURE 6.2 Floating island solar-powered nanobubble generation for water management.

Source: Reprinted from English, 2022.

6.3.2 GROWTH AND PERFORMANCE OF CULTURED SPECIES

A crucial factor for successful fish cultivation is maintaining a well-oxygenated environment. NB technology offers a solution by increasing the DO concentration in the water used for cultivation (Figure 6.3) (Marcelino et al., 2022; Mahasri et al., 2018). The application of oxygen NBs in closed or open-circulating aquaculture systems has shown several beneficial effects. The utilization of NBs in aquaculture holds great promise for improving the growth and performance of cultured fish. With their small size and large surface area, NBs have the ability to efficiently supply oxygen and other gases to the aquatic environment. NBs have been demonstrated to increase DO levels and reduce nitrogen content while also converting nitrite to nitrate and decomposing nitrogen, carbon dioxide, and hydrogen sulfide (Moleaer, 2020). This increased availability of DO is essential for supporting the metabolic processes and overall health of fish, leading to enhanced growth rates, improved feed conversion efficiency, and greater survival rates. By creating a stable and optimal environment, NBs play a significant role in promoting the well-being and productivity of cultured fish in aquaculture systems. These include maintaining high levels of DO, inhibiting the growth of anaerobic and facultative bacteria, and promoting faster growth of aquaculture species (Nghia et al., 2021).

FIGURE 6.3 Nanobubble generator (picture on the left) being used to increase the concentration of dissolved oxygen in the water for fish cultivation.

Source: https://thefishsite.com/articles/a-breath-of-fresh-air-how-nanobubbles-can-make-aquaculture-more-sustainable-dissolved-oxygen

Fish perform better in terms of growth when adequate oxygen levels are maintained in the water owing to the inclusion of nano-oxygen bubbles (Rahmawati et al., 2021). There are specific instances of how the use of NBs increased the growth of fish and prawns. For example, it has been demonstrated that NBs enhance the utilization of DO in white-leg prawns (*Litopenaeusvannamei*) (Galang et al., 2019). According to studies by Ebina et al. (2013), rainbow trout (*Oncorhynchus mykiss*) and sweet fish (*Plecoglossusaltivelis*) cultivated in water treated with air NBs grew better than those in untreated water. Koi fish (*Cyprinus carpio*) and red seabream (*Pagrus major*) have benefited from aeration with micro-NBs as well (Stander, 2018). While the private sector and various organizations have reported early applications of NB technology in the cultivation of oysters and prawns, there is a scarcity of scientific research focusing on this specific aspect (Foundas et al., 2022). Additionally, NBs can help with waste processing in cultivation containers for both organic and inorganic waste. NBs encourage the activity of the bacteria responsible for digesting organic waste by giving DO to the water.

6.3.2.1 DISEASE CONTROL

Due to ignorance of the benefits of this technology, the use of NB-O3 (nanobubbles containing ozone) in aquaculture is now restricted. NB-O3 has been shown to have a number of positive effects on water quality, including an improvement in DO levels without having a detrimental effect on the health of aquaculture species, according to earlier research (Mahasri et al., 2018). It has been discovered that short-term exposure to NB-O3 substantially promotes immune-related genes involved in Nile tilapia's innate defence system. These genes, notably the bacteriolytic enzyme C-type lysozyme (LYZ), heat-shock proteins (HSP70, HSP90), and pro-inflammatory cytokines (IL-1, IL-2, and TNF-), play critical roles in triggering the immune system and facilitating early responses to infections (Linh et al., 2021). In order to lower the danger of infectious diseases in fish and prawns, some studies have reported the possible uses of NB-O3 in pathogen disinfection in aquaculture water (Imaizumi et al., 2018; Jhunkeaw et al., 2021; Kurita et al., 2017). Here, a 1 L/min oxygen flow from the oxygen concentrator is supplied to the ozone generator. After being diffused with water inside the NB generator, the produced ozone is subsequently returned to the tank (Figure 6.4) (Jhunkeaw et al., 2021). Additionally, recent research has shown an additional benefit of NB-O_3 in modulating the innate immune defence system of Nile tilapia to combat pathogens like *Streptococcus agalactiae* (Linh et al., 2021). Furthermore, studies have demonstrated that NB-O_3 treatment can enhance the survivability of Nile tilapia challenged with drug-resistant *Aeromonashydrophila*, *S. agalactiae*, and *Vibrio parahaemolyticus* (Thanh Dien et al., 2021; Linh et al., 2021; Nghia et al., 2021).

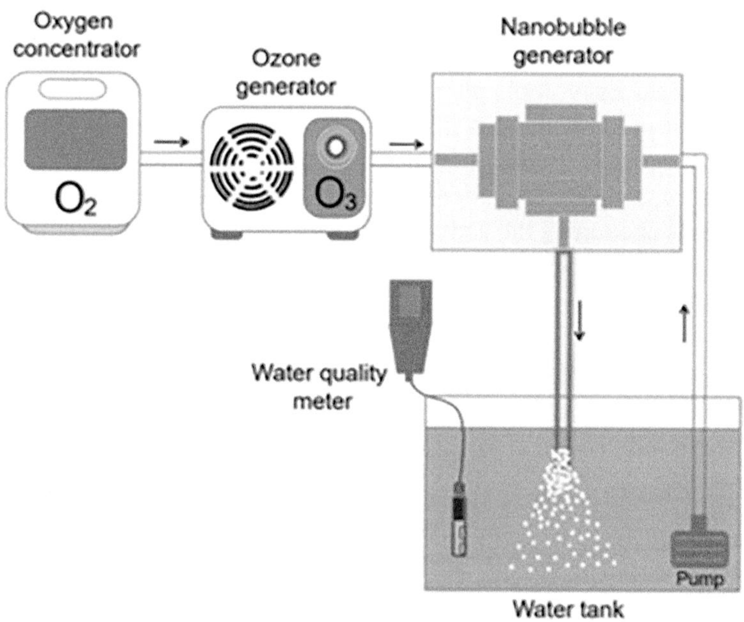

FIGURE 6.4 Laboratory setup of NB-O_3 system.

Source: Reprinted from Jhunkeaw et al. (2021) https://creativecommons.org/licenses/by/4.0/

6.3.3 *EFFECT ON FISH HEALTH*

NBs have been found to play a beneficial role in aquaculture by enhancing nutrient delivery due to the ability to deliver materials in a controlled way. The small size and large surface area of NBs allow for increased contact with water, leading to improved nutrient absorption and availability for aquatic organisms (Favvas et al., 2021). Several research studies have demonstrated the positive effects of NBs on nutrient delivery. Fish health may be enhanced by higher levels of DO during and after treatments in the NB-O_3-treated groups by preserving or enhancing normal physiological functioning. According to studies, fish with high oxygen levels have better immune systems (Thanh Dien et al., 2021). According to Romano et al. (2017), sea bass (*Dicentrarchuslabrax*) performed better in their immunological response when exposed to 12–13 mg/L oxygen. A combination of the synergistic effects of reduced bacterial load, increased DO, and stimulation of the fish immune response may also be accountable for the enhanced survivability of fish subjected to NB-O_3 treatment.

6.3.4 AQUAPONICS

Hydroponics is known for its efficient water usage and faster plant growth (Sambo, 2019). Aquaponics, on the other hand, is a system that combines aquaculture with the principles of recirculation, integrating the production of hydroponic plants or vegetables. The use of NBs in hydroponics can positively impact plant growth promotion, influenced by various factors. Research has demonstrated that NBs can regulate the DO levels in the solution, leading to improved leaf and stem growth (Kobayashi and Yamaji, 2022). Compared to traditional aeration methods, NBs have been shown to elicit better morphological and physiological responses in plants. With their high oxygen-carrying capacity, NBs can significantly increase the DO levels in the water, which is vital for the health and growth of both fish and plants within an aquaponics system.

6.4 CONCLUSION AND FUTURE PERSPECTIVE

The exceptional stability, excellent surface-to-volume ratio, quick gas dissolution, and capacity to produce free radicals are distinctive characteristics of NBs that present substantial potential to alter the physical characteristics and environmental factors of aqueous solutions. These properties make NBs highly relevant and promising for applications in aquaculture. However, to fully understand the underlying principles of NBs and their interactions with complex culture systems, further research is needed. Research should focus on sustainable production methods, expanding the range of applications, and scaling up the technology for large-scale industrial operations. To comprehensively explore the potential applications of

NBs in the fisheries and aquaculture industry, it is essential to conduct additional research. Despite the promising and diverse applications of NB technology in aquaculture, there is currently a limited number of ongoing projects in this field. However, in the next ten years, there is a good chance that NBs will find commercial application because the probability of them moving from the lab to the market is high enough.

KEYWORDS

- **aquaponics**
- **disease control**
- **fish health**
- **nanobubble**
- **water management**

REFERENCES

Demangeat, J. L. (2015). Gas nanobubbles and aqueous nanostructures: the crucial role of dynamization. *Homeopathy, 104*(02), 101–115.

Ebina, K., Shi, K., Hirao, M., Hashimoto, J., Kawato, Y., Kaneshiro, S., Morimoto, T., Koizumi, K., & Yoshikawa, H. (2013). Oxygen and air nanobubble water solution promote the growth of plants, fishes, and mice. *PLoS One, 8*(6), e65339.

English, N. J. (2022). Environmental exploration of ultra-dense nanobubbles: Rethinking sustainability. *Environments, 9*(3), 33.

Favvas, E. P., Kyzas, G. Z., Efthimiadou, E. K., & Mitropoulos, A. C. (2021). Bulk nanobubbles, generation methods and potential applications. *Current Opinion in Colloid & Interface Science, 54*, 101455.

Foundas, A., Kosheleva, R. I., Favvas, E. P., Kostoglou, M., Mitropoulos, A. C., & Kyzas, G. Z. (2022). Fundamentals and applications of nanobubbles: A review. *Chemical Engineering Research and Design, 189*, 64–86.

Galang, D. P., Ashari, A. K., Sulmatiwi, L., Mahasri, G., & Prayogoand Sari, L. A. (2019). The oxygen content and dissolved oxygen consumption level of white shrimp *Litopenaeusvannamei*in the nanobubble cultivation system. *IOP Conference Series: Earth and Environmental Science, 236*, 012014.

Ghadimkhani, A., Zhang, W., & Marhaba, T. (2016). Ceramic membrane defouling (cleaning) by air Nano Bubbles. *Chemosphere, 146*, 379–384.

Hansen, H. H., Cha, H., Ouyang, L., Zhang, J., Jin, B., Stratton, H., Nguyen, N. T., & An, H. (2022). Nanobubble technologies: Applications in therapy from molecular to cellular level. *Biotechnology Advances, 63*, 108091.

Haris, S., Qiu, X., Klammler, H., & Mohamed, M. M. (2020). The use of micro-nano bubbles in groundwater remediation: A comprehensive review. *Groundwater for Sustainable Development, 11*, 100463.

Imaizumi, K., Tinwongger, S., Kondo, H., & Hirono, I. (2018). Disinfection of an EMS/AHPND strain of *Vibrio parahaemolyticus* using ozone nanobubbles. *Journal of Fish Diseases, 41*(4), 725–727.

ISO. (2017). Fine bubble technology–General principles for usage and measurement of fine bubbles–Part 1: Terminology. International Organization for Standardization, Geneva.

Jhunkeaw, C., Khongcharoen, N., Rungrueng, N., Sangpo, P., Panphut, W., Thapinta, A., Senapin, S., St-Hilaire, S., & Dong, H. T. (2021). Ozone nanobubble treatment in freshwater effectively reduced pathogenic fish bacteria and is safe for Nile tilapia (*Oreochromis niloticus*). *Aquaculture, 534*, 736286.

Kobayashi, N., & Yamaji, K. (2022). Leaf lettuce (*Lactuca sativa*L.'L-121') growth in hydroponics with different nutrient solutions used to generate ultrafine bubbles. *Journal of Plant Nutrition, 45*(6), 816–827.

Kurita, Y., Chiba, I., & Kijima, A. (2017). Physical eradication of small planktonic crustaceans from aquaculture tanks with cavitation treatment. *Aquaculture International, 25*, 2127–2133.

Linh, N. V., Panphut, W., Thapinta, A., Senapin, S., St-Hilaire, S., Rodkhum, C., & Dong, H. T. (2021). Ozone nanobubble modulates the innate defense system of Nile tilapia (*Oreochromis niloticus*) against *Streptococcus agalactiae*. *Fish & Shellfish Immunology, 112*, 64–73.

Mahasri, G., Saskia, A., Apandi, P. S., Dewi, N. N., & Usuman, N. M. (2018). Development of an aquaculture system using nanobubble technology for the optimation of dissolved oxygen in culture media for nile tilapia (*Oreochromis niloticus*). *IOP Conference Series: Earth and Environmental Science, 137*(1), 012046.

Marcelino, K. R., Ling, L., Wongkiew, S., Nhan, H. T., Surendra, K. C., Shitanaka, T., Lu, H., & Khanal, S. K. (2022). Nanobubble technology applications in environmental and agricultural systems: Opportunities and challenges. *Critical Reviews in Environmental Science and Technology, 53*(14), 1378–1403.

Moleaer. (2020). Nanobubbles case studies. https://www.nanobubblesystems.com/case-studies

Nazari, S., Hassanzadeh, A., He, Y., Khoshdast, H., & Kowalczuk, P. B. (2022). Recent developments in generation, detection and application of nanobubbles in flotation. *Minerals, 12*(4), 462.

Nghia, N. H., Van, P. T., Giang, P. T., Hanh, N. T., St-Hilaire, S., & Domingos, J. A. (2021). Control of *Vibrio parahaemolyticus* (AHPND strain) and improvement of water quality using nanobubble technology. *Aquaculture Research, 52*(6), 2727–2739.

Phan, K. K. T., Truong, T., Wang, Y., & Bhandari, B. (2020). Nanobubbles: Fundamental characteristics and applications in food processing. *Trends in Food Science & Technology, 95*, 118–130.

Qadri, H., Bhat, R. A., Mehmood, M. A., & Dar, G. H. (2020). In *Fresh water pollution dynamics and remediation*. Springer, Singapore.

Rahmawati, A. I., Saputra, R. N., Hidayatullah, A., Dwiarto, A., Junaedi, H., Cahyadi, D., Saputra, H. K. H., Prabowo, W. T., Kartamiharja, U. K. A., Shafira, H., & Noviyanto, A. (2021). Enhancement of *Penaeus vannamei* shrimp growth using nanobubble in indoor raceway pond. *Aquaculture and Fisheries, 6*(3), 277–282.

Romano, N., Scapigliati, G., & Abelli, L. (2017). Water oxygen content affects distribution of T and B lymphocytes in lymphoid tissues of farmed sea bass (*Dicentrarchuslabrax*). *Fishes, 2*(3), 16.

Sambo, P., Nicoletto, C., Giro, A., Pii, Y., Valentinuzzi, F., Mimmo, T., Lugli, P., Orzes, G., Mazzetto, F., Astolfi, S., & Terzano, R. (2019). Hydroponic solutions for soilless production systems: Issues and opportunities in a smart agriculture perspective. *Frontiers in Plant Science, 10*, 923.

Stander, H. (2018). Nanobubble technology in aquaculture production systems. https://agriabout.com/nanobubble-techn ology-in-aquaculture-production-systems/

Tan, B. H., An, H., & Ohl, C. D. (2021). Stability of surface and bulk nanobubbles. *Current Opinion in Colloid & Interface Science, 53*, 101428.

Thanh Dien, L., Linh, N. V., Sangpo, P., Senapin, S., St-Hilaire, S., Rodkhum, C., & Dong, H. T. (2021). Ozone nanobubble treatments improve survivability of Nile tilapia (*Oreochromis niloticus*) challenged with a pathogenic multi-drug-resistant *Aeromonas hydrophila. Journal of Fish Diseases, 44*(9), 1435–1447.

Wang, Q., Zhao, H., Qi, N., Qin, Y., Zhang, X., & Li, Y. (2019). Generation and stability of size-adjustable bulk nanobubbles based on periodic pressure change. *Scientific Reports, 9*(1), 1118.

Zhu, Z., Sun, D. W., Zhang, Z., Li, Y., & Cheng, L. (2018). Effects of micro-nano bubbles on the nucleation and crystal growth of sucrose and maltodextrin solutions during ultrasound-assisted freezing process. *LWT, 92*, 404–411.

CHAPTER 7

Incorporating Virtual Reality in Fisheries Education

S. ABUTHAGIR IBURAHIM[1,*], B. CHANIKYA NAIDU[1], P. S. ANANTHAN[1], T. PAUL NATHANIEL[1], KAMIL AKAMAD[1], KARANKUMAR RAMTEKE[1], and R. BHARATHI RATHINAM[2]

[1]*ICAR-Central Institute of Fisheries Education, Mumbai, Maharashtra, India*

[2]*Central Institute of Fisheries Education, Versova, Mumbai, India*

*Corresponding author

ABSTRACT

The way students learn is evolving. There have been significant changes to the classroom environment during the past few years. The evolution of whiteboards from its chalkboard predecessors to the cutting-edge technology of today's interactive models is indicative of the rapid pace of change in educational technology. Using virtual reality (VR) in the classroom allows teachers to give their pupils an experience that is so real and so realistic that it sticks with them long after the class is over. These VR activities not only help students learn more effectively, but also develop crucial interpersonal abilities including empathy, teamwork, and social skills. The use of VR in the classroom is revolutionary and holds great promise for the future of fisheries education. There are various VR models that have to be developed in fisheries and the fisheries education institutions of India have a big task in front of them in the coming future.

Recent Advances in Sustainable Aquaculture and Fisheries. Shabir Ahmad Dar, Imtiyaz Qayoom, Gohar Bilal Wani, Adnan Abubakr, and Seemab Zehra (Eds.)

DOI: 10.1201/9781998511372-7

7.1 INTRODUCTION

The idea of virtual reality (VR) was first presented in 1965, when Ivan Sutherland, an inventor, aimed at providing a virtual realistic environment to the user through a head-mounted device, which he called as the "window into a virtual world." VR can be defined as follows "real-time interactive graphics with three-dimensional models, combined with a display technology that gives the user the immersion in the model world and direct manipulation" (Fuchs and Bishop, 1992). It enables the users to interact, sense, and experience the digital simulation created inside the VR. In recent times, this technology has been increasingly adopted in a plethora of sectors ranging from education, entertainment, and services. VR is used in a variety of contexts, from giving medical and sports training (Harrington et al., 2018; Panchuk, et al., 2018) to teaching languages and about cultural traditions (Ibañez-Etxeberria, et al., 2020; Parmaxi, 2020) worldwide. In particular, it has gained popularity in the classrooms that can be attributed to several factors, most notably its ability to bring users to a new world and encourage creative problem solving (Gavish et al., 2015). While omitting certain components that could impede learning, Blascovich et al. (2002) note that its execution allows the student to be placed in numerous contexts with a realism that could never be reached with a textbook. In the classroom, it can transport students to any location or time period (Martínez, 2011). According to Cuesta Cambra and Mañas Viniegra (2016), this technology can be used to overcome the traditional spatial and temporal limitations of the classroom, making it possible for students to engage in meaningful, hands-on learning. Immersion and presence are two fundamental ideas in VR. Despite their frequent synonymy, "immersion" refers formally to the usage of immersive technology (Jensen and Konradsen, 2018) and "presence" refers formally to the user's response to the technology (Slater, 2003). Thus, immersion is regarded as the standout feature of VR among the two because of its uniqueness over available softwares (Mikropoulos and Strouboulis, 2004; Kalyvioti and Mikropoulos, 2014; Webster, 2016). However, VR comes in many forms. The degree of immersion a user has in a VR experience will depend on the features and capabilities of the technology behind the experience (Bowman and Mcmahan, 2007). Systems with a high degree of immersion in VR are called "immersive VR" (IVR). The technology provides a greater realistic experience as if the user is within the virtual mode itself. It is realistic, with constant upgradation of scenes in response to the user's head and body movement. In contrast, semi-immersive VR systems allow users to feel as though they are only partially involved in the virtual world. It accomplishes this goal by enhancing either the sensory inputs or the users' active participation in the virtual world (di Natale et al., 2020). Finally, the nonimmersive VR systems are reality generated solely in an electronic device such as a computer or laptop, where the user can interact using a mouse or keyboard. These desktop VR systems are also referred to as 3D worlds (Freina and Ott, 2015).

7.2 LATEST TRENDS IN VR

The VR and augmented reality (AR) markets are set to grow exponentially in the coming years as per many reports in different sectors. The VR and AR market has increased at a CAGR of 77% from 2019 to 2023, as per international data corporation (IDC) research. Increase in cheaper internet connectivity and mobile gaming will boost the VR market value up to 571 billion USD by 2025, with a CAGR of 63.3% (valuates report). VynZ Research estimated a CAGR of 48.8% having 161.8 billion USD revenue during 2020–2025 and concluded Asia-Pacific as the fastest growing market during the period, with North America having majority of shares in the sector. Major players in the VR and AR market are oculus VR, Alphabet Inc., Microsoft Corporation, LLC, Intel Corporation, Qualcomm Technologies Inc., Samsung Electronics Co. Ltd., Himax Technologies Inc., and Sony Corporation. The market will continue to grow in producing AR and VR headsets, their chips and variety of contents for VR/AR headsets, smartphones, and other devices that can support these contents. This report (by all the research) segments the contents into industry-based 360° videos having application in training, monitoring, and 3D modeling; games; and end-user applications in aerospace, healthcare, defense, consumer industry, education, and so forth. Oculus by Meta (previously known as Facebook) is the segment leader in manufacturing VR headsets in the world. In-build app and content store makes it easy to get along as we do in a smartphone, except the fact that apps will follow VR-based interface.

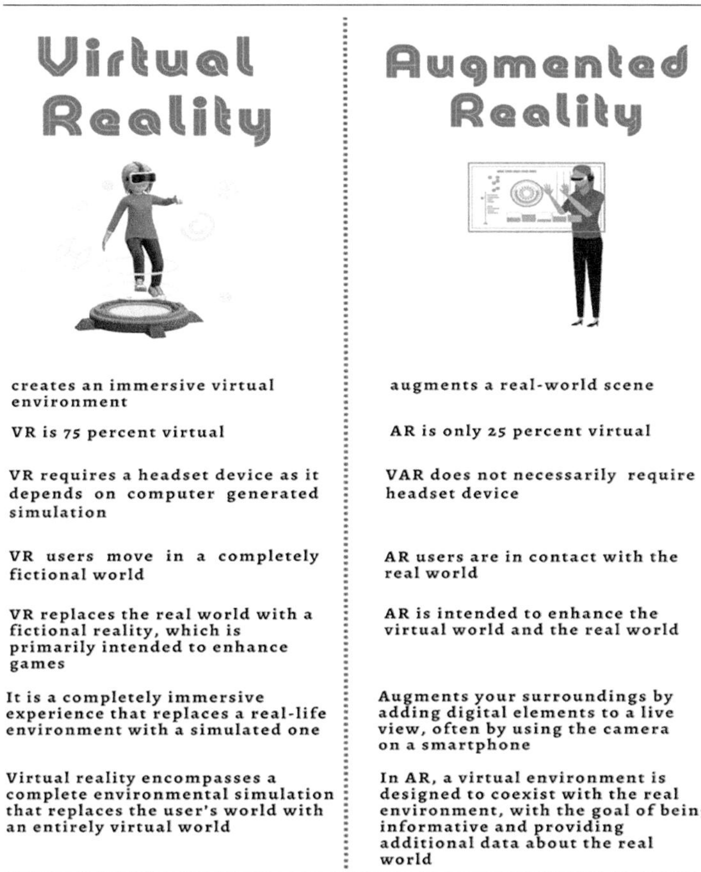

FIGURE 7.1 Difference between virtual reality and augmented reality.

7.3 STATUS OF FISHERIES EDUCATION IN INDIA

In comparison to veterinary and agricultural education, professional fisheries education in India got established later. The central institute of fisheries education in Mumbai started in 1961 and it was in 1969, under the auspices of the University of Agricultural Sciences, Bengaluru, that the first fisheries college in Mangalore opened, marking a new era in the professional fisheries education in India at the state agricultural/veterinary universities. There are more than 32 fisheries colleges and 3 universities are offering fisheries education in India for bachelors, master, and PhD degree programs. The present annual intake capacity of the B.F.Sc., M.F.Sc., and

TABLE 7.1 Virtual Reality Application in Various Domains.

S. No.	Domain	Application	Source
1	Agriculture	VR-based simulation of the natural environment for cows to improve quantity and quality of milk.	Keene, 2020
2	Medicinal science	Application of VR to understand the complex treatment procedure before carrying out the treatment to the patients.	Thomas, 2021
3	Remote sensing and GIS	VR-based interaction with real objects using 3D visualization.	Singla, 2021
4	Oceanography	VR-based creation of plankton zoo and 3D visualization of plankton to learn about phytoplankton.	Walcutt et al., 2019
5	Oceanography	VR application in mimicking natural habitat to elicit behavioral response (e.g.: camouflage) in *Loligo opalescens*	Jaffe et al., 2011; Josef, 2018
6	Oceanography	Creation of immersive virtual aquarium with real-walking navigation by using VR	Jung et al., 2013
7	Fishing areas	Application of VR to map the fishing location present in the US country to know about the different fishing locations.	Rollan et al., 2018
8	Engineering and design	VR-based creation of prototype machinery to examine, modify any defects in the system rather than making a physical prototype.	Bharath et al., 2022
9	Military	Application of VR for realistic military training to combat high-stress situations and improve skills over handling weapons and range of communications.	Fleming et al., 2020
10	Sports	VR sports entertainment accelerates the athlete's training regimen, giving them a chance to run unlimited reps to perform better in a true event.	Derek Belch et al., 2018
11	Manufacturing	VR-based manufacturing will allow the worker to interact with the inside of an engine to create, repair, and to maintain the machinery.	Paolo Samontanez, 2020
12	Veterinary medicine	Application of VR to understand the anatomy of dog, cow, and equine.	Alison Elward, 2023
13	Tourism	VR technology implementation of smart tourism to provide information about destinations and attractions while showing its potential to become a new tourism service.	Pestek and Sarvan, 2020
14	Entertainment	VR-based games, theatre, museum, amusement park, gallery, live music concerts, and hobby lessons.	Anastasia, 2017

TABLE 7.1 *(Continued)*

S. No.	Domain	Application	Source
15	Archaeology	A prototype of pleito cave was reconstructed by using VR which leads to understanding the real-time information of the caves.	Cassidy et al., 2019
16	Education	VR-based learning is a promising tool to learn the concepts clearly and also engages multiple senses.	Christou, 2010
17	Forestry	Application of VR to map the presence of jaguars throughout the Peruvian amazon for the effective conservation.	Bednarz et al., 2016
18	Environment	Application of VR to know the individual carbon print in their daily activities will make them think in a different way to reduce carbon footprints.	Ligia Noronha, 2020
19	Horticulture	VR-based learning of plant physiology can make the individual to grow the plant and can make them to tackle the arising problem effectively.	Ai-guo et al., 2011
20	Medicine	VR technology employed in the drug discovery by engaging different scientist in a place for real time modification or identification of molecules or drugs.	Laura French, 2021

PhD program is 1500, 425, and 185, respectively, while the annual out turn is about 80–85% of intake (Kumar et al., 2018). Universities and colleges have always been at the forefront of new technologies, driving innovation, changing industries, and training the next generation of scientists, developers, and business owners. Right now, VR and AR technologies are at the cutting edge of progress, and things are changing quickly.

7.4 POSSIBLE VIRTUAL REALITY APPLICATION IN FISHERIES EDUCATION

IVR application allows the students to reach to any possible environment without being actually present.

7.5 DISTANCE LEARNING/VIRTUAL CLASSROOMS

Students who are able to think creatively have a better shot at learning the problem-solving, teamwork, and efficiency skills they will need in college and beyond. Teachers can employ exploration-based learning, an active learning strategy that encourages students to learn through curiosity and

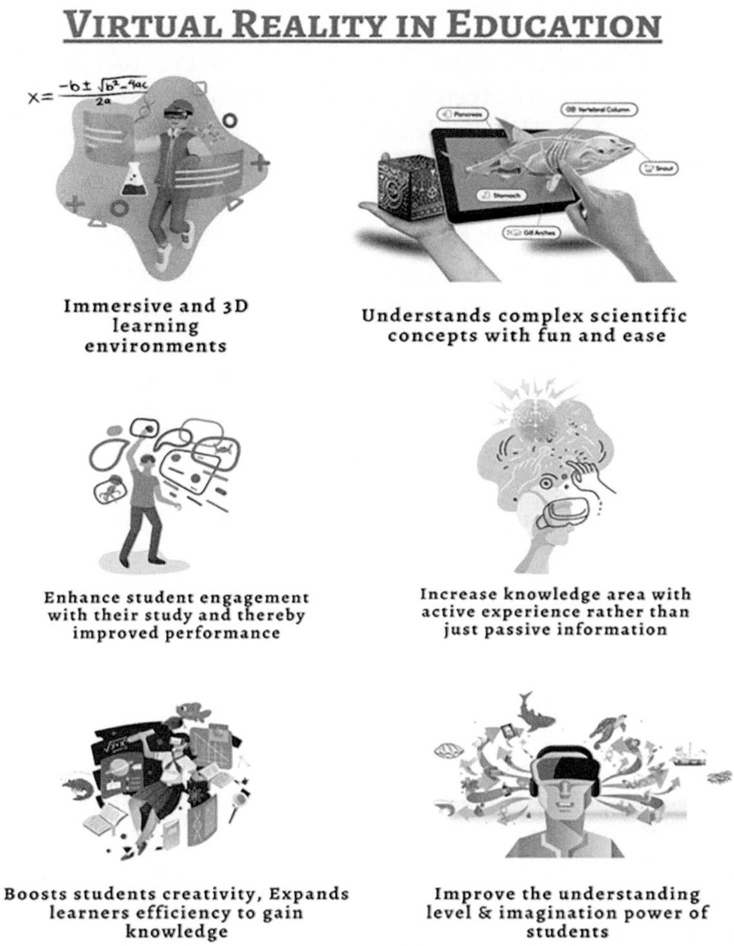

FIGURE 7.2 Virtual reality benefits in education.

enquiry, to assist students to acquire confidence and make their education more meaningful (Figure 7.2). The concept of distance learning is growing up after the COVID-19 lockdown as people get used to online learning platforms. One of the major problem in online learning platforms is physical engagement, which can be sorted out by VR. VR classroom in many examples showed significantly better learning motivation, learning outcomes, and positive impacts on learning students' achievement scores (Liou and Chang, 2018).

7.6 VIRTUAL FIELD VISITS

Visiting a real workplace, especially one that could put students in harm's way, can be prohibitively expensive and time-consuming for those pursuing a practical education. The simulations on underwater environment such as trench and coral reef areas; real-time fishing in ocean and freshwater environments; research cruise; aquaculture farms; virtual museums; climate change studies; and so forth can be created which can enable the student to visit these areas from their convenient places and cut downs the traveling cost and time. VR creates psychological presence, a sense of being there, due to the immersive VR tracking system that detects the user's body movements and enables them to feel as though their body is moving in the virtual world and the virtual world is reacting to their movement. VR is an effective treatment for some of the phobias such as aquaphobia (Morina et al., 2015), and anxiety disorder (Opriş et al., 2012) due to its high level of presence. VR meetings such as conferences, workshops, and symposia are gaining popularity all over the world. One of the best examples in the fisheries is 13th AFAF (Asian Fisheries Aquaculture Forum) held in virtual mode in 2022.

7.7 VIRTUAL FISH AND FISH SYSTEMS

Traditionally, textbooks, live animals, and cadavers are used in fisheries anatomy and biology classes to develop a solid foundation of knowledge. The practical connectivity of students in anatomy of finfish and shellfish, biology of finfish, and shellfish systems are difficult in the textbook education. Three-dimensional (3D) software programs provide a platform for deeper examination of anatomical structures but are unable to deliver a fully immersive experience. The use of VR provides a new type of learning environment and avoid sacrificing live organisms (Figure 7.3). Dissection and identification of vital organs by the students themselves will be very easy in the VR environment as the users can zoom in and zoom out the parts and they can know the functions of the organs and systems.

7.8 VIRTUAL LABS

As the accreditation criteria of many accrediting bodies emphasize, acquiring practical skills is a critical component of learning outcomes in any fisheries degree. It will be especially difficult to ensure efficient achievement of these skills with limited physical delivery. One possible solution to this problem is to use virtual lab experiments and simulations to help students understand the concepts, important relationships between variables, and potential impact on experimental rig operation before physically carrying out an experiment in a laboratory within a much shorter timeline. Field of cell biology, toxicology, pharmacology, histology, and pathology studies in aquatic organisms involves hazardous or toxic chemicals in laboratories, which can be avoided in VR education. The wastage of costly chemicals also can be overcome in VR education and the experiment for teaching can be repeated a number of times.

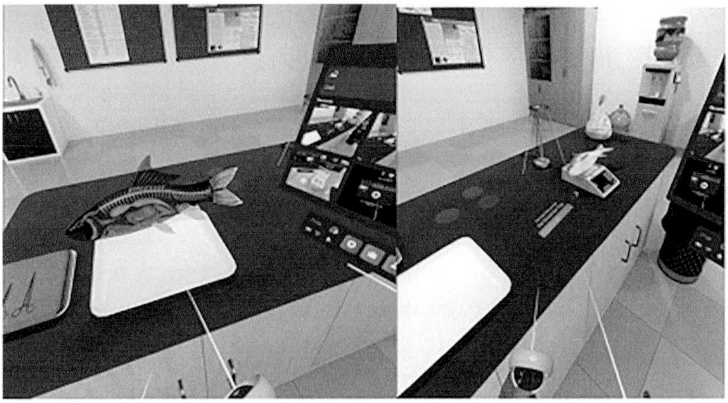

FIGURE 7.3 Glimpse of the virtual fish laboratory developed under NAHEP.

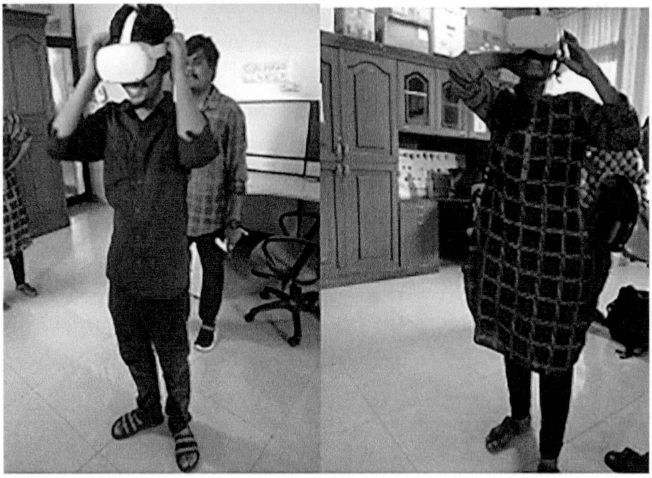

FIGURE 7.4 Students of ICAR-CIFE enjoying the virtual reality programme.

7.9 CONCLUSION

Those interested in the new teaching technique should not miss this chance to learn more. VR has the potential to completely transform education at all levels by bringing a new dimension to the learning process. AR promotes student learning and comprehension through experimentation. We are currently only witnessing the beginnings of a paradigm shift in education brought about by virtual technologies. Increasing accessibility is another key component of the future of VR in the classroom. It benefits both education and technology. AR is gaining popular as application development incorporates new technology.

As VR hardware and software continue to decrease in price, it will soon permeate all facets of the educational system. VR solutions have been proven effective across a variety of educational settings, and they are generally well received by viewers. AR can transform human existence due to its many benefits. We must assess unsolved research problems about VR future barriers and potential benefits. Recent advances in telecommunication systems, especially the roll out of the 5G network in big economies such as China, India, and the United States will boost VR-based markets. The web 3.0 is the next big revolution that is happening in the world today. Ongoing, COVID-19 pandemic pushed the world to adapt and evolve to challenging situations to keep the economy running. Every aspect of life, be it education, shopping, running business, banking, and attending office became online. Being online is the new normal these days. This is why VR is going to have a huge impact in all the sectors. VR is increasingly used in education and training in academia, healthcare, tourism, shopping, and in automobile and space industries.

From an educational point of view, VR will revolutionize learning experience. Students can immerse and understand subjects that are challenging to grasp via conventional learning. Increase in educational content in the VR arena will certainly take education to the next level in the coming generations. Medical students can have real-time experience in complex surgeries via pre-recorded real-time surgery videos captured directly through the eyes of the surgeon himself and an engineering student can have hands-on training in 3D modeling of complex machineries. A fishery student in relative terms will get to experience real-time immersive knowledge in a variety of aspects such as underwater oceanic observations, craft and gear handling at the sea, fish processing in big industries, aqua feed production in feed manufacturing plants, and hatchery operations to which the students have limited accessibility in many grad schools. In limited-resource settings, VR contents can be crafted to aid the students' needs. For example, it is not always necessary for all students to dissect a fish to learn different parts of its body; every time the student will not get exposure to study taxonomy of diverse fish groups within a semester, as fish availability depends on the season and fishing effort, which varies substantially across the country.

VR can solve these problems, by creating interactive contents to give students a real-time immersive learning experience around the year and across the country. Not only does this adaptation make learning wholesome, it cuts down costs and resources associated with it.

ACKNOWLEDGMENT

The authors would like to express their gratitude to ICAR-NAHEP for providing financial assistance for the procurement of VR headsets for ICAR-CIFE, Mumbai. This support greatly facilitated the authors' exploration of the potential applications of VR in fisheries.

KEYWORDS

- **education**
- **fisheries**
- **labs**
- **systems**
- **virtual reality**

REFERENCES

Ai-guo, Z., Ping, S., & He, G. (2011). The technology of VR and its application in the experimental teaching of plant physiology. In *2011 International Conference on E-Business and E-Government (ICEE), Shanghai, China* (pp. 1–2). IEEE.

Bednarz, T., Kim, J., Brown, R., James, A., Burrage, K., Clifford, S., Davis, J., Mengersen, K., Peterson, E., Psaltis, S., & Vercelloni, J. (2016). Virtual reality for conservation. In *Proceedings of the 21st International Conference on Web3D Technology, Web3D 2016* (pp. 177–178).

Blascovich, J., Loomis, J., Beall, A. C., Swinth, K. R., Hoyt, C. L., & Bailenson, J. N. (2002). Immersive virtual environment technology as a methodological tool for social psychology. *Psychological Inquiry*, *13*(2), 103–124.

Bowman, D., & McMahan, R. P. (2007). Virtual reality: How much immersion is enough? *Computer*, *40*(7), 36–43.

Cassidy, B., Sim, G., Robinson, D. W., Gandy, D. (2019). A virtual reality platform for analyzing remote archaeological sites. *Interacting with Computers*, *31*(2), 167–176.

Christou, C. (2010). Virtual reality in education. In *Affective, interactive and cognitive methods for E-learning design: Creating an optimal education experience*, IGI Global, Hershey (pp. 228–243).

Cuesta Cambra, U., & Mañas Viniegra, L. (2016). Integration of immersive virtual reality in communication degrees. *Revista ICONO 14. Revista científica de Comunicación y Tecnologías emergentes, 14*(2), 1–12.

Di Natale, A. F., Repetto, C., Riva, G., Villani, D. (2020). Immersive virtual reality in K-12 and higher education: A 10-year systematic review of empirical research. *British Journal of Educational Technology*, *51*(6), 2006–2033.

Freina, L., & Ott, M. (2015). A literature review on immersive virtual reality in education: State of the art and perspectives. In *The 11th International Scientific Conference ELearning and Software for Education* (pp. 133–141).

Fuchs, H., and Bishop, G. (1992). *Research directions in virtual environments*. Chapel Hill, United States: University of North Carolina.

Gavish, N., Gutiérrez, T., Webel, S., Rodríguez, J., Peveri, M., Bockholt, U., & Tecchia, F. (2015). Evaluating virtual reality and augmented reality training for industrial maintenance and assembly tasks. *Interactive Learning Environments*, *23*(6), 778–798.

Harrington, C. M., Kavanagh, D. O., Quinlan, J. F., Ryan, D., Dicker, P., O'Keeffe, D., Traynor, O., & Tierney, S. (2018). Development and evaluation of a trauma decision-making simulator in Oculus virtual reality. *The American Journal of Surgery*, *215*(1), 42–47.

Ibañez-Etxeberria, A., Gómez-Carrasco, C. J., Fontal, O., & García-Ceballos, S. (2020). Virtual environments and augmented reality applied to heritage education. An evaluative study. *Applied Sciences*, *10*(7), 2352.

Jaffe, J. S., Laxton, B., Zylinski, S. (2011). The sub sea holodeck: A 14-megapixel immersive virtual environment for studying cephalopod camouflage behavior. In *Oceans 2011 Spain* (pp. 1–6). IEEE.

Jensen, L., Konradsen, F. J. E., & Technologies, I. (2018). A review of the use of virtual reality head-mounted displays in education and training. *Education and Information Technologies*, *23*(4), 1515–1529.

Josef, N. (2018). Cephalopod experimental projected habitat (CEPH): Virtual reality for underwater organisms. *Frontiers in Marine Science*, *5*, 73.

Jung, S., Choi, Y. S., Choi, J. S., Koo, B. K., & Lee, W. H. (2013). Immersive virtual aquarium with real-walking navigation. In *Proceedings of the 12th ACM SIGGRAPH International Conference on Virtual-Reality Continuum and its Applications in Industry*, pp 291–294. New York, NY, USA: Association for Computing Machinery.

Kalyvioti, K., & Mikropoulos, T. A. (2014). Virtual environments and dyslexia: A literature review. *Procedia Computer Science*, *27*, 138–147.

Kumar, M., Gurjar, U. R., Keer, N. R., & Kumar, S. (2018). Professional fisheries education in India: History, current status and future-A review. *International Journal of Current Microbiology and Applied Sciences, 7*(6), 3395–3409.

Liou, W. K., & Chang, C. Y. (2018, February). Virtual reality classroom applied to science education. In *2018 23rd International Scientific-Professional Conference on Information Technology (IT), Zabljak, Montenegro* (pp. 1–4). IEEE.

Martínez, J. P. (2011). Presente y futuro de la tecnología de la realidad virtual. *Creatividad y sociedad, 16*, 1–39.

Mikropoulos, T. A., & Strouboulis, V. (2004). Factors that influence presence in educational virtual environments. *Cyberpsychology and Behavior*, *7*(5), 582–591.

Morina, N., Ijntema, H., Meyerbröker, K., & Emmelkamp, P. M. (2015). Can virtual reality exposure therapy gains be generalized to real-life? A meta-analysis of studies applying behavioral assessments. *Behaviour Research and Therapy, 74*, 18–24.

Opriş, D., Pintea, S., García-palacios, A., Botella, C., Szamosközi, Ş., & David, D. (2012). Virtual reality exposure therapy in anxiety disorders: A quantitative meta-analysis. *Depression and Anxiety, 29*(2), 85–93.

Panchuk, D., Klusemann, M. J., & Hadlow, S. M. (2018). Exploring the effectiveness of immersive video for training decision-making capability in elite, youth basketball players. *Frontiers in Psychology*, *9*, 2315.

Parmaxi, A. (2020). Virtual reality in language learning: A systematic review and implications for research and practice. *Interactive Learning Environments*, *31*(1), 172–184.

Pestek, A., & Sarvan, M. (2020). Virtual reality and modern tourism. *Journal of Tourism Futures*, *7*(2), 245–250.

Singla, J. G. (2021). Virtual reality based novel use case in remote sensing and GIS. *Current Science*, *121*(7), 958.

Slater, M. (2003). A note on presence terminology. *Presence Connect*, *3*(3), 1–5.

Keene, M. (2020). Smart farming is ready for augmented and virtual reality. https://www.agritechtomorrow.com/article/2020/11/smart-farming-is-ready-for-augmented-and-virtual-reality/12516sports.

Thomas, L. (2021). Applications of virtual reality in medicine. https://www.news-medical.net/health/applications-of-virtual-reality-in-medicine.aspx.

Walcutt, N. L., Knörlein, B., Sgouros, T., Cetinić, I., & Omand, M. M. (2019). Virtual reality and oceanography: Overview, applications, and perspective. *Frontiers in Marine Science*, *6*, 644.

Webster, R. (2016). Declarative knowledge acquisition in immersive virtual learning environments. *Interactive Learning Environments*, *24*(6), 1319–1333.

CHAPTER 8

Phytoremediation: Aim for a Cleaner Environment

FARAH HANIEF[1], MONISA MALIK[1,*], ADNAN AMIN[1], and ADNAN ABUBAKR[2]

[1]*Division of Aquatic Environmental Management, Faculty of Fisheries, Sher-e-Kashmir University of Agricultural Sciences and Technology, Rangil, Ganderbal, Jammu and Kashmir, India*

[2]*Division of Aquatic Environmental Management, Faculty of Fisheries, Sher-e-Kashmir University of Agricultural Sciences and Technology, Kashmir, SKUAST-K, India*

*Corresponding author

ABSTRACT

Pollution is a global concern which is increasing with each passing day and needs a practical approach. Various technologies have been used time and again to combat this issue, each proving its efficiency. Phytoremediation is one such technique which efficiently removes the contaminants from polluted water and soil. Phytoremediation is an eco-friendly technique which removes, decomposes, or converts toxic heavy metals, toxicants, or pollutants to less harmful form thereby conserving the ecological status of the natural resources. It is a collective and structural approach with visible results. There are various types of techniques involved in phytoremediation such as phyto-extraction, phyto-degradation, phyto-stabilization, phyto-volatilization, rhizo-filtration, rhizo-degradation, and phyto-hydraulics. The plants used for phytoremediation have an inherent ability to tolerate toxic concentrations of heavy metals and other contaminants. These plants gradually adsorb, absorb, and precipitate these toxic compounds into the other less

Recent Advances in Sustainable Aquaculture and Fisheries. Shabir Ahmad Dar, Imtiyaz Qayoom, Gohar Bilal Wani, Adnan Abubakr, and Seemab Zehra (Eds.)

DOI: 10.1201/9781998511372-8

harmful or harmless compounds. Phytoremediation has been seen as a very effective tool for reducing the pollution without harming the ecosystem.

8.1 INTRODUCTION

The term "phytoremediation" is defined as the process of decontaminating or detoxifying the physical components of the atmosphere, that is, air, water, and soil by employing plant-based process. In other words, in phytoremediation, plants are used to remove the contamination of soil, water, or air. Pollution is the major global concern in the present times. The United Nations Environment Program defined phytoremediation as "the efficient use of plants to remove, detoxify or immobilize environmental contaminants" (UNEP, 2019). The level of pollutants in the atmosphere has already increased beyond permissible limits. The uncontrolled use of nonbiodegradable substances definitely adds to the whole misery. Therefore, to combat the increasing menace of pollution and environmental degradation, a number of techniques are used, phytoremediation being one of them (Ali et al., 2020). Phytoremediation has proven as an effective method of cleaning the polluted environment as it makes the pollutants harmless and nonactive. It is an effective way which can be used as a potential method in cleaner technology for environment.

The term phytoremediation has been derived from a combination of Greek word "*phyto*" meaning plant based and Latin word "*remediar*" meaning remedy. Phytoremediation has been proven to be a better innovation for cleaning the polluted environments as it not only detoxifies the physical components of environment, but also helps in mobilizing the toxic substances and converts them to less toxic forms.

Phytoremediation has been found to be very effective as it is a biological process and all the biological processes derive their energy from sun because of which this technique becomes affordable to all for usage. Moreover, it can used on a wide area without any concern as it is economic and environment friendly. Phytoremediation works on both organic and inorganic compounds which makes it preferred over other technologies.

Phytoremediation is a conventional method and is preferred over other methods used for environmental cleaning as it not only reduces the environmental damage but also is hassle free as there is no need to dispose of the contaminated material. The plant species which are usually selected for phytoremediation have an inherent tendency to attract the specific contaminants (Favas et al., 2018). Globally, there are about 350 species found suitable for phytoremediation with ease and efficiency.

8.2 PHYTOREMEDIATION OF AQUATIC WATER BODIES

Water bodies are a main source of waste discharge. Waste from all sources, whether of point or nonpoint origin, finds its way into the water bodies causing pollution of the existing water resources to such an extent that in near future no fresh water would be available. These contaminated water bodies disturb the inhabiting aquatic life, which in turn affects the terrestrial ecosystem. The main causes of the water pollution fall under major groups such as agricultural runoff containing washed off fertilizers, pesticides, insecticides, domestic sewage, and wastes from chemical or petrochemical industries. These wastes after entering the aquatic ecosystem lead to eutrophication due to the release of different ionic forms of nitrogen and phosphorous. These pollutants have been reported to contain a major portion of heavy metals such as cadmium (Cd), lead (Pb), chromium (Cr), copper (Cu), zinc (Zn), nickel (Ni), vanadium (V), and selenium (Se). Majority of these heavy metals have been identified as concerning agents for human health in drinking water by World Health Organization (WHO) in 1984. Therefore, it is of extreme importance that timely remedies are incorporated by government, including awareness at the local level to all for conserving the ecological status of waterbodies.

In this regard, phytoremediation can serve a great deal it being an eco-friendly approach to combat this issue. Phytoremediation is technique which removes accumulated heavy metals and related toxicants from the polluted water bodies. The process of phytoremediation helps in decomposing the heavy metals from waterbodies and converting them into less toxic forms. Since recent times, various researches have been conducted on this technique at different sites of pollution which have proven to be extremely efficient (Favas et al., 2018). Various aquatic plants have been found to act as great phytoremediation plants acting both on organic and inorganic pollutants (Prasad, 2006). However, it is very important to have a thorough knowledge of the plants and their efficiency as phytoremediation candidates (Stefani et al., 2011). Selecting a candidate specie for phytoremediation also depends on the level of pollution that a water body is subjected to.

There are various types of phytoremediation techniques which are employed time and again to treat the contaminated environments. The type of mechanism employed for a particular contaminated area depends on the type of contaminant, bioavailability of that particular contaminant, and on the soil properties. More often, the plants selected for phytoremediation should have roots with large surface area which facilitates more uptake of material from the soil, in turn quickening the process of phytoremediation. The various types of phytoremediation techniques are listed as follows:

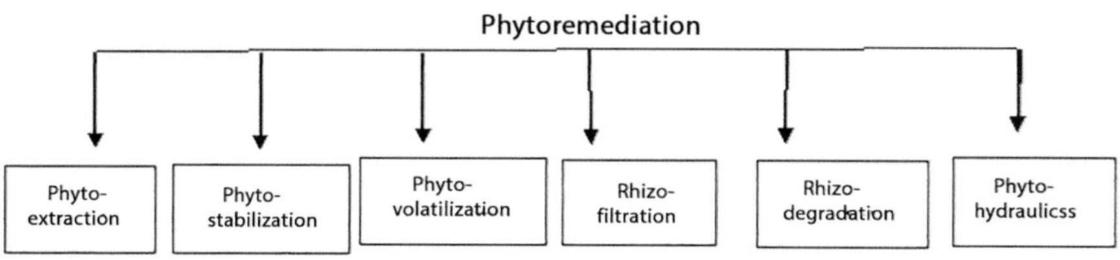

FIGURE 8.1 Types of phytoremediation

8.2.1 PHYTO-EXTRACTION

It is also known as "phyto-accumulation" and involves using the roots of the plants. Plants are grown in contaminated soil where its roots develop fully. The roots of the plants have the capacity to absorb the contaminants present in the soil. These absorbed contaminant compounds are gradually transported to the stem and leaves of the plant. In this way, the soil becomes free from all contaminants in a very natural way.

Phyto-extraction has been used in a large scale to remove metal contaminants such as arsenic, cadmium, lead, mercury, so forth, in an efficient way. The process can be employed in two ways:

1. **Natural phyto-extraction or continuous phyto-extraction:** In this method, the plants take up the contaminants naturally on their own throughout the growth cycle of their life. The roots require no external support for performing the process of phyto-extraction from contaminated soil. For the first time ever, the plants of Brasicaceae family were found to act as natural phyto-extractors. *Thalaspi arvense* was found to be the phyto-extractor of zinc (Zn), and *Brassica juncea* was found to be the phyto-extractor of lead (Pb), zinc (Zn), and cadmium (Cd).

   One the most important example here is that of the *Salicornia* (pickleweed). It is known to act as a phyto-extractor of salt from saline soils. It tends to accumulate salt in vacuole form. Later on, this salt can be extracted from this plant and can be used as table salt.

2. **Assisted phyto-extraction or induced phyto-extraction:** In this method, certain chelating agents are used which enhance the ability of the roots for accelerating the process of uptake of contaminants. Sometimes, certain hyper accumulators are also used which also accelerate the uptake of contaminants from the soil. The most commonly used hyper accumulators for phyto-extraction is *Thlaspi caerulescens.* EDTA is added to soil most commonly which increases the heavy metal extraction by plants, particularly for lead (Pb), cadmium (Cd), so forth.

This method is very effective as it as helps to remove the contaminants from the soil permanently. This method disposes of the waste material almost up to 95%, and even the waste material can be recycled in some special cases.

8.2.2 PHYTO-STABILIZATION/PHYTO-SEQUESTRATION

This is a very interesting method of cleaning the soil of any contaminants. The contaminants present in the root zone of the plants are immobilized by the release of certain biochemical substances from the roots. After immobilization, the contaminants are then stabilized to less toxic forms.

8.2.3 PHYTO-VOLATILIZATION

In this method, all the volatile contaminants present in the soil are absorbed by the roots. These volatile contaminants are then transported to the leaves of the plant. From the leaves, they are gradually released in the air. This method provided better results as it leaves no trace of contaminants behind.

8.2.4 RHIZO-FILTRATION/PHYTO-STIMULATION

In this process of phytoremediation, roots play the major role. The metal contaminants present in the soil or aquatic ecosystems are adsorbed or precipitated on the roots of the plant. The contamination of groundwater and surface water occurs through roots of the plant (Dushenkov et al., 1995). The plants used for this process enhance the process of degradation of contaminants by releasing certain organic compounds. These organic compounds attract the surrounding microbial community to the specific site of release of these compounds where the process of degradation and decomposition takes place. This specific method has been used extensively to treat waste water (Ali et al., 2020).

8.2.5 RHIZO-DEGRADATION

This type of phytoremediation involves the use of micro-organisms. The roots of the plants secrete certain biochemical substances which help in enhancing growth of various micro-organisms specifically bacteria in the root zone. These bacteria in turn enhance the degradation process down in the soil.

8.2.6 PHYTO-HYDRAULICS

This type of phytoremediation deals with ground water. The plants perform their function of phytoremediation by controlling the ground water movement. This process is based on the ability of the plants to capture the contaminants in the groundwater, transport those captured contaminants to upper parts of the plants, and finally transpiring them along with water from the leaves into the air. This method has proven to be very useful where Poplar trees were used to treat groundwater contaminated with methyl-tert-butyl-ether (MTBE).

Out of the abovementioned types of phytoremediation, some are used to treat contaminations of organic origin while some work on metal-contaminated areas.

Organic Contaminants ↓	Metal Contaminants ↓
1. Phytodegradation	1. Phyto-extraction
2. Phyto-stimulation	2. Rhizo-filtration
3. Phyto-volatilization	3. Phyto-stabilization

FIGURE 8.2 Organic and metal contaminants

8.3 ROLE OF AQUATIC MACROPHYTES AS PHYTOREMEDIATION

A lot of research has been conducted in this field since the last decade. Various aquatic plants have been reported to have enormous ability to degrade pollution and heavy metals in water bodies. The normal physiology of the plants is such that they require a certain level of heavy metals to perform their regular metabolic activities. But beyond the requisite concentration, these heavy metals become toxic for the surroundings. Researches have shown that toxic concentrations of heavy metals can lead to the generation of reactive oxygen species (ROS) which impact the normal physiology of the aquatic plants and other surrounding living organisms.

However, various plant species have this innate ability to tolerate the heavy metal concentration beyond permissible limit. These tolerant plants have certain anthocyanins, thiols, and antioxidants due to which they can survive in high concentration of heavy metals (Leao et al., 2014). Prasad and Freitas (2003) have conducted an extensive research on the ability of phytoremediation of plants and have found that there are more than 400 aquatic macrophytes which can serve as candidate species for phytoremediation. The most commonly used aquatic macrophytes are *Lemna, Wolffia, Azolla, Hydrilla, Eichornia, Typha, Pistia,* so forth. Every macrophyte performs a different function when it comes in contact with the contaminants. The list of macrophyte along with their response when they come in contact with any contaminant is given below:

1. *Eichhornia crassipes* shows reduced growth, decreased height of stem, and length of root which is followed by wilting of leaves and chlorosis. Ultimately it leads to the death of the plant.
2. *Ipomonea aquatica* shows a peculiar phenomenon when it comes in contact with the contaminant. The length of the roots shows a decrease but the girth of the roots increases.
3. *Lemna minor* shows reduced activity of enzymes which impacts the photosynthetic ability of the plants. There is a gradual decrease in chlorophyll content in these plants ultimately impacting the growth of roots and shoots.
4. *Pistia stratiotes* suffers cell damage and increased enzymatic activity, particularly of superoxide dismutase (SOD), catalase (CAT), ascorbate peroxidase (APX), and peroxidase (POX).
5. *Isoetes taiwanensis* suffers from degenerative root and shoot growth in presence of heavy metals.

The positive effects of many aquatic weeds, for example, water hyacinth (*Eichhornia crassipes*), water lettuce (*Pistia stratiotes*), duckweed (*Lemna, Spirodela, Wolffia, Wolfiella*), bulrush (*Typha*), common reed (*Phragmites australis*), and vetiver grass (*Chrysopogon zizanioides*) in the elimination of contaminants, have been studied by various researchers (Mkandawire and Dudel, 2007). *Wolffia globosa* has been found to effectively remove the arsenic from waste water. It can eliminate up to 400 mg of arsenic from waste water present in water bodies. Water hyacinth has also been found to be efficient in removing nitrogen from waste water up to 60–80% and potassium up to 69%. The conducted studies have shown that aquatic macrophytes can serve as efficient phytoremediation candidate species. The increasing anthropogenic activities are worsening the already degrading quality of water bodies. In such circumstances, phytoremediation is such a technique which will not only decrease the pollution but also is eco-friendly.

8.4 ADVANTAGES OF PHYTOREMEDIATION

1. The whole operation can be performed in *in situ* as well as in *ex situ* conditions with very little cost input.
2. The metals present in the soil as contaminants can be recovered as re-used without any problem.
3. The overall operation is driven by solar energy, hence, there is no need of any other source of energy. Moreover, the whole process can be monitored very easily.
4. This method is literally harmless for the environment as it uses normal biological process for overall working.
5. Phytoremediation enhances soil health, reduces soil erosion, and maintains soil fertility in the long run.
6. It helps in conserving natural resources by decreasing the loss of environment.
7. It can be applied to a variety of contaminants both from organic and inorganic origin.
8. There is easy acceptance of this technology by people as it is environmentally friendly and esthetically pleasing to eyes.

8.5 DISADVANTAGES OF PHYTOREMEDIATION

1. As the whole operation is plant based, sometimes the contaminants are leached into the ground water.

2. The time requirement is very much as the process is slow and time taking.
3. Bioaccumulation of toxic contaminants can happen in the food chain if the plants are consumed by primary level consumers.
4. It is limited to shallow waters or sediments as roots of plants need soil for growth.
5. This technology has a major impact on the climate. Moreover, in winters, this technology does not work at all.
6. Harvesting of plants that have contaminants accumulated within leads to pollution again.

Phytoremediation can be a boon to aquatic water bodies. Nowadays, the pollution in aquatic ecosystems is increasing at an exponential scale. Wetland technology has helped to reduce this menace to a lot of extent but the more variants of contaminants, the more difficult it is to handle. Phytoremediation can play a crucial role here. The water bodies or aquaculture practices can be decontaminated by using plants which will also facilitate the production of oxygen and removal of carbon dioxides simultaneously.

KEYWORDS

- **aquatic macrophytes**
- **contaminants**
- **heavy metals**
- **phyto-extraction**
- **phytoremediation**
- **rhizo-filtration**

REFERENCES

Ali, S., Abbas, Z., Rizwan, M., Zaheer, I. E., Yavas, I., Unay, A., Kalderis, D., et al. (2020). Application of floating aquatic plants in phytoremediation of heavy metals polluted water: A review. *Sustainability, 12*, 1927.

Dushenkov, V., Kumar, P. B. A. N., Motto, H., & Raskin, I. (1995). Rhizofiltration: The use of plants to remove heavy metals from aqueous streams. *Environmental Science & Technology, 29*, 1239–1245.

Favas, P. J. C., Pratas, J., Rodrigues, N., D'Souza, R., Varun, M., Paul, M. S. (2018). Metal(loid) accumulation in aquatic plants of a mining area: Potential for water quality biomonitoring and biogeochemical prospecting. *Chemosphere, 194*, 158–170.

Leao, G. A., de Oliveira, J. A., Felipe, R. T. A., Farnese, F. S., & Gusman, G. S. (2014). Anthocyanins, thiols, and antioxidant scavenging enzymes are involved in *Lemna gibba* tolerance to arsenic. *Journal of Plant Interactions, 9*, 143–151.

Mkandawire, M., & Dudel, G. S. Are *Lemna* sp. effective phytoremediation agents. *Bioremediation, Biodiversity and Bioavailability, 1*, 56–71.

Prasad, M. N. V., & Freitas, H. M. D. O. (2003). Metal hyperaccumulation in plants - Biodiversity prospecting for phytoremediation technology. *Electronic Journal of Biotechnology, 6*, 3.

Prasad, M. N. V. (2006). Aquatic plants for phytotechnology. In S. N. Singh, R. D. Tripathi (Eds.), *Environmental bioremediation technologies* (pp. 259–274). Springer.

Stefani, G. D., Tocchetto, D., Salvato, M., & Borin, M. (2011). Performance of a floating treatment wetland for in-stream water amelioration in NE Italy. *Hydrobiologia, 674*, 157–167.

UNEP. (2019). Phytoremediation: An environmentally sound technology for pollution prevention, control and remediation. An introductory guide to decision-makers. Newsletter and Technical Publications Freshwater Management Series No. 2 United Nations Environment Programme Division of Technology, Industry, and Economics. http://www.unep.or.jp/Ietc/Publications/Freshwater/FMS2/1.asp (assessed August 18, 2019).

WHO. (1984). Guideline for drinking water quality recommendations. World Health Organization, Geneva.

CHAPTER 9

Commonly Used Software in Fisheries Science and Research for Statistical Analysis

JAHANZAIB KHAN[1,*], AIMEN FIRDOUS[2], TABINDA SHABIR[2], and BURHAN ELLAHI[1]

[1]*Division of Fish Nutrition and Biochemistry, Faculty of Fisheries, Sher-e- Kashmir University of Agricultural Sciences and Technology, Ganderbal, Jammu and Kashmir, India*

[2]*Division of Post-Harvest Technology, Faculty of Fisheries, Sher-e- Kashmir University of Agricultural Sciences and Technology, Ganderbal, Jammu and Kashmir, India*

*Corresponding author

ABSTRACT

This chapter offers a thorough and valuable exploration of the statistical analysis software utilized routinely in the domain of fisheries science and research. It underscores the significance of statistical analysis in comprehending and overseeing fish populations, along with the development of statistical theory over time. This chapter commences with an introduction to STATISTICA, a comprehensive suite of analytical tools renowned for its proficiency in data administration, analysis, visualization, data mining, and machine learning. It emphasizes STATISTICA's adaptability, as it offers both a command-line interface and a user-friendly graphical interface for effective data analysis. Furthermore, the chapter delves into ARTFISH, a standardized statistical approach developed by FAO-FIDI for sample-based fishery surveys, and examines the crucial role

Recent Advances in Sustainable Aquaculture and Fisheries. Shabir Ahmad Dar, Imtiyaz Qayoom, Gohar Bilal Wani, Adnan Abubakr, and Seemab Zehra (Eds.)

DOI: 10.1201/9781998511372-9

of its core components, namely, ARTPLAN, ARTBASIC, and ARTSER, in enhancing national fisheries statistical programs. FishStatJ, Java-based software tailored for statistical time series in fishery and aquaculture, with an emphasis on its capabilities in data storage, processing, and analysis, is also covered. ParFish, another software tool utilizing Bayesian statistics and decision theory to estimate participatory fisheries stocks, is discussed alongside SPSS, a popular statistical program known for its extensive statistical capabilities and user-friendly interface, especially in academic settings. This chapter concludes with an exploration of Statistical Analysis System (SAS), a robust statistical analysis framework with a wide array of applications. Overall, it underscores how these tools empower researchers to conduct thorough statistical analyses, enhance data management, and make informed decisions for the sake of sustainable fisheries management. Researchers and practitioners in fisheries science will find this chapter instrumental in comprehending and effectively utilizing appropriate statistical analysis software for their research pursuits.

9.1 INTRODUCTION

In the world today, fish stands as a crucial source of protein, supporting the livelihoods of millions and contributing substantially to a nation's foreign exchange earnings. In Efron's discerning observation from 1998, the year 1925 marked the maturation of statistical theory, transforming statistics from a collection of ingenious techniques into a coherent discipline. In this context, it is intriguing to note that fisheries science and statistical sciences share a similar historical timeline.

The integration of statistics into fisheries science has closely mirrored the progression of statistical theory, methodologies, and computational tools, facilitating both straightforward and intricate analyses. Specialized statistical analysis software packages have been developed to empower users to undertake complex statistical analyses while also offering features for data organization, interpretation, and presentation. These software solutions typically incorporate data management functionalities alongside their statistical analysis capabilities. Over the past few decades, the utilization of statistical software has evolved through various stages (Muenchen, 2010–2019), reflecting its growth and diversification to address the evolving needs of fisheries science.

9.2 COMMONLY USED SOFTWARES FOR RESEARCH

9.2.1 STATISTICA

Originally crafted by StatSoft, STATISTICA stands as a multifaceted suite of analytical software tools that has traversed intriguing ownership transitions over the years. In 2014, Dell acquired STATISTICA, marking a significant chapter in its journey. However, another milestone occurred in 2017 when TIBCO entered into an agreement to acquire the software suite. STATISTICA's prominence in the realm of data management and analysis is noteworthy. It encompasses a wide spectrum of capabilities, including data visualization, data mining, and machine learning. The suite's adaptability is evident in its various product categories, catering to diverse user needs. These categories encompass Enterprise, designed for broad organizational deployment; Web-Based, tailored for server-based usage through web browsers; Concurrent Network Desktop, facilitating collaborative analysis within networks; and Single-User Desktop, intended for individual users seeking robust analytical tools.

As STATISTICA continues to evolve under new stewardship, its enduring reputation for proficiency in data-related tasks persists. Its comprehensive feature set makes it an invaluable asset for organizations and individuals alike, serving as a pivotal tool in the pursuit of data-driven insights and informed decision-making.

9.2.2 STATA

The name "Stata" is a fusion of the words "statistics" and "data," reflecting its core mission. This powerful statistical software emerged in 1985, and a significant milestone occurred in 2003 when it introduced a user-friendly graphical interface option. Stata has since become a favored tool among researchers in diverse fields such as economics, medicine, and political science. Stata's appeal lies in its robust capabilities, which empower users to not only analyze but also effectively manage and visualize data. What sets Stata apart is its dual-mode usability, offering both a command-line interface for seasoned users and an intuitive graphical user interface (GUI) for those who prefer a more user-friendly approach. Moreover, Stata stands out by accommodating both conventional and unconventional data analysis methods, courtesy of its incorporation of a potent programming language. This versatility allows researchers to tailor their analyses to specific research objectives. Nevertheless, it is worth noting that some of these advanced features may present a steeper learning curve compared to other software applications (Rabe-Hesketh and Everitt, 2019).

9.2.3 ARTFISH

ARTFISH, an acronym for "Approaches, Rules, and Techniques for Fisheries Statistical Monitoring," stands as a testament to FAO-FIDI's commitment to advancing fisheries science. This suite of standardized statistical methodologies was meticulously developed to facilitate the planning and execution of sample-based fishery surveys. The multifaceted outputs of ARTFISH encompass a rich array of resources, including training kits, comprehensive manuals, insightful guidelines, illuminating case studies, informative documents, and invaluable computer software. The primary objective behind the creation of ARTFISH was to empower stakeholders in the fisheries sector to establish cost-effective fishery statistics systems with minimal external assistance. The hallmark of ARTFISH lies in its dedication to providing customers with reliable, user-friendly, and error-free computer software solutions. This approach not only fosters self-sufficiency but also promotes the generation of accurate and insightful fisheries data. In a field where precision and data integrity are paramount, ARTFISH emerges as a vital tool, equipping fisheries professionals with the means to monitor and manage their resources effectively.

The core ARTFISH components are of three modules as described below.

9.2.3.1 ARTPLAN

1. ARTPLAN serves as a tool for survey planning and training within fisheries.
2. Utilizing user-defined parameters, this module generates simulated fishery scenarios for testing and evaluating various sampling options.
3. It aids in foreseeing common challenges related to mobility, cost-benefit tradeoffs, sample size frequency, site accessibility, and sampling schedules.

9.2.3.2 ARTBASIC

1. ARTBASIC comes into play during data sampling and estimation phases.
2. Its primary objective is to establish standardized database structures, adhering to system standards and classifications. It encompasses primary sample data and provides estimations of overall effort, catch, and values.
3. ARTBASIC operates within the context of a calendar month, logical strata, and specific boat/gear types, facilitating decentralized data entry and supporting multilingual production estimates.

9.2.3.3 ARTSER

1. ARTFISH's reporting component is known as ARTSER.
2. ARTSER component is used for creation of reports and graphics.
3. With the help of this component, monthly estimates from ARTBASIC can be integrated into standardized tables that display monthly estimates for catch, CPUEs, effort, pricing, and values over the course of a reference year.
4. Additionally, it offers the capacity to execute data mining and analysis in multiple languages.

The overarching objective of ARTFISH is to enhance the methodological and operational facets of national fisheries statistical programs, thereby fostering development in the field of fishery statistics. This endeavor encompasses several key aspects:

1. It makes use of technology that shortens training periods and improves user cognition.
2. It conducts comprehensive statistical diagnostics to assess data quality.
3. It facilitates a deeper understanding of the fisheries for which data is collected.

9.2.4 FISHSTATJ

FishStatJ is a versatile software application designed for Windows and MAC platforms, accessible to users of all backgrounds. This Java-based program provides a valuable resource for individuals interested in fisheries and aquaculture statistical time series analysis. With a user-friendly interface and a repository of fishery statistical datasets, FishStatJ simplifies the storage and processing of data with a time series structure.

Key Features:

1. Browsing, Data Mining, and Reporting: FishStatJ empowers users to navigate through datasets effortlessly. It facilitates data mining, allowing users to unearth valuable insights. Furthermore, it streamlines the process of generating reports for a comprehensive understanding of the data.
2. Charting, Filtering, Grouping, Data Querying, and Aggregation: The software offers robust visualization capabilities, enabling users to create charts and graphs to visualize trends and patterns. Advanced filtering, grouping, querying, and aggregation tools make it easy to extract specific information from the data, facilitating in-depth analysis.

FishStatJ comprises a core application and various workspaces that host the datasets. The setup module initially includes one dataset, while additional sets can be obtained separately. Datasets are organized within workspace files, and users can install an unlimited number of workspaces on their computer. However, it is important to note that FishStatJ can only operate with one workspace at a time. A range of workspaces is available to cater to diverse analytical needs, including the Global Production Workspace, Global Fish Trade Workspace, Global Fish Processed Products Production Workspace, Fishery Food Balance Sheets Workspace, Regional Workspace, and Border Rejection Workspace.

9.2.5 PARFISH—PARTICIPATORY FISHERIES STOCK ASSESSMENT

This specialized software package harnesses the power of Bayesian statistics and decision theory to assess and manage fisheries populations effectively. At its core, this software relies on a logistical biomass growth model, demanding four critical inputs: current biomass utilization, unexploited biomass, catchability, and growth rate.

Key Features:

1. Estimating Limit and Target Control Levels: By employing Bayesian statistical techniques, the software calculates limit and

target control levels that are instrumental in fisheries management. These levels serve as crucial benchmarks for sustainable resource utilization.

2. Assessing Fisheries Population Condition: The software facilitates an in-depth evaluation of the fisheries population's condition. Through Bayesian analysis, it provides insights into the health and dynamics of the population, aiding decision-makers in ensuring its sustainability.
3. Integration of Fisher Interview Data: Fisher interview data is leveraged to establish "priors" for model parameters. These priors, coupled with additional information, contribute to the generation of robust estimations. The incorporation of real-world insights from fishermen enhances the accuracy of the assessments.
4. Preference-Based Control Level Determination: The software incorporates preference information gathered from fishermen. This invaluable data helps in determining the present stock level and defining control levels that align with fishermen's preferences for catch rates. This collaborative approach ensures that management decisions are not only sustainable but also responsive to the needs and preferences of stakeholders.

9.2.5.1 PRINCIPLES

1. Identify measurable indicators related to policy.
2. Identify state of exploited populations (reference points).
3. Identify controls on fishing.
4. Identify and deal with uncertainty.

9.2.5.2 TYPES OF INDICATORS

1. Stock size.
2. Catch/Landings.
3. Effort/vessels/days-at-sea.
4. Fishing mortality.
5. Employment.
6. Profit/economic rent.
7. Nontarget catch.
8. Interactions/illegal activity, so forth.

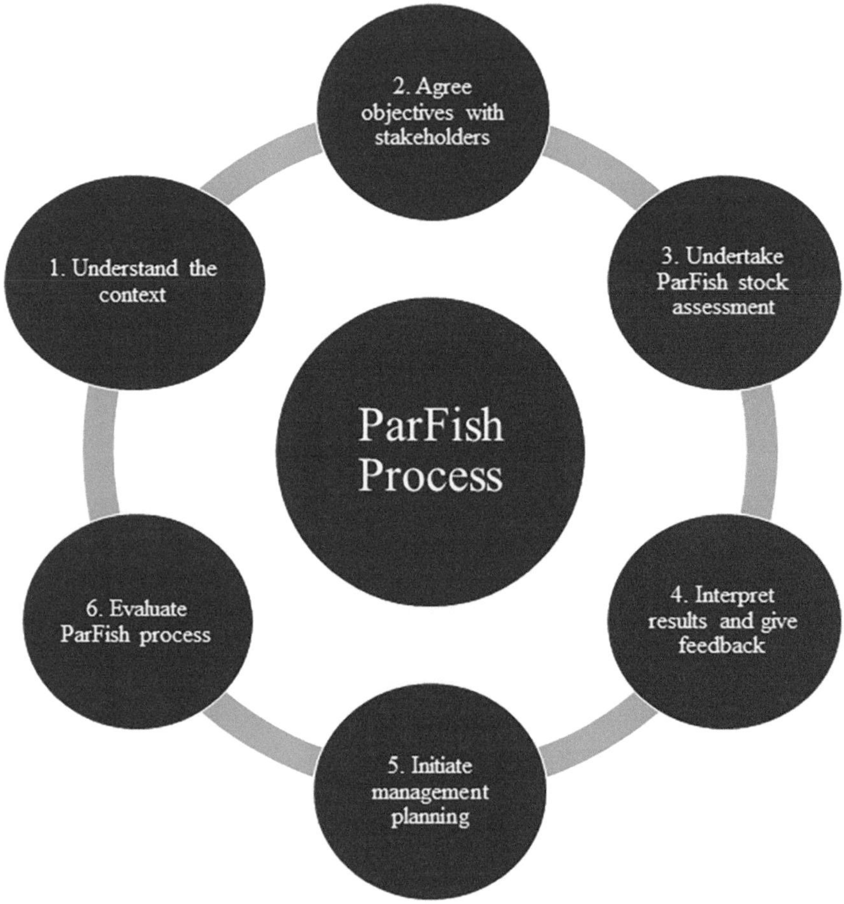

FIGURE 9.1 Process of ParFish.

Example

1. Policy statement: "sustainable utilization maximising economic benefits."
2. Interpret: "…maintain stock size above MSY point, balancing employment and economic rent."
3. Indicators and reference points: Stock size and MSY point Total employment and current employment Vessel profits (catch rates) and break-even point.

9.2.5.3 DATA AND ANALYSIS

1. Stock size: Total Catch and Stock size index.
2. Employment: Number of people employed by sector.
3. Vessel profits: Economic inputs—Vessel, gear, fuel costs, and Economic outputs—Landings and prices.
4. Co-management.

9.2.5.4 REFERENCE POINTS

Impact of fishing on populations

1. Link fishing activity to depletion.
2. Link stock size to productivity.

Use models to interpret data

1. Simple indicators.
2. Complex reference points.

9.2.5.5 UNCERTAINTY

1. Sources of uncertainty.
    - Observation error (sampling).
    - Process error (time series).
    - Structural error (models).
2. Presentation of uncertainty.
3. Making decisions under uncertainty.

9.2.5.6 SUMMARY

1. Interpret independent information in relation to policy aims, for example, indicators and reference points reduction.
2. Address uncertainty.
3. Provide simple understandable advice.
4. Promote management action.

9.2.6 SPSS

SPSS, originally named the Statistical Package for the Social Sciences and later rebranded as Statistical Product and Service Solutions, is a widely embraced software tool in the academic realm and beyond. This Windows-based program serves as a robust platform for data entry, analysis, and the creation of tables and graphs. Its flexibility extends to various analysis types, data transformations, and output formats, catering to the multifaceted needs of researchers. The evolution of SPSS reflects a commitment to blending feature richness with user-friendliness. In the realm of quantitative data analysis, precision and data representation are paramount, driving the software's designers to craft a tool that excels in both accuracy and usability. Over time, SPSS has transcended its roots in the social sciences, becoming an essential data analysis tool across diverse fields, including medicine, engineering, business, education, and government. SPSS empowers professionals across industries, including business managers and analysts, to address critical research and operational challenges. Its comprehensive statistical techniques enable users to derive valuable insights, while additional modules such as SPSS Data Collection and SPSS Modeler cater to specific needs such as survey design and text analytics.

Notably, SPSS transitioned into the IBM family in 2009 and is officially known as IBM SPSS Statistics. Its depth of analysis sets it apart from conventional spreadsheet software, databases, and multidimensional tools. SPSS excels at making predictions and inferences, and its efficiency in tasks such as data manipulation and statistical operations is unrivaled. Users can conduct complex studies and visualize results using the program's extensive chart and graphing capabilities. The program's user-friendly design, supplemented by online guides and readily available documentation, makes SPSS one of the easiest statistical analysis software platforms to learn. Consequently, it enjoys widespread adoption in educational and noneducational institutions alike, serving as an indispensable tool for data-driven decision-making and research across numerous disciplines.

9.2.6.1 ADVANTAGES OF SPSS

1. User-Friendly Interface: SPSS is renowned for its GUI, making it exceptionally user-friendly. This feature makes it accessible to individuals with varying levels of technical expertise.

2. Extensive Statistical Capabilities: SPSS boasts a wide array of statistical capabilities, allowing users to perform complex analyses effortlessly. It can also generate new variables from existing data, enhancing analytical flexibility.
3. Built-in Advanced Statistical Tests: The software includes a comprehensive set of sophisticated statistical tests, eliminating the need for additional tools or add-ons.
4. Versatile File Compatibility: SPSS supports multiple file formats, including Excel, SAS, Stata, and tab-delimited formats, facilitating seamless data import and export.
5. Platform Compatibility: SPSS is compatible with Windows, macOS, and LINUX platforms, ensuring researchers can work on their data regardless of their operating system.
6. Cluster Analysis: SPSS provides cluster analysis capabilities, enabling researchers to identify patterns and groupings within their data.
7. Outlier Detection and Handling: The software offers statistical adjustments to address outliers and automated methods to detect anomalies in the data.
8. Efficient Results Visualization: SPSS excels in creating tables and visualizing results, enhancing the presentation of research findings.
9. Predictive Modeling: SPSS can forecast target variable values based on predictor variables, allowing for the modeling of both linear and nonlinear relationships.
10. Quantitative Data Analysis: SPSS offers a comprehensive suite of tools for quantitative data analysis, catering to the needs of social scientists and researchers (IBM, 2020).

9.2.6.2 DISADVANTAGES OF SPSS

1. Commercial License Requirement: A major drawback is the requirement for a commercial license to use SPSS. While discounts may be available, it is not a free software program.
2. Usability Variation: Some users, especially those familiar with MS Excel, may find the usability of SPSS to be different, which can be a learning curve for new users.

3. Graph Editing Limitations: Editing graphs and customizing their appearance can be challenging, and the graphing options in SPSS are somewhat limited.
4. Visual Appeal: The visual appeal of SPSS-generated results may be less impressive compared to other data visualization tools.
5. Performance Variability: SPSS's performance may vary depending on the system it is installed on, potentially leading to slower operation, particularly with large datasets.
6. Complex Modules: Some modules within SPSS can be more complex than necessary for certain tasks, potentially complicating the analysis process.

Despite these disadvantages, SPSS remains a powerful and indispensable tool for quantitative data analysis in various fields, particularly in the social sciences. Its extensive statistical capabilities, ease of use, and comprehensive features make it a valuable resource for researchers and analysts, often considered essential in certain research contexts.

9.2.7 SAS

The SAS System stands as a robust framework for statistical analysis that goes beyond traditional statistical tasks. In addition to its statistical prowess, SAS empowers programmers and users to delve into report writing, graphics, business planning, forecasting, quality improvement, project management, and much more. Its versatility and capability to handle large datasets and perform complex analyses make it a preferred choice for intermediate and advanced users. Established as a business entity in 1976, SAS has evolved into a leading provider of statistical software, garnering widespread adoption across academic, commercial, and governmental organizations. Over the years, SAS has expanded its analytical software offerings to include data and text mining, forecasting, research, and econometrics. Building on these foundations, SAS now provides specialized business analytics tools tailored to industry-specific challenges, such as customer intelligence, fraud prevention, and risk management. SAS primarily operates through programming syntax rather than point-and-click menus, which necessitates some programming language experience. This approach can be intimidating for newcomers, requiring them to learn how to write code and apply appropriate methods.

However, SAS offers a solution through Enterprise Guide, which offers a user-friendly point-and-click interface, enabling users to select options from menus and find solutions without the need to write scripts.

SAS's evolution includes transitioning from its original PL/I and Fortran roots to a portable system capable of running on various platforms, including PCs. The introduction of the Output Delivery System (ODS) enhanced reporting flexibility by providing complete control over tabular result structures. Furthermore, SAS introduced ODS Statistical Graphics, incorporating graphics into statistical analyses. SAS offers comprehensive support for both Windows and UNIX/Linux operating systems, ensuring that users can work on their preferred platforms.

Key Features of SAS:

1. Statistical Versatility: SAS encompasses a wide range of statistical techniques, including analysis of variance, mixed models, regression, categorical data analysis, Bayesian analysis, multivariate analysis, survival analysis, psychometric analysis, cluster analysis, and survey data analysis.
2. Graphical Capabilities: SAS has advanced tools for graphing main effects and interactions, with PROC GLM offering robust analysis of variance tools and interaction dissection.
3. Platform Support: SAS is fully compatible with both Windows and UNIX/Linux operating systems, staying current with advancements in both platforms.
4. Comprehensive Data Management: SAS provides a full suite of integrated tools and solutions for data access, management, and analysis.
5. Business Analytics: SAS offers specialized tools for addressing industry-specific challenges, making it an invaluable resource for sectors such as customer intelligence, fraud prevention, and risk management.

For those seeking more information, the SAS website (http://www.sas.com/) offers comprehensive resources. In academic settings, access to SAS is often facilitated through institutional agreements, ensuring that students and researchers can leverage its extensive capabilities for data analysis and beyond.

9.2.8 LFDA

LFDA, or Local Fisher Discriminant Analysis, represents a powerful dimensionality reduction technique that combines the strengths of local-preserving projection and Fisher discriminant analysis. Its primary aim is to enhance between-class separability while minimizing within-class separability and preserving the local data structure. Researchers find LFDA particularly useful for various applications, including the estimation of growth parameters and death rates from fish length frequency distributions.

9.2.8.1 KEY FEATURES AND IMPLEMENTATION

1. Native R Package: LFDA is the first package to implement such methods in the native R language, providing a familiar environment for researchers to work with dimensionality reduction and discriminant analysis.
2. Multiple Variants: LFDA offers multiple variants of Fisher discriminant analysis, including local, kernel, and semi-supervised local Fisher discriminant analysis. These variations allow researchers to experiment with different approaches tailored to their specific research needs.
3. Nonlinear Dimensionality Reduction: LFDA's capabilities extend to nonlinear dimensionality reduction scenarios through the use of kernel techniques, enabling the analysis of complex data structures.
4. Visualization Tools: The package includes visualization tools that facilitate the exploration of dimensionality reduction effects. Researchers can visualize results in 2D using tools such as ggfortify, in 3D with rgl, or create interactive visualizations using autoplotly.
5. Growth Parameter Estimation: LFDA is equipped to estimate growth parameters, including those of the von Bertalanffy growth curve, which comes in nonseasonal and seasonal forms. Researchers can utilize LFDA's techniques for parameter determination and analysis.

6. Shepherd's Length Composition Analysis: LFDA incorporates Shepherd's Length Composition Analysis method for further analysis and insights into length frequency data.
7. Simulation Capabilities: The package includes a program that allows users to simulate length frequency data under various hypotheses, enabling researchers to assess different scenarios and test hypotheses.

KEYWORDS

- **ARTPLAN**
- **ParFish**
- **SAS**
- **STATISTICA**
- **statistical tools**
- **statistics**

REFERENCES

Efron, B. (1998). R. A. Fisher in the 21st century. *Statistical Science, 13*, 95–122.

Muenchen, R. A. (2010–2019). The popularity of data analysis software. http://r4stats.com/articles/popularity/ 2010–2019

Rabe-Hesketh, S., & Everitt, B. (2006). In *A handbook of statistical Analyses using Stata* (Vol. 10, p. 2019). Chapman & Hall/CRC.

CHAPTER 10

Role of Cell Line in Conservation of Gene Pool of Declined Native Fish Species of Kashmir

KEEZIA KHURSHID[1,*], FEROZ AHMAD SHAH[1], and MEHAK HAFEEZ[2]

[1]*Division of Aquatic Animal Health Management, Faculty of Fisheries, Sher-e-Kashmir University of Agricultural Sciences and Technology of Kashmir (SKUAST-K), Rangil, Ganderbal, Jammu and Kashmir, India*

[2]*Division of Fish Genetics and Biotechnology, Faculty of Fisheries, Sher-e-Kashmir University of Agricultural Sciences and Technology of Kashmir (SKUAST-K), Rangil, Ganderbal, Jammu and Kashmir, India*

*Corresponding author

ABSTRACT

Cell culture techniques are beginning to take the place of conventional research methods for more accurate and productive research. Research on the establishment of cell lines and their use in several fisheries-related domains, including virology, has been started in the field of fisheries science. L-15 medium was used to standardize the primary cell culture by experimenting with different fetal bovine serum (FBS) concentrations and temperature ranges. It was shown that the tissue preparation method of explanting was more dependable than the enzymatic dissociation method. The conception of repositories and gene banks may offer opportunities for collaboration among the top researchers in the fields of fish ecology, physiology, and cryobiology in order to develop efficient strategies for the conservation of fish genetic resources. The number of fish cell lines that have been developed worldwide over the past ten years has increased to

Recent Advances in Sustainable Aquaculture and Fisheries. Shabir Ahmad Dar, Imtiyaz Qayoom, Gohar Bilal Wani, Adnan Abubakr, and Seemab Zehra (Eds.)

DOI: 10.1201/9781998511372-10

783. These 280 new fish cell lines were created from 20 different tissue types of 44 different fish families. This chapter focuses on the various factors that contribute to the decline in fish biodiversity, suggests solutions, and promotes cell line repositories as an alternative tool for protecting fish germplasm.

10.1 INTRODUCTION

Fisheries is an emerging sector in Jammu and Kashmir (J&K) economy which has been registering consistent growth during the past few decades. The significant fish fauna of Kashmir water bodies comprises of about 22 species, including indigenous schizothoracids *(Schizothorax* spp.)*, Bangana diplostoma, Crossocheilus diplochilus, and* exotic carp (*Cyprinus carpio*). *C. carpio* is represented by two phenotypes, *C. carpio* var. *communis* and *C. carpio* var. *specularis*. Schizothoracid form the main indigenous group with a wide commercial importance in the valley (Shah and Balkhi, 2020). It inhabits both lentic and lotic water bodies of Kashmir (Ahmad et al., 2017). Schizothoracids are highly valued and preferred table fish in Kashmir valley. It consumes debris, attached plants (including algae), coatings on stones and other rocks, and the invertebrate fauna that is present there. The *Schizothorax* species of Kashmir are mainly represented by *S. niger (Ale gad)*, *S. curvifrons (Satter gad), S. plagiostomus (khont), S. esocinus (chirruh),* and *S. labiatus* (chush) (Yousuf et al., 2006; Khan and Ali, 2013).

10.2 FACTORS AFFECTING NATIVE SPECIES

The decrease in fish stock is caused by a number of human actions, including construction of dams and barrages, siltation, agricultural run-off, and many more, in addition to the natural pressure. Overfishing, water pollution, flow alteration, habitat destruction or degradation, and invasion by exotic species are the five main challenges to the biodiversity of fish worldwide, and fish stock is declining as a result of interacting and combining effects. Intensification of aquaculture in cold water sector has resulted in disease outbreaks such as parasitic, bacterial, fungal, and predominantly of viral etiology leading to the mass mortality of fishes and, therefore, has been regarded as one of the most important obstacles to the development of sustainable aquaculture. In an aquatic ecosystem, disease occurs due to the complex interaction between the host, pathogen, and the environment (Engering et al., 2013). This might be due to the adoption of intensive farming practices, extensive use of antibiotics, unregulated use of inputs, and inbreeding in hatcheries. In the intensive system, fish are farmed in high stocking density which in turn results in exposure of animals to opportunistic pathogens. Disease outbreak causes economic loss to farmers due to lower production and harvests of cultured fish. Moreover, pathogen infestation has been increased due to unplanned and unregulated expansion of fish farms. Recent observations suggested that degradation in ecosystem and introduction of exotics from last decade has resulted in drastic decline in fish fauna along with emergence of diseases (Qureshi et al., 2016).

10.3 CONSERVATIVE STRATEGY BY USING CELL LINES

The necessity for conservation action across species, the scope of its application, and the results it has had in averting species extinctions must be better understood in order to increase the efficiency of conservation action. Numerous conservation measures have been created to address the situation, which involves major risks to both biodiversity and the stability of aquatic ecosystems across the world (Goswami et al., 2016). Effective analysis, careful planning, and active resource management are required for the conservation of fish genetic resources. One of the vital living resources for the preservation of fish genetic diversity is acknowledged to be fish cell lines. Cell culture is essentially the process of cells surviving and reproducing in a synthetic environment. The development of piscine cell lines started with the establishment of rainbow trout gonad (RTG-2) cell lines (Wolf and Quimby, 1962). As primary cell cultures or cell lines, almost 20,000 different species of bony fish have been isolated. More than 80% of the world's production, or over 60% of cell lines, are produced in Asia. These cell lines come from 19 different freshwater fish species, 22 marine fish species, and 11 brackish water species (Lakra et al., 2010).

First-time successive initiative was made on established and characterized continuous cell line (SISK) from kidney of *Lates calcarifer*, which is India's first marine fish cell line (Hameed et al., 2006). Using a mechanical method or an enzymatic process, the cells are separated from the sample organism and then placed in a medium for cell culture (Goswami and Lakra, 2012) that contains all nutrients and energy sources required for its survival. The majority of fish cell lines were derived from healthy tissues, such as the skin, gills, heart, liver, kidney, spleen, swim bladder, brain, so forth. Due to its remarkable propensity to regenerate, the fin is the second most-often utilized tissue for culture after the ovary (Fryer and Lannan, 1994). During the process of cell cultivation, cells reach the stage of confluence, that is, when they proliferate and occupy the whole available substrate on which these are grown. The cells must be a passage to a different medium and sub-cultured at this stage, known as primary culture, in order to give them more room to develop (Khurshid et al., 2022). After the successful completion of first sub-culture, the primary cell culture is called a cell line, characterized of being finite in nature and having limited life span. The population exhibits some degree of genotypic and phenotypic homogeneity due to the predominance of cells with the highest growth ability. Cloning or other positive selection techniques can turn a cell line into a cell strain by selecting a subpopulation from the culture. Following the start of the parent line, a cell strain frequently develops new genetic modifications. Expanding fish cell culture research in the nation has been greatly aided by the Department of Biotechnology (DBT), Government of India. The production of cell lines in this nation has been a success, which has increased the necessity for cell line preservation in one place. The National Bureau of Fish Genetic Resources (NBFGR), Lucknow, is home to a cutting-edge laboratory for the creation and preservation of cell lines.

Cell lines have a basic and practical importance in determining basic cell physiology offering wide range of applications, viz., biomedical research; gene regulation, expression, and gene transfer studies; pathological studies (Villena, 2003); and conservation of germplasm (Kumar et al., 2020). Cell culture technique has low ethical constraints, and thus widespread application in life sciences, cytogenetics, physiology, toxicology, host-pathogen interactions, and storage of cell line repositories. Individual cells of a complex tissue can be used to know the pharmacodynamics and pharmacokinetic effects of various drugs and toxins under different conditions. Fish cell culture technique has significant importance in isolation of fish viruses that are agents of epizootics in economically significant farmed fish species (Grant and Smail, 2003; Crane et al., 2005). Fish cell lines are essential in the development of vaccines for optimizing two fundamental parameters: adjuvant selection and delivery method, in addition to the understanding and examination of immune responses in cell lines to different vaccine components. The advancement of our knowledge of human biology as well as ground-breaking applications in cell therapy and regenerative medicine are all made possible by cell culture. The fish germplasm would be preserved by using the cell line repository as a biobank.

KEYWORDS

- **characterization**
- **cryopreservation**
- **germplasm**
- ***in vitro***
- **primary cell culture**
- **virus isolation**

REFERENCES

Ahmad, I., Sheikh, Z. A., & Ahmad, I. (2017). Current status of fish fauna of river Jhelum and Dal Lake of Kashmir valley. *Bulletin of Pure and Applied Sciences- Zoology, 36A*(2), 85–92.

Crane, M. S. J., Young, J., & Williams, L. M. (2005). Epizootic hematopoietic necrosis virus (EHNV): Growth in fish cell lines at different temperatures. *Bulletin-European Association of Fish Pathologists, 25*(5), 228–231.

Engering, A., Hogerwerf, L., & Slingenbergh, J. (2013). Pathogen-host- environment interplay and disease emergence. *Emerging Microbes and Infections, 2*(1), 1–7.

Fryer, J. L., & Lannan, C. N. (1994). Three decades of fish cell culture: A current listing of cell lines derived from fishes. *Journal of Tissue Culture Methods, 16*, 87–94.

Goswami, M., & Lakra, W. S. (2012). In *Fish cell and tissue culture* (1st Ed., p. 14). Narendra Publishing House

Goswami, M., Mishra, A., Ninawe, A. S., Trudeau, V. L., & Lakra, W. S. (2016). Bio-banking: An emerging approach for conversation of fish germplasm. *Poultry Fisheries & Wildlife Sciences, 4*(1), 1–8.

Grant, R., & Smail, D. A. (2003). Comparative isolation of infectious salmon anaemia virus (ISAV) from Scotland on TO, SHK-1 and CHSE-214 cells. *Bulletin-European Association of Fish Pathologists, 23*(2), 80–85.

Hameed, A. S., Parameswaran, V., Shukla, R., Singh, I. B., Thirunavukkarasu, A. R., & Bhonde, R. R. (2006). Establishment and characterization of India's first marine fish cell line (SISK) from the kidney of sea bass (*Lates calcarifer*). *Aquaculture, 257*(1-4), 92–103.

Khan, I., & Ali, M. (2013). Current status of fish fauna of river Jhelum, Kashmir, J&K. *Open Access Scientific Reports, 2*(3), 1–3.

Khurshid, K., Qadiri, S. S. N., Shah, F. A., Bhat, F. A., Bhat, B. A., Abubakr, A., Dar, S. A., & Wali, A. (2022). Establishment of a primary cell culture system from heart of *Schizothorax esocinus*. *The Pharma Innovation, 11*(5), 330–335.

Kumar, M. S., Soni, P., Singh, N., Kumar, R., Srivastava, S., Mishra, A. K., Singh, V. K., & Kushwaha, B. (2020). Establishment and characterization of eye muscle cell line from snowtrout, *Schizothoraxrichardsonii* (Gray, 1832), a vulnerable coldwater fish, for *in vitro* studies. *Fisheries and Aquatic Sciences, 21*(2), 95–105.

Lakra, W. S., Swaminathan, T. R., Rathore, G., Goswami, M., Yadav, K., & Kapoor, S. (2010). Development and characterization of three new diploid cell lines from *Labeo rohita* (Ham.) *Biotechnology Progress, 26*(4), 1008–1013.

Qureshi, N. W., Krishnan, M., & Chandrasekaran, S. (2016). 'Fish for all' verses 'Fish of choice'- growth, instability and stakeholders' responses for enhancing fish production in major lakes of Kashmir. *Current Science, 110*(8), 1495–1504.

Shah, F. A., & Balkhi, M. H. (2020). In *Handbook on fish disease and diagnosis* (pp. 1–13). Narendra Publishing House, New Delhi, India.

Villena, A. J. (2003). Applications and needs of fish and shellfish cell culture for disease control in aquaculture. *Reviews in Fish Biology and Fisheries, 13*, 111–140.

Wolf, K., & Quimby, M. C. (1962). Established eurythermic line of fish cells in vitro. *Science, 135*(3508), 1065–1066.

Yousuf, A. R., Bhat, F. A., & Mahdi, M. D. (2006). Limnological features of river Jhelum and its important tributaries in Kashmir Himalaya with a note on fish fauna. *Journal of Himalayan Ecology and Sustainable Development, 1*, 37–50.

CHAPTER 11

Biochar for Sustainable Aquaculture

TAO KARA[1], CHALUNGAL THAMPI AMAL[1], VIDYA SHREE BHARTI[1], SOIBAM NGASOTTER[1], and SOIBAM KHOGEN SINGH[2,*]

[1]*ICAR-Central Institute of Fisheries Education, Mumbai, Maharashtra, India*

[2]*College of Fisheries, Central Agricultural University (Imphal), Tripura, India*

ABSTRACT

Due to the intensification of aquaculture in recent years, farmers are using chemicals like fertilizers, pesticides, and antibiotics in the culture system to maximize production within minimum time periods. These chemicals are enhancing production but degrading the environment in the long run. In such a scenario, biochar from agro-waste emerges as a sustainable option to incorporate in aquaculture systems to maximize production without affecting the environment. Biochar is a predominantly stable, recalcitrant organic carbon compound, which can be obtained when biomass is heated to a temperature usually between 300 and 1000°C under low (preferably zero) oxygen concentrations. Biochar has properties like high cation exchange capacity, high surface area, high micro- and macropores, lower bulk density, and so on, so it can be used in aquaculture for various purposes. For instance, when used as a feed additive, biochar improves the feed digestibility, growth, and survival rate in fish and other terrestrial animals. Other uses of biochar in aquaculture include water quality management, pond sediment enhancement, and mineral supplementation in aquaculture systems like inland saline systems, which lack potassium,

Recent Advances in Sustainable Aquaculture and Fisheries. Shabir Ahmad Dar, Imtiyaz Qayoom, Gohar Bilal Wani, Adnan Abubakr, and Seemab Zehra (Eds.)

DOI: 10.1201/9781998511372-11

calcium, and magnesium. Biochar also helps in acid neutralization due to its alkaline nature, thus reducing the need for lime application in aquaculture systems. Furthermore, it has recently been found that biochar can remove antibiotics from the system, which will help mitigate antimicrobial resistance in bacteria. In this context, this chapter will provide an overview of biochar from agricultural waste for application in aquaculture to enhance fish production by maintaining sustainability.

11.1 INTRODUCTION

To enhance aquaculture production, farmers are adopting intensive farming systems. Intensive fish culture is farming the aquatic organisms at higher stocking densities in a confined environment. With increased stocking density and intensification, there is a high chance of disease outbreaks, which may cause huge losses for the farmers. To overcome these issues, farmers are forced to use antibiotics, fertilizers, and other chemicals in aquaculture systems. These chemicals help achieve high production, but they degrade the environment in the long run, like soil losing fertility, and water and soil getting polluted at the same time. Due to the overuse of antibiotics, many bacteria are becoming resistant. Microbes are getting resistant to antibiotics because it has been reported that more than 75% of antibiotics given in drugs are released from the body of the fish as secretions, as nonmetabolites, into the aquatic environments (Lalumera et al., 2004).

Therefore, sustainable, environment-friendly, cost-effective technologies are needed to improve aquaculture production systems' efficiency. In this context, incorporating biochar into aquaculture systems can be the best option while considering the novel properties of biochar. The pollutants from sediments and aquatic environments can be removed with the help of biochar. Biochar can act as an adsorbent for the inorganic and organic contaminants from the system due to its highly porous nature. Biochar can also be used as feed additives to improve fish growth, sediment, and water quality. When used as a feed additive, it decreases feed efficiency ratio (PER) (feed conversion ratio [FCR]), increases length and weight gain, increases digestive enzymes, reduces stress enzymes, enhances digestibility, and improves hematological parameters. When incorporated into sediments, it enhances the sediment's structure and texture; at the same time, it can act as a home for thousands of beneficial microbes. It can also be used to remove or remediate the antibiotics from sediment and aquatic environments. Biochar is a solid carbonaceous material that is produced at high temperatures in the absence of oxygen or low oxygen. It can be produced from all sorts of biomass. In a broader sense, the preparations and utilizations of biochar in various fields can act as a carbon sink because it does not mineralize as fast as the original biomass and has a very long residence time. Therefore, biochar can also be considered a carbon-negative technology.

11.2 BIOCHAR AND ITS HISTORY

The term biochar generally refers to the carbon-rich solid product when biomass is thermally decomposed in the absence of oxygen (Lehmann et al., 2006). In China, black pottery mixed with charcoal was discovered in relics unearthed from the Hemudu site, which is more than 7000 years ago. It was mixed with charcoal to minimize cohesion and to speed up the finished products. In 1971, charcoal was found in the Han Dynasty, the Tomb owner was buried more than 2100 years ago, but still, the body was well preserved due to the special type of charcoal found in the Tomb. In the 1960s, Wim Sombroek discovered soil rich in black matter in the Amazon Basin known as *Terra Preta*. This soil contains organic matter, nitrogen, phosphorus, potassium, and other plant nutritional elements (Tenenbaum, 2009; Harder, 2006). *Terra Preta* was believed to contain three times higher soil organic matter and nutrient levels (Glaser, 2007). Archaeological evidence suggests that biochar production and utilization started 2500 years in the Amazon basin (Glaser, 2007). Biochar can be produced from any type of biomass residue. Figure 11.1 depicts the biochar produced from paddy straw.

11.3 METHODS OF BIOCHAR PRODUCTION

Biochar is produced under anaerobic or limited oxygen conditions. Production methods can be broadly divided into, that is, traditional and modern methods based on their advancement and modernization. The traditional method mostly includes pyrolysis. Further, it is divided into fast and slow pyrolysis based on the heating rate, temperature, and residence

FIGURE 11.1 Biochar produced from paddy straw.

time, and modern methods into gasification, Torre faction, flash pyrolysis, hydrothermal carbonization, and vacuum pyrolysis. The electric kiln used in biochar preparation is shown in Figure 11.2.

FIGURE 11.2 Electric kiln used in biochar preparation.

11.3.1 TRADITIONAL PYROLYSIS

11.3.1.1 SLOW PYROLYSIS

In this process, it takes more time to produce biochar as a main product. It is also called a traditional process, in which biomass is charred at 300–600°C with a charring rate of 5–7°C/min. In slow pyrolysis, biochar yield is 35–45%, along with 25–35% other products and 20–30% syngas (Lai et al., 2013; Méndez et al., 2015). A continuous screw/Auger pyrolizer reactor is employed in slow pyrolysis.

11.3.1.2 FAST PYROLYSIS

Fast pyrolysis is a highly efficient thermochemical process where biomass-derived biofuels are produced. Short retention time and a high product recovery rate are advantages of fast pyrolysis. However, the major product of fast pyrolysis is biofuel and syngas rather than biochar (Huang et al., 2016). The fast pyrolysis process is usually carried out at a high temperature (>500°C) with a charring rate of more than 300°C/min in anoxic conditions. As the name suggests, in fast pyrolysis, biochar is produced within no time, that is, it takes a few seconds to complete the process. The syngas and bio-oil are the major main product of this process, that is, 60% bio-oil, 20% syngas, and the remaining 20% biochar (Dai et al., 2017; Huang et al., 2016).

11.3.2 MODERN METHODS OF BIOCHAR PRODUCTION

11.3.2.1 GASIFICATION

Gasification is a common method for producing syngas from different solid fuel resources. Compared with other traditional methods, it will yield greater syngas volume and lower levels of emissions (Gabhane et al., 2020). The major product in the gasification is hydrogen, though it generates some sort of biochar. The biochar produced during the gasification is assumed to be waste. Still, it has many uses, such as dye removal from wastewater, adsorption of chemicals, carbon sequestration, and soil amendment (González et al., 2008).

Gasification uses temperatures over 700°C where organic carbon-containing biomass is transformed into H_2, CO_2, and CO, which is carried out under a controlled environment of steam and oxygen. At the time of the gasification process, heat transfer occurs within a particle which increases the temperature of the biomass and that leads to the removal of water. Further, it removes the pyrolytic volatiles, that is, a precursor of the primary volatile compound: hemicellulose, cellulose, lignin, and permanent gas species are CO_2, CO, and methane (Gabhane et al., 2020).

11.3.2.2 TORREFACTION

Torrefaction is the pretreatment for raw material or biomass where physical, chemical, and biochemical characteristics are improved for better combustion and gasification at relatively low temperatures (230–300°C) (Tumuluru et al., 2011; Almeida et al., 2010). Torrefaction is the thermochemical process of treating raw material, performed only under atmospheric pressure under anoxic conditions. Mostly fixed bed reactor is needed to carry out the torrefaction process. Some studies suggest that when they carried out light torrefaction at 240°C, it was found to have a significant effect on hemicellulose, while it had no significant effect on lignin and cellulose. Further, they have suggested that light torrefaction is better for the pretreatment of biomass for producing fuels because 60% of biomass is preserved in light torrefaction.

11.3.2.3 FLASH PYROLYSIS

Flash pyrolysis is an improvised and reconstructed form of fast pyrolysis (Gabhane et al., 2020). Sometimes flash pyrolysis is also termed as very fast pyrolysis, characterized by a rapid heating rate of more than 1000°C/s and a very high reaction temperature rate (900–1200°C) (Demirbas and Arin, 2002). Flash pyrolysis yields high bio-oil due to a high heating rate with high temperature and low residence time; however, it reduces biochar yield. A fluidized bed and twin-screw mixing reactor are used to carry out flash pyrolysis. It has limited industrial application due to difficulties in constructing the reactor, which needs a very high temperature and heating rate.

11.3.2.4 VACUUM PYROLYSIS

It is a thermal decomposition of biomass carried at low pressure in anoxic conditions. The temperature and pressure must be controlled between

450 and 600°C and 0.05 and 0.2 MPa, respectively (Tripathi et al., 2016). Though the heating rate and heating condition are identical to slow pyrolysis, end products are greatly different from one another due to the effective elimination of vacuum during pyrolysis. In vacuum pyrolysis, only vacuum or low pressure is required to eliminate the vapor generated during pyrolysis, which shows a significant positive impact on the product quality and yield due to the prevention of devolatilize inorganic (Roy and Chaala, 2001). Vacuum pyrolysis produces very high-quality biochar with high porosity and thus is very useful in the adsorption of minerals and nitrogen; hence, it is used in soil amendment (Uras et al., *2012*).

11.3.2.5 HYDROTHERMAL CARBONIZATION

It is an advanced and prominent technique for the preparation of biochar. It has advantages over conventional charring because microwave generates thermal energy through dielectric heating and there is no contact between the energy source and reaction mixture. It decreases the required temperature by 200°C to achieve a similar result (Patil and Yadav, 2018; Huang et al., 2016). Microwave pyrolysis is selective charring of material and is also more rapid than the conventional technique. It is drawing more attention in the academic and industrial fields because of its rapid, selective, and uniform heating and decrease in the sintering temperature that enhances the gasification.

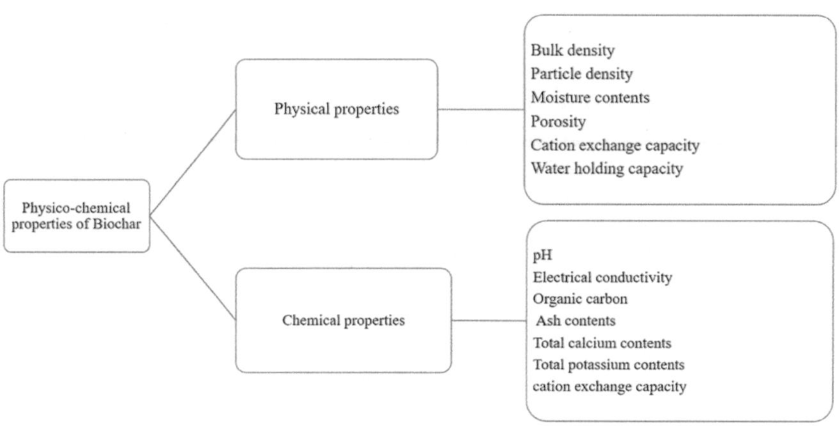

FIGURE 11.3 Physicochemical properties of biochar.

TABLE 11.1 Types of Biochar and Its Properties.

Feed Stocks	Pyrolysis Temp. (°C)	pH	EC (dS/m)	Carbon (%)	Reference
Eucalyptus	300	5.94	0.68	37.25	Dejene and Tilahun (2019)
Acacia	300	8.07	3.79	65.50	
Rice straw	300	6.40	4.29	40.90	
Orange peel	300	12.1	1.2	60.4	Oh et al. (2012)
Wood	500	10.8	0.8	67.7	
Sludge	500	6.6	0.4	7.8	
Coconut coir	400	8.04	–	49.73	Suman and Gautam (2018)
Peanut shell	300	7.8	–	68.3	Ahmad et al. (2012)
Wheat straw	400	9.1	–	65.5	Kloss et al. (2012)
Areca catechu dust	450	8.9	–	54.03	Bardalai et al. (2017)
Poultry litter	450	9.9	5.4	38	Chan et al. (2008)
Sugarcane bagasse	500	8.4	0.18	65	Quirk et al. (2012)
Sugarcane bagasse	600	11.7	4.65	66.90	Haghighatjou and Shirvani (2020)

11.4 APPLICATIONS OF BIOCHAR

Carbon in biochar is highly stable; therefore, it was initially used for soil amendment to fix carbon in the soil and hence helps in carbon sequestration (Vaccari et al., 2011; Novak and Johnson, 2019). However, biochar has multifunctional applications beyond carbon fixation. As biochar obtained from organic residues is a rich source of nutrients (Novak et al., 2019), it is used to improve the soil's physiochemical and biological properties and productivity. However, its contents change with types of biochar and pyrolysis conditions. Due to the highly porous nature of biochar, it acts as a home for thousands of useful microbes. Biochar is also used to mitigate contaminants like heavy metals and organic and inorganic pollutants from the sediments. It is known to improve soil enzyme activities because it can harbor microbes. When biochar is used as feed additives, it improves animal health by improving nutrient digestibility, and absorption, thus producing biochar. Therefore, animal husbandry regularly uses biochar as a feed additive (Man et al., 2021). Similarly, there are reports of using biochar as a feed additive in aquaculture (Thu et al., 2010; Khaki, 2018).

Biochar contains surface functional groups such as amines (NH_2^+) and hydroxyl (OH^-), making biochar an excellent adsorbent. Hence, biochar is widely used to remove pollutants and contaminants from the system, such as heavy metals, ammonia, and phosphate, and helps mitigate greenhouse gases (GHGs). Apart from all these applications, it has also been found to be effective in removing antibiotics from the system. Some of the important physiochemical properties of biochar have been shown in Figure 11.3. The properties of biochar, like pH, carbon contents, and electrical conductivity, change with raw materials. Types of biochar and their properties are shown in Table 11.1. In the following sections, we have elaborated on the applications of biochar in these fields.

11.4.1 APPLICATIONS OF BIOCHAR IN SEDIMENT

11.4.1.1 NUTRIENTS SOURCE

Biochar is rich in various micro and macro minerals that plants and animals require for their growth and maintenance. These essential nutrients can be met by adding biochar to the sediments. Some of these essential nutrients in the soil are phosphorus, magnesium, nitrogen, potassium, calcium, and other trace elements, which are naturally present in biochar (Quilty and Cattle, 2011). These mineral contents of the biochar rely on the raw materials and the temperature at pyrolysis process is carried out. Therefore, biochar can be used as fertilizers for improving sediment quality in inland saline soil, which is lacking in potassium and leads to an imbalance of calcium and magnesium ratio, which are critical for the shrimp culture system (Dahlawi et al., 2018). Applications of biochar made from sugarcane bagasse biochar and banana peduncle in inland saline sediment improve the availability of potassium, calcium, and magnesium (Jateen et al., 2023). The mineral composition in biochar may vary with raw materials and pyrolysis temperature. For instance, biochar from poultry litter, rice straw, and chicken manure contains higher potassium than apple branches, rice husks, and corn stalks. Potassium contents of biochar are positively correlated with pyrolysis temperature. The potassium contents of chicken manure increased from 4.16% to 5.93% when the pyrolysis temperature was increased from 250 to 500°C (Xiao et al., 2018). Biochar can act as a nutrient sink because it reduces the leaching loss of nutrients when applied to the soil. Its capacity to minimize nutrient loss is due to high porosity, high surface area, high CEC, organic coating, and other factors like pH

(Zheng et al., 2013; Conte and Laudicina, 2017). Therefore, the administration of biochar in ponds reduces the loss of nutrients via seepage and leaches into the ground due to increased water holding capacity.

11.4.1.2 SOIL HEALTH

Soil is a finite and fragile resource that can be easily degraded by some natural calamities and anthropogenic activities. Healthy soil can support the rich biodiversity of the microorganisms, which are essential for recycling essential nutrients, enhancing the soil structure, and ultimately leading to higher crop production (FAO, 2011). Biochar has all the required characteristics and properties to maintain soil health. As it can act as a home for thousands of microbes, it shall help recycle important nutrients in the system. Applying biochar to the soil reduces the soil's bulk density due to biochar's porous nature (Burrell et al., 2016; Sorrenti et al., 2016). Many studies suggest that biochar increases sediment's water-holding capacity when applied to the sediment system (Asai et al., 2009; Bordoloi et al., 2018; Sun et al., 2018). Another reason for an increase in water holding of sediment is due to the organic carbon contents of biochar, which helps in the aggregation of biochar (Dahlawi et al., 2018). Biochar contains the highest percentage of organic carbon, which plays a very crucial role in maintaining the physical structure of soil by increasing CEC, sediment aggregation, and releasing phosphorus into sediment solution by increasing sediment porosity (Lashari et al., 2013; Sharma et al., 2016; Zheng et al., 2013). Biochar is known for its lower bulk density. When applied in the soil, it reduces the bulk density of the same, and a decrease in bulk density improves the soil structure (Sharma et al., 2016). When biochar is applied to soil, it reduces the EC; the reason for the decrease in EC of the soil is due to an increase in soil porosity and hydraulic conductivity which helps in the retention of salts in pores of biochar (Thomas et al., 2013). Biochar application in inland saline pond soil reduces the soil's pH because of increased cations such as K, Ca, and Mg (Hasan, 2018). Biochar is alkaline in nature; therefore, applying it in aquaculture ponds will reduce the lime requirements due to its alkaline nature and good buffering capacity (Schmidt and Wilson, 2012). The application of biochar decreases the nitrogen available from normal and salt-affected soil; this is because when biochar is administered in the soil, it enhances the microbial population of the system, as nitrogen is required to carry out metabolism activities (Norton and Firestone, 1996).

11.4.1.3 IMMOBILIZATION OF CONTAMINANT AND TOXIC SUBSTANCES

Biochar is known for removing almost all types of pollutants from the system. It can fix potentially toxic elements in biochar. This is called the immobilization of the pollutants. The efficiency of the biochar to immobilize the potentially toxic elements depends on the raw materials, biochar dose, pyrolysis conditions, soil texture, and other properties of soil. Nowadays, organic contaminants like pesticides, persistent organic pollutants, endocrine disrupters, and pharmaceutical compounds are mostly found in water and soil. Biochar is one of the best options for immobilizing these compounds, so they will not be bioavailable to plants and animals. This property is due to high surface area, micropores, high specific surface area, basicity, and highly aromatic. The soil properties like pH, clay contents, organic matter, and microbial communities also play important roles. The most important factor responsible for immobilizing the organic compound is aromaticity and surface polarity, which depend on raw materials, pyrolysis temperature, and modification methods. Different types of modification like acid modification, base modification, and doping of nonmetallic elements with biochar can extend these qualities. It has been used to remove some organic pollutants like carbendazim, imidacloprid, atrazine, polychlorinated biphenyl, and polycyclic aromatic hydrocarbons (PAHs) from the sediments system. The removal mechanism of organic pollutants by biochar is through various processes like pore diffusion, electrophilic interaction, electrostatic attraction/repulsion, and hydrophobia through various functional groups present in the biochar. Inorganic pollutants like heavy metals are usually not degraded by the activities of microorganisms and remain in sediment for a longer time causing severe health problems to humans and animals. They can easily enter animals and human beings via the food chain. It has been shown that biochar applications reduce the bioavailability and phyto availability of Pb, Cu, Cu, and Cr in the sediments (Meng et al., 2018; Zhang et al., 2019). Some studies say that biochar produced at a lower temperature is more suitable for the remediation of inorganic pollutants (<500°C).

Biochar has very important properties responsible for the remediation of inorganic pollutants due to high organic carbon contents, high specific surface area, and many functional groups, which interact with heavy metals in many ways. However, sorption is the most prominent mechanism for eliminating heavy metals. Other mechanisms like adsorption at the macropore, nanopore, and micropore make the heavy metal more stable. Some studies show that heavy metal removal by biochar occurs through complexation and surface interaction between heavy metal and functional groups present in biochar. The interaction of biochar with heavy metal depends on pyrolysis temperature, types of biomasses, rate of application, and pH of the biochar and sediments. At lower temperatures, the removal occurs by sorption, and at higher temperatures, removal occurs via complexation and electrostatic interaction. The removal efficiency increases with an increase in pH, surface area, and biochar application rate. Hardwood biochar more efficiently eliminates heavy metals like Cd^{2+}, As^{3+}, and Cu^{2+} via metal complexation. Biochar also helps reduce As^{5+} to As^{3+} and immobilize As^{3+} in sediments.

Nowadays, pesticides and herbicides are more frequently used in agricultural and aquaculture systems, the residues remain in the soil for a longer time, and these pollutants need to be removed from the system. In this context, biochar can be used to eliminate these contaminants from the system.

11.4.1.4 SOIL ENZYMES ACTIVITIES

Soil enzymes play a crucial role in maintaining the soil microbial activities, which helps maintain the biogeochemical cycle and soil ecosystem. Therefore, it is very important in nutrient release, organic matter conversion, and soil fertility maintenance. Soil enzymes can also indicate various characteristics of soil, like fertility levels, the status of soil pollution, soil chemical properties, and microbiological characteristics. When administered in soil, it increases enzymatic and microbial activities and improves the C/N ratio due to an increase in carbon content. Studies show that biochar improves the all-important enzymatic activities of soil, like urease, catalase, phosphatase, and sucrase (Jiang et al., 2021). The main reason for the increase in soil enzyme activities is that biochar increases the organic carbon and nitrogen pool, which supplies energy to microbes and supports the growth of microbes, increasing the system's enzymatic activities. Other factors responsible for increased enzymatic activity are N, P, soil moisture, and microbial biomass carbon. It is known that phosphorus is a limiting nutrient in the lake, river, and pond system because the phosphorus cycle is not available in gas form (follows sedimentary cycles). Therefore, it is found only in sediment and aquatic environment. Hence, maintaining phosphorus is very important. As mentioned above, biochar can act as a home for thousands of microbes. These microbes help the solubilization of phosphorus compounds via enzyme activities and the production of inorganic and organic acid, production of CO_2, and H_2S. The role of biochar in maintaining soil fertility and the biogeochemical cycle is shown in Figure 11.4.

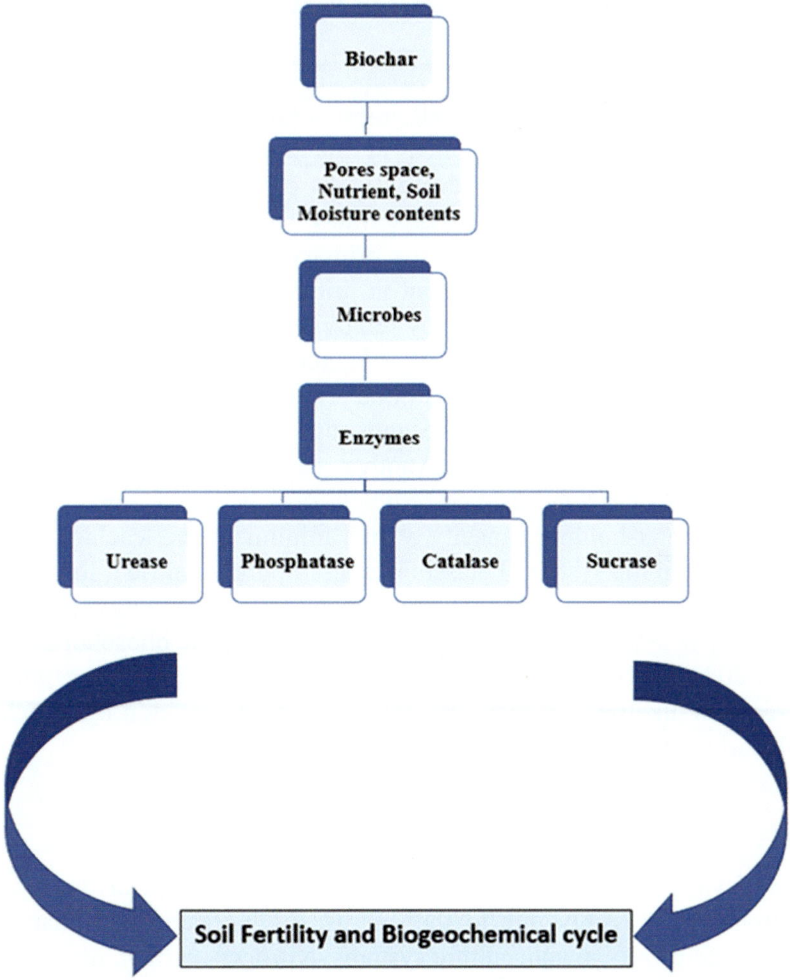

FIGURE 11.4 Role of biochar in maintaining soil fertility and biogeochemical cycle.

Recent studies also confirm that biochar-amended sediment in inland saline shrimp culture systems had significantly low dehydrogenase activity compared to the control sediment sample, where no such amendment has been done. Reduced dehydrogenase activity corresponds to low organic matter availability in the system; thus, compared to control sediment, the sediment amended with biochar is healthier and is conducive for successful aquaculture.

11.4.2 APPLICATIONS OF BIOCHAR IN WATER

11.4.2.1 PRIMARY PRODUCTIVITY

Biochar is known for the overall improvement of the system's water quality. When biochar is applied in the aquaculture system, it will positively impact primary productivity. As plankton are very important components of any cultural system. Study shows that biochar application in water and feed increases the primary productivity of the system (Lan et al., 2016). Its improvement in primary productivity may be due to enhancement in water quality by removing pollutants from the system and supplementing nutrients like N, P, K, and Mg. By applying sugarcane bagasse biochar in inland saline pond systems, the potassium content significantly increased in water because sugarcane bagasse biochar is rich in potassium (Raul et al., 2021). Another reason for the increase in nutrients in the biochar-administered sediment may be due to organic carbon and good water-holding capacity, which minimizes the leaching loss of nutrients from the system. Potassium is an important nutrient for primary productivity (Wakeel, 2013). The phytoplankton concentration positively correlates with the alkalinity, pH, nitrate, and available phosphorus (Raul et al., 2021). If biochar is applied, it will increase the alkalinity, and the increase in alkalinity triggers the growth of phytoplankton by enriching with carbonate and calcium to the system.

11.4.2.2 WATER QUALITY IMPROVEMENT

Nowadays, eutrophication is one of the key challenges in the aquatic environment, possibly due to the overuse of inorganic phosphate and nitrogen fertilizers in aquaculture systems. Nitrogen, carbon, magnesium, phosphorus, and potassium in water are the main factors for eutrophication. The porous nature, high surface area, and surface functional group in biochar are responsible for removing contaminants from water, especially heavy metals, organic pollutants, and other pollutants. The removal efficiency of biochar depends on the raw materials, pyrolysis temperature, pH, and biochar application rate. It has been noticed that phosphate adsorption capacity correlated with Mg and Ca contents of biochar. Biochar is being used to remove heavy metals from the aquaculture system. It will improve the fish meat, growth performance, immunity against diseases, and hatchability. Some studies show that applying biochar in shrimp and fish culture systems gives better immunity and weight gain results than in a system without biochar. Biochar can also be used for filtration media. When biochar is used as filtration media, it will reduce the heavy metal from the pond system and reduce the chance of bioaccumulation in fish meat. Due to its highly porous nature and high surface area, biochar will adsorb the pollutants from the water. Therefore, it is one of the most suitable materials to be used as filtration media in recirculatory aquaculture systems for total ammonical nitrogen (TAN) and unionized ammonia since TAN and unionized ammonia is one of the major concerns in recirculatory aquaculture systems. Biochar can also be used in biofloc technology as an alternative carbon source because it is found to enhance the biofloc system's water quality (Abakari et al., 2020).

11.4.3 BIOCHAR FOR IMPROVING GROWTH AND HEALTH OF FISH

11.4.3.1 BIOCHAR IN FEED

Charcoal has a long history of being used for treating digestive problems like diarrhea for humans as well as livestock. Regular application of charcoal as feed additives started from the 19th to 20th century for improving and growth performance of the animal, which is very similar to biochar. Recently, people began using biochar instead of charcoal because it can be produced from any type of biomass. Many studies show that feeding biochar as feed additives have various positive benefits like an increase in weight gain, PER, specific growth rate, feed intake, survival rate, decrease in FCR, improved digestibility, aids in detoxification of toxic substances, enhances hematological parameters, and maintains good water qualities (Najmudeen et al., 2019; Khaki, 2018; Khalid et al., 2022; Prasai et al., 2016; Saroeun et al., 2018). It has been seen that when biochar was fed to striped catfish (*Pangasius hypophthalmus*), the growth rate was increased by 36%. At the same time, it reduced TAN, phosphate, nitrite, and chemical oxygen demand in the tank (Lan et al., 2016). When biochar is used in the pokkali system, it significantly improves the growth of fish and paddy compared to the control, where no biochar is used (Najmudeen et al., 2019). The improvements in the growth of the fish may be due to various reasons like biochar added to feed increases the dissolved oxygen level of water, pH, alkalinity improves digestibility and enhances the immunity of fish. Some other reasons are high surface area, which harbors the microbes, and those microbes help in the digestion of solid feed in the digestive tract of fish by forming a biofilm. The biochar-enriched fish feed is shown in Figure 11.5.

It has also been found that when Nile tilapia was fed with activated charcoal through feed as feed additives, it increased the length and number of goblet cells, which play a key role in the absorption of digested food from digestive tracks (Boonanuntanasarn et al., 2014). Another reason for better growth in biochar-incorporated feed is that biochar slows down the passage of digested particles in the gastrointestinal tract, which increases the villi function and reduces the pathogenic bacteria in the guts.

It has been noticed that feeding biochar reduces the stress enzyme and increases the digestive enzymes. Feeding biochar increases the microbes in the guts of the fish, which are responsible for the increase in digestive enzymes in the gut system.

FIGURE 11.5 Fish feed enriched with biochar.

Microbes like *Acinetobacter* found in Atlantic salmon and seabass larvae, which are responsible for the production of amylase, also improve the hematological parameters and health status of fish and other terrestrial animals. At present, only a few studies are there which investigate the effect of biochar feeding in fish concerning hematological parameters, but many studies have been done in veterinary science (Mabe et al., 2018; Dim et al., 2022; Elghalid, 2022).

11.4.3.2 BIOCHAR IN SOIL

Nowadays, biochar is commonly used in soil to improve sediment qualities. When it is used in the soil, it improves the soil's properties and structure at the same time. As mentioned in the above section, it also removes organic and inorganic pollutants from the system. Besides improving soil quality, it enhances growth parameters, survivability, enzymatic activities, blood and serum parameters. Applying biochar to the shrimp culture system increases the specific growth rate and weight gain. It also reduces the FCR and increases the PER. The reasons may be due to enhanced digestive enzyme activities. Hemocyanin plays a fundamental role in the osmoregulation and molting of shrimp. It has been seen that application of biochar increases the concentration of hemocyanin in shrimp. Incorporation of biochar in a culture system has been found to have many other advantages like metabolic enzyme activities (alanine transaminase and aspartate transaminase) and an increase in amylase, protease, and lipase (Jateen et al., 2023). Administration of biochar in soil has also been shown to boost fish's growth performance, specific growth rate, and weight gain increase when biochar is incorporated into the sediment bottom of the culture system (Najmudeen et al., 2019).

11.4.4 BIOCHAR FOR THE REDUCTION OF GREENHOUSE GAS

In aquaculture, fish pond activities generate emissions, principally CO_2, O_2, NH_3, CH_4, N_2O, and N_2 (Ewoukem et al., 2013). This is mostly due to the high-protein-rich diet provided to fish, which was either poorly digested or used, resulting in increased excretion into water. Aquaculture systems have substantially higher biological density, productivity, and enrichment from fertilizer and feeds than natural water bodies (Ye et al., 2022). These circumstances encourage rapid respiration and methanogenic rates, resulting in significant CO_2 and CH_4 emissions (Yang et al.,

2021). By 2030, aquaculture might contribute 6% of anthropogenic N_2O emissions if it grows at the current rate, with an average yearly growth of roughly 8% (Hu et al., 2012). Therefore, aquaculture is a practice that contributes to the greenhouse effect through the fundamental processes in the soil–water–animal system. Depending on the management techniques, it can either act as a source or drain these gases.

Biochar and pyrogenic carbon capture and storage are recognized for the first time as promising negative emission technologies in the 15th IPCC (Intergovernmental Panel on Climate Change) special report on global warming of 1.5°C. Both agricultural and environmental benefits can be achieved by applying them to soil (Sohi, 2012). By carbon sequestration and control over GHG fluxes in soil, biochar amendment has been widely employed to combat climate, because of its ability to affect the pH of the soil, adsorb dissolve organic carbon (DOC) and NH_4^+/NO_3^-, and alter the microbial population, biochar can reduce GHG emissions from the soil (Van Zwieten et al., 2014). Using biochar in rice fields might greatly reduce GHG emissions (GoI MAFW, 2017).

By changing soil parameters like soil pH, Eh, or other contributing variables, biochar addition may indirectly lower CH_4 emissions, according to research by Pratiwi and Shinogi (2016). Soil DOC concentration and soil organic matter were adversely linked with CH_4 emission (Qin et al., 2016). In addition, the recalcitrance of biochar limits the availability of DOC for methanogens, which may hinder CH_4 synthesis. The negative association between CH_4 emission and DOC indicates the improvement of re-oxidation capacity in the rhizosphere after adding biochar. Yang et al. (2017) found that biochar addition increased methanotroph activity while decreasing CH_4 emissions. An increasing pH may limit methanogenic activity and result in lower CH_4 emissions. The cascading use of biochar in animal farming systems also lowers the damaging environmental losses of ammonia through volatilization or nitrate through leaching (Liu et al., 2018; Borchard et al., 2019; Sha et al., 2019) and has the potential to lower GHG emissions like nitrous oxide N_2O (Kammann et al., 2017; Borchard et al., 2019) or CH_4 (Jeffery et al., 2016).

11.4.5 REMOVAL OF ANTIBIOTICS

The prevalence of antimicrobial resistance (AMR) is one of the key issues in the aquaculture system in the present scenario. The AMR in aquaculture can be easily transferred in the environment through horizontal gene transfer, affecting the whole ecosystem. Nowadays, most ornamental and cultured fish are facing the problem of multiple antibiotic resistances. We can say that aquaculture and fish farm have become a hotspot for AMR genes.

Approximately, 51 antibiotics are used in the aquaculture system, which is important for human medicine. Of these, six antibiotics are recommended by the World Health Organization as critically important for human beings (Done et al., 2015). In the present scenario, almost all the aquaculture-producing countries use antibiotics in their culture system. One of the studies says that in Thailand, 600 mt of antibiotics are used in shrimp farms, which may be the major reason for the development of AMR because more than 70% of antibiotics used in feed are excreted in the water or remain in an aquatic environment. These antibiotics are responsible for antimicrobial gene resistance. We cannot ignore the significant increase in the growth of the aquaculture sector due to the use of antibiotics. Still, it has negative consequences on the environment and human health. Some studies confirmed that antibiotics concentrate in the sediments and environment and accumulate in tissues of cultured species which has a high chance of transferring to human beings and other carnivorous animals. Therefore, we need to remove or remediate the antibiotic from the system. In this context, biochar can play a crucial role; it will help to remediate or remove the antibiotic from the system.

The removal mechanism of antibiotics by biochar is mainly based on degradation and adsorption pathway. As we know that biochar is highly porous and can adsorb antibiotics. Using biochar to remove antibiotics has many advantages, less cost of production, is easily available, and has a simple design. There are three stages in the adsorption process: particle diffusion, external diffusion, and surface reaction. The absorption mechanism is due to pore filling, electrostatic attraction, surface complexation, π–π interaction, and hydrophobic interaction (Cheng et al., 2022). The adsorption capacity of biochar changes with types of biochar and biomass and pyrolysis temperature because pore structure changes with biomass, pyrolysis temperature, and residence time. It has been shown that biochar produced at higher temperatures is more efficient in removing antibiotics because it has high pore space. When biochar is applied in sediment or aquaculture systems, it will absorb the antibiotic from the culture system and make it unavailable to animals and plants to uptake, reducing the risk of entering into humans and higher animals via the food chain. The advantages of using biochar for adsorbents are that it is a cheaper replacement for activated charcoal due to its low cost of production ($350–$1200 per ton biochar vs. $1100–$1700 per ton activated charcoal) and it can be produced from any types of biomasses (Thompson et al., 2016). Another mechanism of antibiotic removal by biochar is degradation via a persulfate-based advanced oxidation process (AOP). The mechanism involved in the degradation of antibiotics is the production of reactive oxygen species. Persulfate-based AOP is the most suitable among other oxidation like fenton oxidation, persulfate oxidation, and ozone oxidation because it can work at a high pH range and selective oxidation. Two pathways work in the degradation of antibiotics in a persulfate-based AOP: free radical transfer and nonradical transfer.

In free radical pathways, three types of radicals are responsible for degradation, that is, OH, SO_4^-, and O_2, which are produced by biochar. SO_4^- is a strong oxidant that can oxidize many organic pollutants, including antibiotics. In addition, SO_4^- has a wide pH range (2.0–9.0). The oxidation process by OH is known to be a safe and clean technology. Due to the strong oxidation effect, it can easily degrade antibiotics, organic pollutants, and its intermediate compound into H_2O and CO_2. O_2^- radical has a relatively weaker oxidizing capacity; therefore, it plays a crucial role in the catalytic process.

Electron transfer and singlet oxygen are responsible for nonradical pathways of persulfate-based AOPs. The main highlight of nonradical pathways is that they can work in all pH ranges, that is, acidic, neutral, and basic conditions. In this process, organic pollutants or antibiotics are electron donors, biochar is an electron conductor and persulfate is an electron acceptor. In this process, adsorbed electrons from antibiotics are transferred to oxidants with the help of the surface of biochar, which restrains the recombination of electron and hole pairing, enhancing the degradation efficiency. Electron transfer capacities are directly proportional to the extent of porosity and graphitization of biochar. Therefore, we can improve the efficiency of the electron transfer capacity of biochar by different techniques of modifications, which shall enhance the porosity and surface area of biochar. Modification can be done by acid treatment (HNO_3), alkali treatment (KOH), changing pyrolysis temperature, and making a biochar composite.

Antibiotics from sediment and the aquatic system can also be remediated with the help of microbes, as it is known that biochar supports the growth of microbes due to high surface area and pore space and improves the soil qualities. Microorganisms are effective in degrading organic pollutants and antibiotics. It has been seen that biochar immobilized with microorganisms are more effective in the degradation of antibiotics than removal. Biochar also helps reduce antibiotic resistance genes (ARGs) transmission in microbes, as *Brevundimonas*, *Rhodanobacter*, and *Proteobacteria* host ARGs. It is because the administration of biochar changes the abundance of *Brevundimonas*, *Rhodanobacter*, and *Proteobacteria* in the system. Some factors crucial for microbial degradation of antibiotics are pH, initial concentration, growth of microbes, and temperature. The pH plays a crucial role in the microbial degradation of antibiotics, and some studies show that the degradation of antibiotics increases with an increase in pH. The maximum degradation of tetracycline was found at pH 9.0, and when pH decreases, the degradation capacity also decreases. More the initial concentration of the antibiotics, the higher is the degradation capacity. More microbes mean more microbes acting on antibiotics and a higher degradation capacity. One of the studies suggests that 40–55°C is most suitable for the degradation of antibiotics by microbes depending on thermophilic or mesophilic.

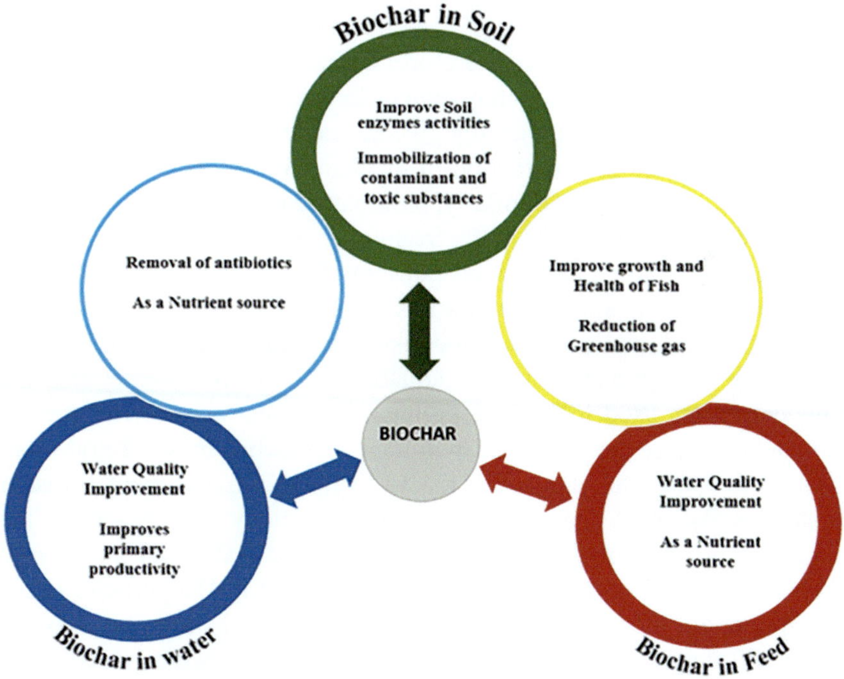

FIGURE 11.6 Interaction of biochar with different components in the culture system.

11.5 LIMITATIONS OF BIOCHAR

Though most of the literature shows biochar's beneficial effects, some limitations in biochar production and application have been briefly discussed in this section. Biochar production is not so popular in developing and underdeveloped countries, so the cost of production may be relatively higher. The production of biochar at large volumes is challenging since the yield of biochar is on the lower side, and getting a huge quantity of raw material is difficult. Biochar may inhibit the aging of the sediment because it is recalcitrant. Fresh biomass may need to be added to the system to maintain nutrient cycling. Some studies show that application negatively affects a very important component of the soil environment, that is, earthworms and fungi. The administration of biochar on soil may have direct and indirect effects on soil microbes, that is, direct effects like microbes may come in contact with heavy metal and organic pollutants trapped inside the biochar and may have detrimental effects on useful microbes in the soil. Indirect effects like biochar application may change the environmental condition like pH, changing acidic pH to alkaline pH; it may hamper the release and activities of enzymes. The effects of biochar on sediment may change from region to region; therefore, we cannot expect the same types of results worldwide, as they may give different results based on different regions. For an instance, it may not give the same results in sandy soil and loamy soil because the interaction of biochar with sandy soil and loamy soil may not be the same. Similarly, pollutant removal capacity in marine or brackish water may differ in the freshwater system.

Biochar is known to provide nutrients like nitrogen and phosphorous to the system. Some biomass is known to have very high contents of nitrogen and phosphorous, and using a large quantity of biochar may cause eutrophication of the aquatic environment. Sometimes the nanocomposites used to remove pollutants may act as a carrier for the pollutants because the adsorbed pollutant in biochar may infiltrate into groundwater and cause a potential risk to an aquatic environment. Some of the studies also show that biochar has higher toxicity on crustaceans; the toxicity effects of biochar are positively correlated to the concentration of PAHs in biochar. It has been seen that the levels of PM10 particles in the atmosphere are increased during the production and handling of biochar which will degrade the air quality. This is because biochar is very light and tends to be easily carried away by natural wind conditions and natural processes and microbial activities can break down aged biochar. It has been mentioned that feeding biochar increases the growth parameters of fish or animals, but sometimes adding biochar in the feed may reduce the feed acceptance in fish.

KEYWORDS

- **circular economy**
- **sustainability**
- **waste valorization**
- **water quality enhancement**

REFERENCES

Abakari, G., Luo, G., Meng, H., Yang, Z., Owusu-Afriyie, G., Kombat, E. O., & Alhassan, E. H. (2020). The use of biochar in the production of tilapia (*Oreochromis niloticus*) in a biofloc technology system-BFT. *Aquacultural Engineering, 91*, 102123. https://doi.org/10.1016/j.aquaeng.2020.102123

Ahmad, M., Lee, S. S., Dou, X., Mohan, D., Sung, J. K., Yang, J. E., & Ok, Y. S. (2012). Effects of pyrolysis temperature on soybean Stover-and peanut shell-derived biochar properties and TCE adsorption in water. *Bioresource Technology, 118*, 536–544. https://doi.org/10.1016/j.biortech.2012.05.042

Almeida, G., Brito, J. O., & Perré, P. (2010). Alterations in energy properties of eucalyptus wood and bark subjected to torrefaction: The potential of mass loss as a synthetic indicator. *Bioresource Technology*, *101*(24), 9778–9784. https://doi.org/10.1016/j.biortech.2010.07.026

Asai, H., Samson, B. K., Stephan, H. M., Songyikhangsuthor, K., Homma, K., Kiyono, Y. et al. (2009). Biochar amendment techniques for upland rice production in Northern Laos: 1. Soil physical properties, leaf SPAD and grain yield. *Field Crops Research, 111*(1–2), 81–84. https://doi.org/10.1016/j.fcr.2008.10.008

Bardalai, M., Bordoloi, N. K., & Mahanta, D. K. (2017). Production and biodegradability analysis of bael shell pyrolysis oil. *Materials Today: Proceedings, 4*(2), 603–610. https://doi.org/10.1016/j.matpr.2017.01.063

Boonanuntanasarn, S., Khaomek, P., Pitaksong, T., & Hua, Y. (2014). The effects of the supplementation of activated charcoal on the growth, health status and fillet composition-odor of Nile tilapia (*Oreochromis niloticus*) before harvesting. *Aquaculture International, 22*, 1417–1436. https://doi.org/10.1007/s10499-014-9756-8

Borchard, N., Schirrmann, M., Cayuela, M. L., Kammann, C., Wrage-Mönnig, N., Estavillo, J. M. et al. (2019). Biochar, soil and land-use interactions that reduce nitrate leaching and N_2O emissions: A meta-analysis. *Science of the Total Environment, 651*, 2354–2364. https://doi.org/10.1016/j.scitotenv.2018.10.060

Bordoloi, S., Garg, A., Sreedeep, S., Lin, P., & Mei, G. (2018). Investigation of cracking and water availability of soil-biochar composite synthesized from invasive weed water hyacinth. *Bioresource Technology, 263*, 665–677. https://doi.org/10.1016/j.biortech.2018.05.011

Burrell, L. D., Zehetner, F., Rampazzo, N., Wimmer, B., & Soja, G. (2016). Long-term effects of biochar on soil physical properties. *Geoderma, 282*, 96–102. https://doi.org/10.1016/j.geoderma.2016.07.019

Chan, K. Y., Van Zwieten, L., Meszaros, I., Downie, A., & Joseph, S. (2008). Using poultry litter biochars as soil amendments. *Soil Research, 46*(5), 437–444. https://doi.org/10.1071/SR08036

Cheng, G., Tan, B., Zhang, Z., Fu, S., Haiyan, W., & Wang, F. (2022). Characteristics of coal-oxygen chemisorption at the low-temperature oxidation stage: DFT and experimental study. *Fuel, 315*, 123120. https://doi.org/10.1016/j.fuel.2021.123120

Conte, P., & Laudicina, V. A. (2017). Mechanisms of organic coating on the surface of a poplar biochar. *Current Organic Chemistry, 21*(6), 559–565.

Dahlawi, S., Naeem, A., Rengel, Z., & Naidu, R. (2018). Biochar application for the remediation of salt-affected soils: Challenges and opportunities. *Science of the Total Environment, 625*, 320–335. https://doi.org/10.1016/j.scitotenv.2017.12.257

Dai, L., Fan, L., Liu, Y., Ruan, R., Wang, Y., Zhou, Y. et al. (2017). Production of bio-oil and biochar from soapstock via microwave-assisted co-catalytic fast pyrolysis. *Bioresource Technology, 225*, 1–8. https://doi.org/10.1016/j.biortech.2016.11.017

Dejene, D., & Tilahun, E. (2019). Characterization of biochar produced from different feed stocks for waste management. *International Journal of Environmental Sciences & Natural Resources, 20*(3), 98–102.

Demirbas, A., & Arin, G. (2002). An overview of biomass pyrolysis. *Energy Sources, 24*(5), 471–482.

Dim, E. C., Akuru, E. A., Mgbor, F. N., Ethelbert, C., Oyeagu, A. B. F., Lewu, F. B., & Onyimonyi, A. E. (2022). *International Journal of Veterinary Science, 11*(3), 315–320. http://www.ijvets.com/pdf-files/Volume-11-no-3-2022/315-320.pdf

Done, H. Y., Venkatesan, A. K., & Halden, R. U. (2015). Does the recent growth of aquaculture create antibiotic resistance threats different from those associated with land animal production in agriculture? *The AAPS Journal, 17*, 513–524. https://doi.org/10.1208/s12248-015-9722-z

Elghalid, O. (2022). Effect of graded levels of biochar supplementation as a growth promoter on productive and physiological performance of broiler chicks. *Egyptian Poultry Science Journal, 42*(3), 243–263. DOI: 10.21608/epsj.2022.263846

Ewoukem, T. E., Hassouna, M., Robin, P., Mikolasek, O., Aubin, J., & Ombredane, D. D. (2013). Estimation test of gaseous emissions from fishponds with contrasted inputs. https://hal.science/hal-01461133/

FAO Save (2011). *Grow: A policymaker's guide to the sustainable intensification of smallholder crop production*. Food and Agricultural Organization of the United Nations.

Gabhane, J. W., Bhange, V. P., Patil, P. D., Bankar, S. T., & Kumar, S. (2020). Recent trends in biochar production methods and its application as a soil health conditioner: A review. *SN Applied Sciences, 2*, 1–21. https://doi.org/10.1007/s42452-020-3121-5

Glaser, B. (2007). Prehistorically modified soils of central Amazonia: A model for sustainable agriculture in the twenty-first century. *Philosophical Transactions of the Royal Society B: Biological Sciences, 362*(1478), 187–196. https://doi.org/10.1098/rstb.2006.1978

González, J. F., Román, S., Bragado, D., & Calderón, M. (2008). Investigation on the reactions influencing biomass air and air/steam gasification for hydrogen production. *Fuel Processing Technology, 89*(8), 764–772. https://doi.org/10.1016/j.fuproc.2008.01.011

Haghighatjou, M., & Shirvani, M. (2020). Sugarcane bagasse biochar: Preparation, characterization, and its effects on soil properties and zinc sorption-desorption. *Communications in Soil Science and Plant Analysis, 51*(10), 1391–1405. https://doi.org/10.1080/00103624.2020.1763383

Harder, B. (2006). Smoldered-Earth policy: Created by ancient Amazonian natives, fertile, dark soils retain abundant carbon. *Science News, 169*(9), 133–133. https://doi.org/10.2307/3982299

Hasan, Ö. Z. (2018). A new approach to soil solarization: Addition of biochar to the effect of soil temperature and quality and yield parameters of lettuce (*Lactucasativa* L. Duna). *Scientia Horticulturae, 228*, 153–161. https://doi.org/10.1016/j.scienta.2017.10.021

Hu, Z., Lee, J. W., Chandran, K., Kim, S., & Khanal, S. K. (2012). Nitrous oxide (N2O) emission from aquaculture: A review. *Environmental Science & Technology, 46*(12), 6470–6480. https://doi.org/10.1021/es300110x

Huang, Y. F., Chiueh, P. T., Kuan, W. H., & Lo, S. L. (2016). Microwave pyrolysis of lignocellulosic biomass: Heating performance and reaction kinetics. *Energy, 100*, 137–144. https://doi.org/10.1016/j.energy.2016.01.088

Jateen, S., Bharti, V. S., Prakash, S., Krishnan, S., Paul, T., & Kumar, S. (2023). Sugarcane bagasse biochar-amended sediment improves growth, survival, and physiological profiles of white-leg shrimp, *Litopenaeusvannamei* (Boone, 1931) reared in inland saline water. *Aquaculture International,* 1–20. https://doi.org/10.1007/s10499-023-01077-9

Jeffery, S., Verheijen, F. G., Kammann, C., & Abalos, D. (2016). Biochar effects on methane emissions from soils: A meta-analysis. *Soil Biology and Biochemistry, 101*, 251–258. https://doi.org/10.1016/j.soilbio.2016.07.021

Jiang, Y., Wang, X., Zhao, Y., Zhang, C., Jin, Z., Shan, S., & Ping, L. (2021). Effects of biochar application on enzyme activities in tea garden soil. *Frontiers in Bioengineering and Biotechnology, 9*, 728530. https://doi.org/10.3389/fbioe.2021.728530

Kammann, C., Ippolito, J., Hagemann, N., Borchard, N., Cayuela, M. L., Estavillo, J. M. et al. (2017). Biochar as a tool to reduce the agricultural greenhouse-gas burden-knowns, unknowns and future research needs. *Journal of Environmental Engineering and Landscape Management, 25*(2), 114–139. https://www.tandfonline.com/doi/abs/10.3846/16486897.2017.1319375

Khaki, N. D. (2018). Innovative approaches in aquaculture for conservation and management of threatened fish species: The case of different strains of brown trout (*Salmo trutta* and *Salmo ghigii*). https://www.repository.unipr.it/bitstream/1889/3539/1/Thesis%20PhDdelsouz%202018.%20submition.pdf

Khalid, M. A., Hussain, S. M., Mahboob, S., Al-Ghanim, K. A., & Riaz, M. N. (2022). Biochar as a feed supplement for nutrient digestibility and growth performance of *Catla catla* fingerlings. *Saudi Journal of Biological Sciences, 29*(12), 103453. https://doi.org/10.1016/j.sjbs.2022.103453

Kloss, S., Zehetner, F., Dellantonio, A., Hamid, R., Ottner, F., Liedtke, V., Schwanninger, M., Gerzabek, M. H., & Soja, G. (2012). Characterization of slow pyrolysis biochars: effects of feedstocks and pyrolysis temperature on biochar properties. *Journal of Environmental Quality, 41*(4), 990–1000.

Lai, W. Y., Lai, C. M., Ke, G. R., Chung, R. S., Chen, C. T., Cheng, C. H. et al. (2013). The effects of woodchip biochar application on crop yield, carbon sequestration and greenhouse gas emissions from soils planted with rice or leaf beet. *Journal of the Taiwan Institute of Chemical Engineers, 44*(6), 1039–1044. https://doi.org/10.1016/j.jtice.2013.06.028

Lalumera, G.M., Calamari, D., Galli, P., Castiglioni, S., Crosa, G., & Fanelli, R. (2004) Preliminary investigation on the environmental occurrence and effects of antibiotics used in aquaculture in Italy. *Chemosphere, 54*(5), 661–668.

Lan, T. T., Preston, T. R., & Leng, R. A. (2016). Feeding biochar or charcoal increased the growth rate of striped catfish (*Pangasius hypophthalmus*) and improved water quality. *Livestock Research for Rural Development, 28*(5), 84. http://www.lrrd.cipav.org.co/lrrd28/5/lan28084.html

Lashari, M. S., Liu, Y., Li, L., Pan, W., Fu, J., Pan, G. et al. (2013). Effects of amendment of biochar-manure compost in conjunction with pyroligneous solution on soil quality and wheat yield of a salt-stressed cropland from Central China Great Plain. *Field Crops Research, 144*, 113–118. https://doi.org/10.1016/j.fcr.2012.11.015

Lehmann, J., Gaunt, J., & Rondon, M. (2006). Bio-char sequestration in terrestrial ecosystems—A review. *Mitigation and Adaptation Strategies for Global Change, 11*, 403–427. https://doi.org/10.1007/s11027-005-9006-5

Liu, Q., Zhang, Y., Liu, B., Amonette, J. E., Lin, Z., Liu, G. et al. (2018). How does biochar influence soil N cycle? A meta-analysis. *Plant and Soil, 426*, 211–225. https://doi.org/10.1007/s11104-018-3619-4

Mabe, L. T., Su, S., Tang, D., Zhu, W., Wang, S., & Dong, Z. (2018). The effect of dietary bamboo charcoal supplementation on growth and serum biochemical parameters of juvenile common carp (*Cyprinus carpio* L.). *Aquaculture Research, 49*(3), 1142–1152. https://doi.org/10.1111/are.13564

MAFW. (2017). Annual report 2016–17, Department of agriculture, cooperation and farmers welfare. Directorate of Economics and Statistics, Ministry of Agriculture and Farmers Welfare, Government of India, New Delhi. https://agricoop.nic.in/Documents/Annual_rpt_201617.pdf (accessed 03/20/2023).

Man, K. Y., Chow, K. L., Man, Y. B., Mo, W. Y., & Wong, M. H. (2021). Use of biochar as feed supplements for animal farming. *Critical Reviews in Environmental Science and Technology, 51*(2), 187–217. https://doi.org/10.1080/10643389.2020.1721980

Méndez, A., Paz-Ferreiro, J., Gil, E., & Gascó, G. (2015). The effect of paper sludge and biochar addition on brown peat and coir based growing media properties. *Scientia Horticulturae, 193*, 225–230. https://doi.org/10.1016/j.scienta.2015.07.032

Meng, J., Tao, M., Wang, L., Liu, X., & Xu, J. (2018). Changes in heavy metal bioavailability and speciation from a Pb-Zn mining soil amended with biochars from co-pyrolysis of rice straw and swine manure. *Science of the Total Environment, 633*, 300–307. https://doi.org/10.1016/j.scitotenv.2018.03.199

Najmudeen, T. M., ArakkalFebna, M. A., Rojith, G., & Zacharia, P. U. (2019). Characterisation of biochar from water hyacinth *Eichhornia crassipes* and the effects of biochar on the growth of fish and paddy in integrated culture systems. *Journal of Coastal Research, 86*(SI), 225–234. https://doi.org/10.2112/SI86-033.1

Norton, J. M., & Firestone, M. K. (1996). N dynamics in the rhizosphere of Pinus ponderosa seedlings. *Soil Biology and Biochemistry, 28*(3), 351–362.

Novak, J. M., & Johnson, M. G. (2019). Elemental and spectroscopic characterization of low-temperature (350 C) lignocellulosic-and manure-based designer biochars and their use as soil amendments. In *Biochar from biomass and waste* (pp. 37–58). https://doi.org/10.1016/B978-0-12-811729-3.00003-0

Novak, J., Ro, K., Ok, Y. S., Sigua, G., Spokas, K., Uchimiya, S., & Bolan, N. (2016). Biochars multifunctional role as a novel technology in the agricultural, environmental, and industrial sectors. *Chemosphere, 142*, 1–3. https://doi.org/10.1016/j.chemosphere.2015.06.066

Oh, T. K., Choi, B., Shinogi, Y., & Chikushi, J. (2012). Characterization of biochar derived from three types of biomass. https://doi.org/10.5109/22049

Patil, P. D., & Yadav, G. D. (2018). Application of microwave assisted three phase partitioning method for purification of laccase from *Trametes hirsuta*. *Process Biochemistry, 65*, 220–227. https://doi.org/10.1016/j.procbio.2017.10.006

Prasai, T. P., Walsh, K. B., Bhattarai, S. P., Midmore, D. J., Van, T. T., Moore, R. J., & Stanley, D. (2016). Biochar, bentonite and zeolite supplemented feeding of layer chickens alters intestinal microbiota and reduces campylobacter load. *PLoS One, 11*(4), e0154061. https://doi.org/10.1371/journal.pone.0154061

Pratiwi, E. P. A., & Shinogi, Y. (2016). Rice husk biochar application to paddy soil and its effects on soil physical properties, plant growth, and methane emission. *Paddy and Water Environment, 14*, 521–532. https://doi.org/10.1007/s10333-015-0521-z

Qin, X., Wang, H., Liu, C., Li, J., Wan, Y., Gao, Q. et al. (2016). Long-term effect of biochar application on yield-scaled greenhouse gas emissions in a rice paddy cropping system: A four-year case study in south China. *Science of the Total Environment, 569*, 1390–1401. https://doi.org/10.1016/j.scitotenv.2016.06.222

Quilty, J. R., & Cattle, S. R. (2011). Use and understanding of organic amendments in Australian agriculture: A review. *Soil Research, 49*(1), 1–26. https://doi.org/10.1071/SR10059

Quilty, J. R., & Cattle, S. R. (2011). Use and understanding of organic amendments in Australian agriculture: A review. *Soil Research, 49*(1), 1–26. https://doi.org/10.1071/SR10059

Quirk, R. G., Van Zwieten, L., Kimber, S., Downie, A., Morris, S., & Rust, J. (2012). Utilization of biochar in sugarcane and sugar-industry management. *Sugar Tech, 14*, 321–326. https://doi.org/10.1007/s12355-012-0158-9

Raul, C., Prakash, S., Lenka, S., & Bharti, V. S. (2021). Sugarcane bagasse biochar: A suitable amendment for inland saline pond water productivity. *Journal of Environmental Biology, 42*(5), 1264–1273. http://www.jeb.co.in/journal_issues/202109_sep21/paper_08.pdf

Roy, C., & Chaala, A. (2001). Vacuum pyrolysis of automobile shredder residues. *Resources, Conservation and Recycling, 32*(1), 1–27. https://doi.org/10.1016/S0921-3449(00)00088-4

Saroeun, K., Preston, T. R., & Leng, R. A. (2018). Rice distillers' byproduct and molasses-urea blocks containing biochar improved the growth performance of local Yellow cattle fed ensiled cassava roots, cassava foliage and rice straw. *Bone, 3*(3), 3. http://lrrd.cipav.org.co/lrrd30/9/saroe30162.html

Schmidt, H. P., & Wilson, K. (2012). 55 uses of biochar. *Ithaka Journal, 1*(May), 286–289. https://www.terrapreta.bioenergylists.org/files/e082012-55-uses-of-bc.pdf

Sha, Z., Li, Q., Lv, T., Misselbrook, T., & Liu, X. (2019). Response of ammonia volatilization to biochar addition: A meta-analysis. *Science of the Total Environment, 655*, 1387–1396. https://doi.org/10.1016/j.scitotenv.2018.11.316

Sharma, I., Uniyal, A., & Rajwar, G. S. (2016). Comparison of biochar and urea amendment on growth and productivity of *Glycine max* (L.) Merr. *Plant Archives, 16*(2), 949–954. http://plantarchives.org/PDF%20162/949-954.pdf

Sohi, S. P. (2012). Carbon storage with benefits. *Science, 338*(6110), 1034–1035. https://www.science.org/doi/abs/10.1126/science.1225987 (accessed 20 Mar 2023).

Sorrenti, G., Masiello, C. A., Dugan, B., & Toselli, M. (2016). Biochar physico-chemical properties as affected by environmental exposure. *Science of the total Environment, 563*, 237–246. https://doi.org/10.1016/j.scitotenv.2016.03.245

Suman, S., & Gautam, S. (2018). Biochar derived from agricultural waste biomass act as a clean and alternative energy source of fossil fuel inputs. *Energy Systems and Environment, 207*.

Sun, J., Yang, R., Li, W., Pan, Y., Zheng, M., & Zhang, Z. (2018). Effect of biochar amendment on water infiltration in a coastal saline soil. *Journal of Soils and Sediments, 18*, 3271–3279.

Tenenbaum, D. J. (2009). Biochar: Carbon mitigation from the ground up. *Environmental Health Perspectives, 117*(2), A70–A73. https://doi.org/10.1289/ehp.117-a70

Thomas, S. C., Frye, S., Gale, N., Garmon, M., Launchbury, R., Machado, N. et al. (2013). Biochar mitigates negative effects of salt additions on two herbaceous plant species. *Journal of Environmental Management, 129*, 62–68. https://doi.org/10.1016/j.jenvman.2013.05.057

Thompson, K. A., Shimabuku, K. K., Kearns, J. P., Knappe, D. R., Summers, R. S., & Cook, S. M. (2016). Environmental comparison of biochar and activated carbon for tertiary wastewater treatment. *Environmental Science & Technology, 50*(20), 11253–11262. https://doi.org/10.1021/acs.est.6b03239

Thu, M., Koshio, S., Ishikawa, M., & Yokoyama, S. (2010). Effects of supplementation of dietary bamboo charcoal on growth performance and body composition of juvenile Japanese flounder, Paralichthys olivaceus. *Journal of the World Aquaculture Society, 41*, 255–262.

Tripathi, M., Sahu, J. N., & Ganesan, P. (2016). Effect of process parameters on production of biochar from biomass waste through pyrolysis: A review. *Renewable and Sustainable Energy Reviews, 55*, 467–481. https://doi.org/10.1016/j.rser.2015.10.122

Tumuluru, J. S., Wright, C. T., Hess, J. R., & Kenney, K. L. (2011). A review of biomass densification systems to develop uniform feedstock commodities for bioenergy application. *Biofuels, Bioproducts and Biorefining, 5*(6), 683–707. https://doi.org/10.1002/bbb.324

Uras, Ü., Carrier, M., Hardie, A. G., & Knoetze, J. H. (2012). Physico-chemical characterization of biochars from vacuum pyrolysis of South African agricultural wastes for application as soil amendments. *Journal of Analytical and Applied Pyrolysis, 98*, 207–213. https://doi.org/10.1016/j.jaap.2012.08.007

Vaccari, F. P., Baronti, S., Lugato, E., Genesio, L., Castaldi, S., Fornasier, F., & Miglietta, F. (2011). Biochar as a strategy to sequester carbon and increase yield in durum wheat. *European Journal of Agronomy, 34*(4), 231–238. https://doi.org/10.1016/j.eja.2011.01.006

Van Zwieten, L., Singh, B. P., Kimber, S. W. L., Murphy, D. V., Macdonald, L. M., Rust, J., & Morris, S. (2014). An incubation study investigating the mechanisms that impact N2O flux from soil following biochar application. *Agriculture, Ecosystems & Environment, 191*, 53–62. https://doi.org/10.1016/j.agee.2014.02.030

Wakeel, A. (2013). Potassium–sodium interactions in soil and plant under saline-sodic conditions. *Journal of Plant Nutrition and Soil Science, 176*(3), 344–354. https://doi.org/10.1002/jpln.201200417

Xiao, R., Wang, J. J., Gaston, L. A., Zhou, B., Park, J. H., Li, R. et al. (2018). Biochar produced from mineral salt-impregnated chicken manure: Fertility properties and potential for carbon sequestration. *Waste Management, 78*, 802–810. https://doi.org/10.1016/j.wasman.2018.06.047

Yang, P., Zhao, G., Tong, C., Tang, K. W., Lai, D. Y., Li, L., & Tang, C. (2021). Assessing nutrient budgets and environmental impacts of coastal land-based aquaculture system in southeastern China. *Agriculture, Ecosystems & Environment, 322*, 107662. https://doi.org/10.1016/j.agee.2021.107662

Yang, X., Lan, Y., Meng, J., Chen, W., Huang, Y., Cheng, X. et al. (2017). Effects of Maize Stover and its derived biochar on greenhouse gases emissions and C-budget of brown earth in Northeast China. *Environmental Science and Pollution Research, 24*, 8200–8209. https://doi.org/10.1007/s11356-017-8500-0

Ye, W., Sun, H., Li, Y., Zhang, J., Zhang, M., Gao, Z. et al. (2022). Greenhouse gas emissions from fed mollusk mariculture: A case study of a *Sinonovacula constricta* farming system. *Agriculture, Ecosystems & Environment, 336*, 108029. https://doi.org/10.1016/j.agee.2022.108029

Zhang, J., Zhang, J., Wang, M., Wu, S., Wang, H., Niazi, N. K. et al. (2019). Effect of tobacco stem-derived biochar on soil metal immobilization and the cultivation of tobacco plant. *Journal of Soils and Sediments, 19*, 2313–2321. https://doi.org/10.1007/s11368-018-02226-x

Zheng, H., Wang, Z., Deng, X., Zhao, J., Luo, Y., Novak, J. et al. (2013). Characteristics and nutrient values of biochars produced from giant reed at different temperatures. *Bioresource Technology, 130*, 463–471. https://doi.org/10.1016/j.biortech.2012.12.044

CHAPTER 12

E-DNA in Aquatic Ecology

SIPAEE VENKATESH GOUD and GOLAM ZIAUDDIN

Department of Fisheries Resource Management, Faculty of Fishery Science, West Bengal University of Animal and Fishery Sciences, Kolkata, India

ABSTRACT

Identifying and monitoring species based on their physical origin is essential for understanding ecological patterns, phylogenetics, and conservation. Traditional species detection methods, particularly in aquatic and marine environments, face significant limitations due to complex life cycles, logistical challenges, and potential ecological harm. Environmental DNA (e-DNA) has emerged as an innovative, non-invasive alternative that overcomes many of these constraints. e-DNA refers to genetic material shed by organisms into the environment through tissues, secretions, or decomposition, which can be collected from samples such as water or soil. Initially applied in the 1980s to identify microbial communities, e-DNA techniques have since evolved to track diverse aquatic species and ecological changes. Its applications now extend to bio surveillance, aquaculture, biosecurity, and conservation, offering high sensitivity and specificity. The method distinguishes between particulate DNA (P-DNA) ↓ and dissolved DNA (D-DNA), enabling the detection of microscopic genetic Often likened to DNA barcoding, e-DNA detection is considered a "mini barcode" method due to its focus on short DNA sequences. This approach holds immense potential for revolutionizing species monitoring in aquatic systems.

Recent Advances in Sustainable Aquaculture and Fisheries. Shabir Ahmad Dar, Imtiyaz Qayoom, Gohar Bilal Wani, Adnan Abubakr, and Seemab Zehra (Eds.)

DOI: 10.1201/9781998511372-12

12.1 INTRODUCTION

It is necessary to gather (detect) and identify individuals based on their physical origins to figure out the patterns, distributions, and ample of species or populations. Therefore, the identification or recognizing of species is vital to the study of phylogenetics, conservation, and ecology. However, it can be very challenging to identify a species, particularly in aquatic and marine habitats where the organisms have complicated life cycles, and it is nearly impossible to directly see the earliest phases of development (Díaz-Ferguson and Moyer, 2014). These habitats have been subjected to typical direct observation techniques (visual or auditory) for species detection. However, traditional detection procedures, such as sea-bottom trawls, rotenone poisoning, and electrofishing, might have logistical limits, be tedious, lavish, and in some instances, be hazardous to the ecosystem (Thomsen et al., 2012). The development of anecdote, forensic, and molecular techniques has produced inventive ways for identifying aquatic and marine creatures that may get around the foregoing restrictions.

e-DNA perception is a new strategy that endeavors to ascertain the existence of life forms and other population factors by collecting genetic data emitted from live and deceased organisms in ambient samples such as water. An unparalleled potential to distinguish the species has been created for bio-surveillance, commercial fishing, bio-security, aquaculture, and aquatic conservation management activities globally thanks to the quick development of tools and technology related to e-DNA.

e-DNA is the hereditary material extricated straight from the environmental prototypes such as water, soil, silt, etc. that manifest no affirmations of biological origin. This genetic material (deoxyribonucleic acid) advances from expelled cell material or tissues such as hair, skin, and urine as well as the deceased one's genetic material, in the case of higher organisms. Despite its reputation as a new technique, e-DNA has been cast off to ascertain bacterial communities in marine deposits dating back to the mid-1980s. In the 1990s, e-DNA technologies were accustomed to track phytoplankton blooms and quantify transposes in bacterial community biomass (Bailiff and Karl, 1991; Weirbauer et al., 1993). Particle size was set for categorizing in this period. Thus, e-DNA developed in clusters larger than 0.2 m auxiliary with the cells (i.e., microbial e-DNA) was referred to as P-DNA or particulate DNA, whereas e-DNA found in smaller aggregates (e.g., dissolved viral-DNA) was referred to as D-DNA or dissolved DNA. e-DNA is thought to be a molecular categorizing tool similar to DNA barcoding bearing the main difference that e-DNA particles are typically microscopic (100 base pairs or less). Because of these resemblances, writers refer to e-DNA detection as mini barcode detection.

12.1.1 APPLICATIONS OF ENVIRONMENTAL DEOXYRIBONUCLEIC ACID (E-DNA)

Thomsen et al. (2012) were the first people to use e-DNA applications for the detection of organisms in the marine environment to monitor marine mammals' genetic diversity. Detection of aquatic invasive species (AIS), population dynamics, trophic interactions, etc. are just a few among the many uses of e-DNA that are available for use in the fields of conservation biology and marine ecology. An overview of each of these applications is provided below.

12.1.1.1 WILDLIFE DNA FORENSICS

To meet the growing demand for DNA forensic aids in wildlife law enforcement, the realm of DNA forensics for wildlife was fabricated. It is a blend of conservation genetics and forensic genetics. The use of e-DNA as a forensic technique seems to hold great promise for addressing fundamental identification problems at the individual, population, or species levels. The recognition of e-DNA fragments can also be used to track illegal fishing goods and offer proof of the illegal wildlife trade.

12.1.1.2 DETECTION OF LOW-DENSITY POPULATION

It seems promising to use e-DNA to identify species or populations that are scarce (such as threatened or endangered taxa) or evasive to the naked eye. The greater susceptibility of e-DNA compared to conventional examining techniques may help in the identification of this species; conversely, the absence of detection may cause the government to declare this species to be considered extinct. However, care is to be taken when using the absence of e-DNA detection to assess the conservation status of a taxon as the likelihood of e-DNA detection differs reliant on the amount of water. Future assessments of the species' conservation status will depend heavily on the development of sampling techniques that take into account the likelihood of e-DNA detection.

12.1.1.3 POPULATION DYNAMICS

The recognition and quantification of e-DNA can be utilized as a benchmark or an auxiliary indicator of population characteristics like abundance, biomass, and distribution. Correlations between e-DNA content and the dispensation of species and biomass in the freshwater ecosystems have been found. Thus, e-DNA concentration was utilized as a surrogate for population dispersion in fish, amphibians, and the reptile species. The link between e-DNA concentration and species distribution, biomass, and abundance has not been studied in marine species. However, during blooming events, the concentration of DNA in coastal waters has been associated with the geographical and temporal oscillations of bacterial and phytoplankton groups (Bailiff and Karl, 1991).

12.1.1.4 DISCOVERING THE AQUATIC INVASIVE SPECIES (AIS)

The molecular techniques for AIS detection have become increasingly popular nowadays. The creation of genetic indicators unique to ambitious species offers conservation managers a new aid to track AIS. Therefore, using e-DNA makes it possible to confirm AIS recognition allowing management to take immediate action to reduce the invader's dissemination and settlement. Additionally, AIS e-DNA detection offers hints to identify the source of the introduction and potential invasion paths.

12.1.1.5 BIODIVERSITY AND COMMUNITY STRUCTURE

Multispecies identification using e-DNA samples is conferred as DNA meta-barcoding. The strategy makes use of Next Gen Sequencing (NGS), which enables the progression of countless base pair reads, as well as the construction of reference databases. As a result, assuming the DNA patterns hoarded in a taxonomic library, an e-DNA meta-barcoding method could recognize the e-DNA of any taxon gathered in a water sample rather than just one species from a single water sample. Meta-barcoding can, therefore, be employed as a technique to provide biodiversity appraisal that is taxonomically thorough, prompter to compile, and less dependent on taxonomic virtuosity.

12.1.1.6 ECOSYSTEM HEALTH

Existing native populations may suffer significant demographic and genetic effects from the presence of AIS as well as newly imported infections (fungal or viral). Managers can indirectly check on the health of the ecosystem by utilizing e-DNA to monitor invasive species or virus concentration. Additionally, using e-DNA to track changes in species diversity and community composition can potentially be used as a stand-in for accessing environmental health. More significantly, modifying species diversity can have a direct or indirect impression on the dynamics of an ecosystem. For example, they can lower water quality and modify nutrient dynamics or change where submerged macrophytes are found. Therefore, e-DNA might prove valuable for future environmental impact analyses or risk-based resolutions for natural resources.

12.1.1.7 HISTORICAL MARKING OF SPECIES DISTRIBUTION

Short DNA sequences can survive for a very long time, especially in cold conditions with little or no light. Studies in ancient sediments and ice cores have shown this to be the case. So, ancient e-DNA in sediments from ice cores or any other environmental sources may contain genetic information that enables the reconstruction of historical ecological processes and community structure.

12.1.1.8 TROPHIC INTERACTIONS AND DIETARY STUDIES

Using stomach contents, faeces, or faecal pellets, traditional investigations enable estimating and evaluating the association between prey and predator relationships. Direct detection of the prey items in the gut or the faeces is frequently challenging, as it can bias results or diminish the taxonomic resolution. Therefore, nutritional and trophic investigations can be conducted without observing or identifying the prey in the stomach or faeces by using e-DNA fragments or a meta-barcoding technique using stomach contents as target DNA.

12.1.1.9 META-BARCODING IN DIETARY STUDIES

Due to the better taxonomic accuracy provided by DNA-based stomach content analysis tools, that can enhance our grasping of trophic linkages which otherwise remain murky using conventional methods, dietary studies have gained more attention. This has been pivotal for species whose diets pose particular assessment challenges, such as those whose highly miscellaneous consumer-resource systems are made up of cryptic species (Leray et al., 2015), those whose prey deteriorate in dietary remains like stomach contents or those that consume food in remote habitats and those species that are difficult to study (Walters et al., 2019).

e-DNA is a hereditary and genetic material isolated from a water sample devoid of any identifying macro-organism traces. The technique makes use of DNA that organisms continuously emit into the environment through mucus, gametes, faeces, blood, and other cells. Based on an environmental sample, it acquires, analyzes, and determines the nucleotide sequence of the DNA. A potentially effective method for accessing the architecture of aquatic communities is e-DNA analysis. Through two methods, e-DNA barcoding and e-DNA meta-barcoding, analysis of e-DNA can provide us with knowledge about the organisms, their abundance, and biomass. In the former, samples are targeted for specific species using quantitative or standard PCR and the conventional Sanger sequencing method. In the latter, NGS and several conserved primers are used to screen the entire community. According to studies, e-DNA meta-barcoding performs better than conventional survey techniques in terms of cost-effectiveness, and non-invasive sampling. The use of e-DNA to enhance current survey techniques is currently gaining more attention.

12.1.2 METHODOLOGY

12.1.2.1 SAMPLE COLLECTION

To increase the dependability of e-DNA detection surveys, it is first important to check defilement during "sample collection" and subsequent e-DNA capture and extraction. Grab sampling and composite sampling are now used in water sampling strategies. Fish e-DNA seasonal and regional variations should be completely taken into account when collecting samples (Handley et al., 2019). The quantity of fish e-DNA present in the sample may be correlated with the sampling volume. Typically, 1–2 L is the usual sampling volume for seas, rivers, lagoons, and streams.

12.1.2.2 E-DNA CAPTURE

The three basic techniques for extracting e-DNA from aquatic environments are centrifugation, precipitation, and filtration. The approach that is most frequently used to collect e-DNA is filtering. Most filtration methods employ membrane filters.

12.1.2.3 E-DNA EXTRACTION

The popular kits for e-DNA extraction after e-DNA capture are the DNeasy Blood and Tissue Kit and Power Water DNA Isolation Kit.

12.1.2.4 PCR SEQUENCING

By using the specific universal primers, DNA molecules are separated and analyzed.

12.1.2.5 BIOINFORMATICS ANALYSIS

Remove all the flaws, overlapped readings, and repetitions from the product.

12.1.2.6 STATISTICAL ANALYSIS

The data obtained is analyzed using the statistical tool R v.3.6.0 using the "exact rank tests" package v.0.8-30 and "vegan" package v.2.5-5.

12.1.3 REFERENCE SEQUENCE DATABASE

DNA primer samples obtained were identified using the universal reference databases established at (i) The National Centre for Biotechnology Information database, (ii) The Barcode of Life Database, (iii) The European Molecular Biology Laboratory database, (iv) The Fish Barcode of Life Initiative (FISH-BOL), and (v) The Mito-Fish database.

12.1.4 ENVIRONMENTAL DNA META-BARCODING

12.1.4.1 OVERVIEW OF SEDIMENT E-DNA META-BARCODING STUDIES

The application of molecular taxonomy as an effective substitute for morpho-taxonomic inventory is now possible thanks to the quick advancement of high-throughput sequencing technologies. The usage of these DNA-based techniques is predicated on the idea that taxa may be accurately recognized based on a specific section of their genome known as a "DNA barcode" region. Meta-barcoding is the process of using high-throughput sequencing of a DNA barcode region to simultaneously identify numerous species from complicated samples. While e-DNA meta-barcoding refers to the examination of environmental materials like sediments or water, bulk meta-barcoding refers to the analysis of samples made up of heterogeneous assemblages of species.

12.1.4.2 CURRENT APPLICATIONS OF SEDIMENT E-DNA TO BENTHIC MONITORING

1. Environmental impact assessment of marine industries:

   Benthic habitats are heavily impacted by marine industries such as coastal aquaculture and oil and gas operations. For these businesses, regular benthic monitoring is required as part of the environmental impact assessment. This data is used to guide management decisions that support the upkeep of healthy ecosystems and the accompanying functions and services. This relies heavily on the employment of macro-invertebrate bio-indicators and severely restricts the scope and scale of monitoring. The data's ability to inform adaptive management of the benthic environment and prompt responses to pollution events is consequently constrained.

12.1.4.3 BIOMONITORING OF FRESHWATER ECOSYSTEMS

With a focus on oligochaetes and nematodes, respectively, freshwater sediment meta-barcoding has also been utilized to analyze macro- and meiofaunal populations. Vivien et al. (2019) compared a genetic index obtained from total sediment, sieved sediments, and mixed specimens to a morphology-based oligochaete index (bulk DNA). In comparison to the e-DNA meta-barcoding of sieved or whole sediments, they discovered that bulk DNA meta-barcoding offered a more accurate assessment of the local biodiversity. Despite differences in species composition and abundance between community structures determined from morphological and meta-barcoding analyses, ecological assessments based on these two methods essentially concurred.

12.1.4.4 BIOSECURITY

Early detection of hazardous organisms and invasive alien species is greatly enhanced by sediment e-DNA meta-barcoding (e.g., toxic algae, pathogenic bacteria, and viruses). Since there is a high chance of non-native species establishing in closed aquatic systems (like ports), this is especially important (e.g., ballast waters and bio-fouling). It is also a major issue in aquaculture, where non-native animals are consciously introduced for cultivation but may escape and spread across the environment with their parasites and other associated organisms. Confident detection and precise species-level identification are essential for bio-security, in contrast to bio-monitoring for the assessment of ecosystem health, because the management response might be very expensive.

12.1.5 BIOTIC AND ABIOTIC FACTORS

The biotic and abiotic factors are illustrated in the figure below.

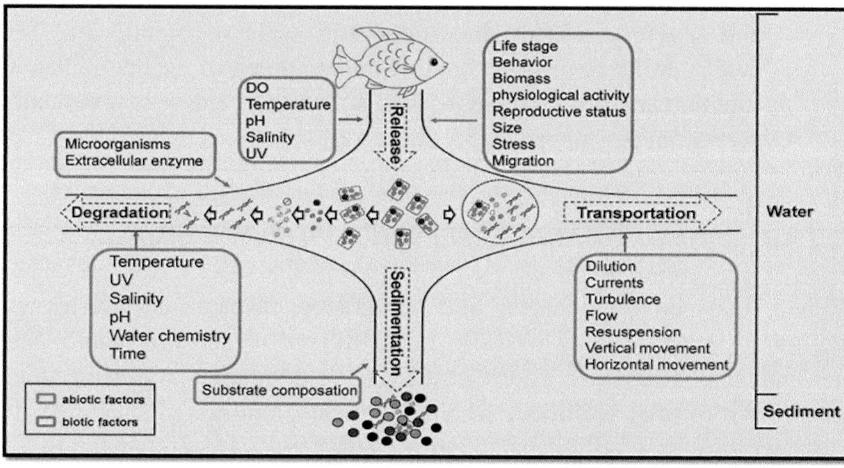

Source: Reprinted with permission from Wang et al., 2021. Copyright © 2020 Elsevier B.V.

12.1.6 ADVANTAGES OF E-DNA

e-DNA analysis method in aquatic ecosystems reduces the cost and time for large-scale identification of the species from DNA sequences that are often simple and accurate. Various aquatic species may be ascertained from a small water sample with the least destructive operation procedures. e-DNA analysis also encounters and monitors (qualitative) the absence/presence of invasive and endangered species and assesses biodiversity. It also discovers organisms for bio-surveillance, bio-security archaeology, ecology, aquaculture, and conservation management activities throughout the world (Kelly et al., 2014).

12.1.7 GAP IN KNOWLEDGE/LIMITATIONS

There are still unknown processes governing the fish e-DNA migration in lotic (rivers, streams, and oceans) and lentic (lakes and ponds) settings. The factors including PCR effectiveness, primer tags, sequencing efficacy limited meta-barcoding, and the absence of reference sources for identification of all the species are still unaccomplished fully. To obtain a more accurate assessment, distribution of fish and its abundance/biomass based on e-DNA, degradation, patterns of release, and diffusion of e-DNA under field circumstances should be taken into account.

12.1.8 THE FUTURE SCOPE OF E-DNA

One of the biggest advantages of e-DNA is that it saves time and money by eliminating many of the requirements for traditional bio-surveys, including manpower, equipment, permits, safety concerns, and organism handling. In addition, it offers a way to do extensive biodiversity comparisons over both time and area. As a result, e-DNA holds the potential to change the fields of conservation, ecology, and archaeology (Kelly et al., 2014). Moving from meta-barcoding methods to real meta-genomics will be the following stage. This might soon be possible given the growing amount of genome data being produced and will enable more accurate identification of the species and quantitative assessments of their abundance in the environment.

KEYWORDS

- **biomass**
- **diversity**
- **environment**
- **fishing**
- **observation**
- **organisms**
- **phytoplankton**
- **predator**

REFERENCES

Bailiff, M., & Karl, D. (1991). Dissolved and particulate DNA dynamics during a spring bloom in the Antarctic Peninsula region, 1986–1987. *Deep Sea Research: Part 1. Oceanographic Research, 38*(8–9), 1077–1095.

Díaz-Ferguson, E. E., & Moyer, G. R. (2014). Invited Review: History, applications, methodological issues and perspectives for the use environmental DNA (e-DNA) in marine and freshwater environments. *Revista de Biología Tropical, 62*(4), 1273–1284.

Handley, L. L., Read, D. S., Winfield, I. J., Kimbell, H., Johnson, H., Li, J., Hahn, C., Blackman, R., Wilcox, R., Donnelly, R., Szitenberg, A., & Hänfling, B. (2019). Temporal and spatial variation in distribution of fish environmental DNA in England's largest lake. *Environmental DNA, 1* (1), 26–39.

Kelly, R. P. et al. (2014). Environmental monitoring: Harnessing DNA to improve environmental management. *Science, 344*, 1455–1456. DOI: 10. 1126/science.1251156.

Leray, M., Meyer, C. P., & Mills, S. C. (2015). Metabarcoding dietary analysis of coral dwelling predatory fish demonstrates the minor contribution of coral mutualists to their highly partitioned, generalist diet. *PeerJ, 3*, e1047. https://doi.org/10.7717/peerj.1047

Thomsen, P., Kielgast, J., Iversen, L., Moller, P., Rasmussen, M., & Willerslev, E. (2012). Detection of a diverse marine fauna using e-DNA from seawater samples. *PLoS ONE, 7*(8), e41732.

Vivien, R., Apothéloz-Perret-Gentil, L., Pawlowski, J., Werner, I., & Ferrari, B. J. (2019). Testing different (e) DNA meta-barcoding approaches to assess aquatic oligochaete diversity and the biological quality of sediments. *Ecological Indicators, 106*, 105453.

Walters, T. L., Lamboley, L. M., López-Figueroa, N. B., Rodríguez-Santiago, Á. E., Gibson, D. M., & Frischer, M. E. (2019). Diet and trophic interactions of a circumglobally significant gelatinous marine zooplankter, Dolioletta gegenbauri (Uljanin, 1884). *Molecular Ecology, 28*, 176–189. https://doi.org/10.1111/mec.14926

Weirbauer, M., Fucks, D., & Peduzzi, P. (1993). Distribution of viruses and dissolved DNA along a coastal trophic gradient in the Northern Adriatic Sea. *Applied Environmental Microbiology, 59*(12), 4074–4082.

Wang, S., Yan, Z., Hänfling, B., Zheng, X., Wang, P., Fan, J., & Li, J. (2021). Methodology of fish eDNA and its applications in ecology and environment. Science of the Total Environment, 755, 142622.

CHAPTER 13

Integrated Multi-Tropical Aquaculture System (IMTA): Toward Sustainable Aquaculture

AITWAR VAIJNATH[1,*], EDIGA ARUN GOUD[2], and SURYAWANSHI UMESH[3]

[1]*Karnataka Veterinary, Animal and Fisheries Sciences University, Bidar, Department of Aquatic Environment Management, College of Fisheries, Mangaluru, Karnataka, India*

[2]*Dr. Balasaheb Sawant Konkan Krishi Vidyapeeth, Dapoli, Department of Aquaculture, College of Fisheries, Shirgaon, Ratnagiri, Maharashtra, India*

[3]*Karnataka Veterinary, Animal and Fisheries Sciences University, Bidar, Department of Fisheries Resource Management, College of Fisheries, Mangaluru, Karnataka, India*

*Corresponding author

ABSTRACT

A novel idea in aquaculture is provided by integrated multi-trophic aquaculture (IMTA). It transforms aquaculture from the farming of a single species to the farming of multiple species in concert with one another. The extraction and conversion of extra nutrients into other crop production by IMTA, which substitutes the dilution of waste materials. The farm replicates natural ecosystem cycles and enables nutrient reuse, decreasing waste and negative environmental impacts. The industry is served by IMTA's waste reduction system. A fully circular system might be created by adding plants or other species as a component to the fish food to promote circularity. This type of IMTA produces more beneficial products while minimizing the ecological effects close to the aquaculture business.

Recent Advances in Sustainable Aquaculture and Fisheries. Shabir Ahmad Dar, Imtiyaz Qayoom, Gohar Bilal Wani, Adnan Abubakr, and Seemab Zehra (Eds.)

DOI: 10.1201/9781998511372-13

13.1 INTRODUCTION

The Indian fisheries industry is surrounded by a distinctive and varied collection of resources, from the vast Indian Ocean to the pristine waters of the Himalayas. Millions of people rely on the fisheries biodiversity of the nation, which includes a wide range of physical and biological elements. Resources for fishing are located in several habitats. The necessity for sustainable development of aquatic resources is now more important than ever due to population growth and rising fish protein consumption. In today's date, people are realizing how crucial the fisheries and aquaculture industries are to the world's nutrition and food security in the 21st century. The advancement of revolutionary reforms in policy, management, innovation, and investment is required for the future expansion of this contribution to achieving long-term and equitable global fisheries and aquaculture. Total fisheries and aquaculture production has increased significantly over the past 70 years, growing at an average annual rate of 3.3%, from 19 million tons in 1950 to nearly 179 million tons in 2018. After a slight decline in production in 2019, it achieves 178 million tons in 2020. The combined value of fisheries and aquaculture production at first sale in 2020 was projected to be USD 406 billion, with aquaculture production accounting for USD 265 billion (FAO, 2022).

In 2013, aquaculture produced approximately 51% of the world's fish, outpacing supplies from catch fisheries. This accomplishment was made possible in large part by the commercialization of farm-raised aquatic species like catfish, shrimp, salmon, bivalves, and tilapia. This industry also benefited from the enormous amount of low-value freshwater species that were produced through integrated farming for domestic consumption. Concerns regarding the effects of aquaculture waste on the environment have increased with the rise of the marine aquaculture industry. Aquaculture species are raised close to one another to take advantage of the interactions between species and to recapture unconsumed feed, microelements, and byproducts of one species and transform them into fertilizer, nourishment, and energy for the other crops (Chopin, 2013).

One of the primary issues for the aquaculture industry's sustainable development is minimizing ecological damage while it expands. Although extensive and semi-intensive farming systems still account for the majority of aquaculture production, the current rise in intensive farming of marine carnivorous feed-species is linked to environmental worries. A useful technology for sustainable mariculture is the incorporation of organisms that produce garbage (are fed), as well as those that clean it up. Aquaculture effluents can be transformed into commercial crops and water quality can be improved in a well-balanced integrated system.

13.1.1 WHAT IS INTEGRATED MULTI-TROPHIC AQUACULTURE SYSTEM?

Several monoculture farming systems, differently farm fed-aquaculture species and organic/inorganic extractive aquaculture species (bivalves, herbivorous fishes, and aquatic plants) in different geographical locations, resulting in a significant shift in environmental processes. IMTA combines fed species with extractive species that grow on inorganic and organic waste from aquaculture. This farming method differs from finfish "polyculture," in which the fish share the same chemical and biological processes, potentially leading to ecosystem shifts. The integration of species from various trophic levels within the same system is referred to as multi-trophic.

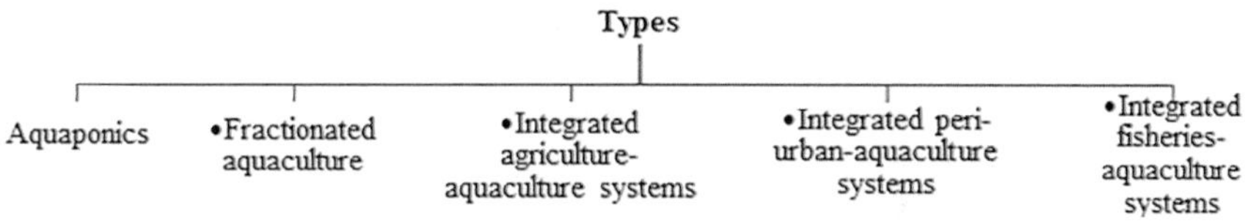

13.2 VARIOUS TYPES OF IMTA CULTURE SYSTEM

13.2.1 DESIGN AND STRUCTURE FOR IMTA SYSTEM

In order to remove both detritus and dissolved solid waste created during culture, a good IMTA operation requires the selection, arrangement, and distribution of numerous components or species. For waste product recapture, the system's chosen species and design should be optimized. Water-soluble organic particles, such as unconsumed feed as well as faecal matter, settle below the cage system and are ingested by detritivores. Simultaneously, filter-feeding animals, that is, bivalves, filter the fine suspended particles out of the water column. To enable them to filter out some of the dissolved inorganic nutrients from the water, such as nitrogen and phosphorus, the seaweeds are positioned a little further from the site in the water flowing direction. IMTA species should be able to be produced commercially as aquaculture products and should be raised at densities that maximize waste uptake and usage throughout the whole manufacturing cycle (Anonymous, 2013).

13.3 SPECIES SELECTION CRITERIA

Because environmental sustainability is a key consideration in IMTA, the criteria influencing species selection are imitating natural ecosystems. Feeding organisms, including such carnivorous fish and shrimp, are nourished by pellets or trash fish. Extractive organisms obtain their food from their surroundings. Bivalves and seaweed are two economically significant cultured groups that belong to this category. Compositions of co-cultured species must be carefully chosen based on a number of conditions and characteristics (Barrington et al., 2009):

1. **Site selection:** When choosing a farm site, the variety of particle sizes, dissolved inorganic nutrients, and particulate organic matter should all be taken into account.
2. **Species suitability:** Utilize species that will cooperate with one another at various trophic levels. For instance, in order for the newly integrated species to enhance the water quality and flourish effectively, they must be able to feed on the waste products of the other species. It is not like all species can always be successfully grown together.
3. **Bio-mitigation:** Utilize species that can generate a lot of biomass. If the organism is to function as a biofilter, taking out the majority of the surplus nutrients from the water, this characteristic is essential. Another choice is the critically important species, which allows for the production of limited proportions.
4. **Adaptability with habitat:** It is possible to use native species that reside within their natural range and have access to technology. As a result, there will be less chance that invasive species could harm the local environment or perhaps other establishments. Native species have also evolved to become highly environment-adaptable.

5. **Market demand and consumer preference:** Utilize species with a recognized or presumed monetary worth. For farmers to raise their income, they must be able to market the substitute species. They should, therefore, develop market buyers before making big investments.

Benefits:

1. **Bio-mitigation:** Effluent reduction through the use of biological filters tailored to the specific habitat of the fish farming site. This has the potential to address a number of the ecological issues raised by monoculture aquaculture.
2. **Diversification:** The cultivated and commercial sold by-products increase the operation's overall economic value.
3. **Economic growth of local people:** Financial growth is achieved through income and employment, as well as product manufacturing and distribution.
4. **Crop reimbursement:** Diversification strategy may provide economic protection and reduce potential losses in the event of market fluctuations or crop loss due to disease or adverse weather.
5. **Disease prevention:** Some seaweeds help prevent or reduce disease in farmed fish because of their antimicrobial property against fish bacterial infections.
6. **Profitability:** Possibility of IMTA products being differentiated through eco-labelling or agricultural certificate programs.

Challenges:

1. **High input cost:** A higher level of technology and engineering complexity, as well as an initial investment, are needed for integrated farming in open waters.
2. **Collaboration problem:** It would necessitate strong coordination and cooperation of management and production activities if implemented by several operators (such as independent fish farmers and mussel farmers) operating together.
3. **Increased need for farmland:** Aquaculture rivals with other users for the limited marine and coastal habitats, even if aquaculture has the ability to relieve strain on fish stocks and IMTA

offers special positive advantages for the businesses and the environment. Conflicts between stakeholders are frequent and might involve anything from site selection to local interests to worries about pollution and its effects on wild fish populations.

4. **Implementation challenges in the absence of open water leasing policies:** Only a few nations have integrated regulation of coastal zones that is well-developed or national aquaculture programmes. Because of political pressures and regional objectives, decisions about site development, authorization, and regulation are frequently made on the fly. Additionally, a number of mariculture sites are in danger from pollution from cities and industries as well as accidental harm as the coastal zone becomes more congested.

13.4 CONCLUSION

Aquaculture's potential for growth requires safe technologies and procedures. Aquaculture that is sustainable should be productive, profitable, environmentally friendly, and good for society. By combining extractive species, which grow by consuming excess nutrients from fed aquaculture in the form of inorganic (such as excess seaweed or other aquatic vegetation) and organic (such as excess suspension- and deposit-feeders) sources, with fed species (such as finfish or shrimp fed sustainable commercial diets), IMTA seems to have the potential to accomplish these goals. In IMTA system, unconsumed feed, nutrients and other by-products that are recognized as loss from the fed component will be recaptured and converted into commercially valuable and usable safe seafood.

KEYWORDS

- **fisheries**
- **IMTA**
- **nourishment**
- **nutrient cycle**
- **organic waste**
- **polyculture**

REFERENCES

Anonymous. (2013). Aquaculture in Canada: Integrated multi-tropic aquaculture (IMTA). *Fisheries and Oceans, Canada*, 1–5.

Barrington, K., Chopin, T., & Robinson, S. (2009). Integrated multi-trophic aquaculture (IMTA) in marine temperate waters—Integrated mariculture: A global review. *FAO Fisheries and Aquaculture Technical Paper, 529*, 7–46.

Chopin, T. (2013). Aquaculture, integrated multi-trophic (IMTA). *Sustainable Food Production,* 184–205. DOI: 10.1007/978-1-4614-5797-8_173

FAO. (2022). *The state of world fisheries and aquaculture 2022: Towards blue transformation* (p. 236). FAO. https://doi.org/10.4060/cc0461en

CHAPTER 14

Fishing Gears in Reservoirs in India: Present Status and Future Prospects

SANDHYA KAVITHA MANDHIR* and REMESAN MELETH PARAMBIL

Fishing Technology Division, ICAR - Central Institute of Fisheries Technology, Kochi, Kerala, India

*Corresponding author

ABSTRACT

This chapter discusses in detail about the present status of fishing gears operated in reservoirs of India. Gillnets are most commonly used in reservoirs due to the presence of underwater obstacles. Hook and lines, cast nets, drag nets, traps, scoop nets, and push nets are other gears used in reservoirs. Major issues in the reservoir fishing are improper design parameters, lack of fishing regulations, inadequate stocking and monitoring, and losses due to extended soaking time. This chapter also briefly discusses the future prospects in exploitation methods to enhance the fish catch from reservoirs in a sustainable manner. Application of newer technologies and improvements in fishing gears suiting to the prevailing conditions were also suggested for augmenting fish catch from reservoirs.

14.1 INTRODUCTION

Reservoirs play a major role in fish production in India and provide nutritional security especially in remote areas that lack adequate supplies of animal protein and sustain the livelihood of fishers. The largest inland waterbodies in the country, covering an area of 3.46 million ha, are often

Recent Advances in Sustainable Aquaculture and Fisheries. Shabir Ahmad Dar, Imtiyaz Qayoom, Gohar Bilal Wani, Adnan Abubakr, and Seemab Zehra (Eds.)

DOI: 10.1201/9781998511372-14

referred to as *Sleeping Giants* due to their untapped fisheries potential (NAAS, 2016; Sugunan and Suresh, 2022). Fisheries exploitation was less in reservoirs compared to other inland waterbodies for several decades when fishing was primarily for subsistence. Over the years, significant technological advancements in fish capturing methods in reservoirs had occurred (Khan et al., 1991). These developments had played a greater role in improving the yield and economic returns from Indian reservoirs.

14.2 FISHING GEARS

Fishing techniques vary according to the topography of the waterbody, ecology, and nature of fish stock in different inland water bodies. Similarly, differences can be seen in fishing gears used in reservoirs, also because the indigenous gears suitable for the riverine environments are not appropriate to the lacustrine condition of reservoirs. Also, the presence of underwater obstacles such as stones, tree trunks, submerged plants, and trenches limit the use of many active fishing gears in reservoirs (George, 2002).

14.2.1 GILLNETS

Gillnets are the most common fishing gear used for commercial fishing in Indian reservoirs. Simple gillnets have a head rope with floats and footrope with sinkers, and some are without the footrope too. In most reservoirs, nowadays gillnets are mostly made of polyamide (PA)/ nylon monofilaments (0.16–0.4 mm diameter) due to their catching efficiency. However, PA multifilaments are also in use (twine size 210 × 1 × 2 to 210 × 6 × 3). Mesh size varies from 20 to 300 mm according to the species targeted. Floats are either thermocol, plastic, or wood. Nowadays, a few empty plastic bottles are also used as floats instead of large numbers of small plastic floats for reducing entanglement while setting and hauling the nets. Sinkers are mostly lead sinkers and their size, shape, and weight depend on the type of net and mode of operation. In addition, burned clay, pieces of concrete or stones are also used as sinkers. Hanging coefficient varies from 0.4 to 0.5. Height of the net varies from 1 to 15 m depending upon the operating conditions and season. Gillnet setting is usually done before sunset, the nets are soaked overnight, and hauling is done during early morning. Triple walled gillnets/trammel nets are also used in some reservoirs for capturing prawns like *Macrobrachium* spp. (Rajeswari et al., 2015). The two outer walls consist of bigger meshes which are 3–5 times larger than the inner wall consisting of smaller meshes and mostly made of PA multifilament webbings.

Two types of gillnets are commonly used in reservoirs based on operation, set gillnets and drift gillnets. Generally, set gillnets are used in reservoirs and are set either in the surface or column layers, anchored to the bottom by means of heavy weights like stone anchors or tied to poles. In drift gillnets, one end of the net is tied to the buoy while the other end drifts due to wind action and is mostly operated in the deeper areas of the reservoir.

14.2.2 HOOK AND LINES

Hook and lines are popular in reservoirs mainly for capturing carnivorous fishes. Lines are made of PA monofilament. The advantage of lines is that they can be used in areas where other gears could not be employed due to submerged obstructions. The simplest form of hook and line includes hand lines having a line with a single hook. Length of the line varies from 5 to 100 m based on the depth and type of the area of operation. Different types of baits are used such as insects, earthworms, small fishes, tadpoles, and mollusc mussels according to the fish targeted. Pole and line consist of a line with a hook attached to a pole. Pole and line has various local modifications and the length of the pole also varies in different reservoirs. Longlines consist of a main line and numerous branch lines with hooks attached to it. Main line is either made of PA monofilament (1.2–2 mm diameter) or polypropylene rope (4–6 mm diameter) and branch line made of PA monofilament (0.6–0.7 mm diameter) with length varying from 1 to 1.5 m, attached to every 2–3 m of interval. About 50–80 hooks are used in longlines to exploit predatory fishes in reservoirs. Vertical long lines are also used in some reservoirs (Pravin et al., 2014). Usually, lines are soaked in reservoirs during the night, fishes are collected in the morning, and catches are mostly used for household consumption.

14.2.3 DRAG NETS

Drag nets are less in use in reservoirs due to the requirement of considerable manpower, plain terrain, and relatively low water level for their operation. The rectangular shaped net is dragged toward shore by a group of 8–10 people. Top of the net is attached with PA head ropes. Foot ropes are made of PA or coir. These PA nets have very small mesh sizes (8–12 mm) or sometimes very fine meshed knotless mosquito nets are also used. The dimension of the net varies according to the area of operation. Catch usually consists of small fishes and prawns. In most of the reservoirs, operation of this type of fine meshed nets is banned due to the capture of juvenile fishes (Bhakta et al., 2016). Stick held dragnets are also in practice in some reservoirs where the webbing is fixed to bamboo sticks at intervals (Rajeswari et al., 2015).

14.2.4 CAST NETS

Cast nets are operated in shallow marginal areas of reservoirs either from shore or from small platforms or boats. Both stringed and stringless cast nets are used for catching small fishes such as *Puntius* spp., *Gudusia chapra*, small indigenous fishes, and small prawns. Cast nets are made of either PA monofilament or multifilament. Mesh size ranges from 20 to 40 mm for stringed cast nets and for stringless, it ranges from 20 to 45 mm (Pravin et al., 2014). Sinker materials are made of lead with a total weight ranging from 3 to 8 kg per net.

14.2.5 TRAPS

Box type traps made of split bamboo are traditionally used to catch fresh water prawns in reservoirs. These square-shaped traps have a vertical inlet at both sides. The inlet structure usually has a "V" shape and the sticks from either end touch together at the middle, which resemble spines and thereby preventing the trapped prawns from moving out. Baits are mostly fresh coconut kernels to lure the prawns to the trap (Rajeswari et al., 2015). Similarly, box-shaped and conical-shaped bamboo traps are used for capturing fishes. Type of baits also varies according to the species and location.

14.2.6 SCOOPNETS AND PUSH NETS

Scoop nets are used for fishing in the shallow areas of the reservoir. The circular nets with small meshes /mosquito nets made of PA webbing are attached to a circular iron frame with a handle. Scoop nets are usually operated either from a boat or from the bank especially near the outgoing canal of the reservoir for catching slow-moving fishes, crabs, and so forth. Sometimes these nets are also used to take out the larger fishes from nets such as dragnets. Push nets have a triangular bamboo frame to which the small meshed nets are attached, mostly for capturing small fishes and prawns, and so forth (Bhakta et al., 2016).

14.3 ISSUES IN RESERVOIR FISHING AND SOLUTIONS

14.3.1 IMPROPER DESIGN PARAMETERS

The catching efficiency of gillnets which are the dominant fishing gear in reservoirs, depends on parameters such as mesh size, color, height, hanging ratio, yarn/twine diameter, and gear material. Inappropriate measurements of these parameters will result in catching of different species and different size groups of the same species. Loosely hung nets catch fishes by entangling than gilling which results in capturing of non-targeted fishes and juveniles also.

Improvements/design modifications for gillnets, have been discussed by Kuriyan (1973), Natarajan (1976), Kartha and Rao (1991), Pravin et al. (2014), and Sundaramoorthy et al. (2013). Modifications were made in gillnets to increase slackness by providing vertical lines or framing. For larger fishes, capturing is mostly by entangling than gilling and for improving the entangling ability, gillnets should have more slackness. This can be achieved by giving vertical lines in gillnets from headrope to footrope. In framed gillnets, slackness is increased by giving both vertical and horizontal lines in the net and dividing the main netting into compartments. Framed gillnets were reported to be more efficient than simple gillnets because their design creates small net bags in which fish become entangled as well as gilled (Khan et al., 1991). Both framed gillnets and vertical line nets are found to be more efficient in reservoirs, particularly where the fish population is low and comprises large-sized fishes (George, 2002).

Monofilament gillnets are mostly used in reservoirs as they are almost invisible in water, and hence catching efficiency is also high. Selection of the right color also yields better catches as studies in reservoirs show that yellow- and orange-colored gillnets are most suited for Gobindsagar reservoir and green- and yellow-colored nets resulted in more catches in Hirakud reservoir (George, 2002). White-colored gillnets yielded better catches in Thirumoorthy reservoir (Velmurugan et al., 2016).

14.3.2 LACK OF FISHING REGULATIONS

Mesh size regulations as well as landing size limits are absent in many reservoirs and fishers also use smaller mesh sizes than the optimum for the target species causing juvenile capture. Since gillnets are highly selective, a particular mesh size is most efficient for a narrow size range of fishes (Thomas, 2019). In addition, gears of very small mesh sizes or mosquito nets like dragnets cause adverse impacts due to the capture of juveniles. Even though control measures exist in many reservoirs, illegal fishing using small-meshed nets is done by fishers and requires strict monitoring also. The reservoir fisheries management in Himachal Pradesh is note-worthy where mesh regulations, closed season, and so forth are strictly monitored which can be adopted by other states for fishery improvement (Thomas, 2015).

14.3.3 LOSSES DUE TO EXTENDED SOAKING TIME

Set gillnets, which are often left in reservoirs for hours, extending 12–24 h, cause spoilage of fish caught in the initial hours of setting. Extended soaking time leads to spoilage/damage to the fish in the nets and also gives a chance for predation by birds/wild fishes. Studies from Bhavanisagar had shown that the average annual loss due to depredation was 2.6% of the total catch (Thomas et al., 2020). Freshwater otters, cormorants, and crocodiles were the main predators, and the problem was severe during summer due to the decline in water level in the reservoir. Reducing the soaking time will help to minimize these types of losses to a greater extent.

14.3.4 INADEQUATE STOCKING AND MONITORING

Commercially important species are less available in many reservoirs due to inadequate stocking. In addition, use of small meshed nets during the stocking period adversely affects the fish stocks. Regular stocking and proper monitoring will help to improve the production.

14.4 FUTURE PROSPECTS

In view of the stagnating marine fish catches and degradation of many inland open water systems due to habitat modifications, reservoir fishery has been given focussed attention due to its untapped potential and vast resource area. Appropriate harvesting methods had greater role in improving the catches through capture of most available species at proper sizes without undermining the sustainability of the fishery. There is always a greater need for the application of newer technologies and improvements in fishing gears suiting to the prevailing conditions to enhance the fish catches from reservoirs in a sustainable manner. For developing appropriate and sustainable fishing methods in reservoirs, the following points can be considered.

1. Baseline information on fishing gears in different reservoirs needs to be updated for evaluating the current status and for developing modification/improvements.
2. Optimum mesh size estimates for commercially important species and standardization of various gear parameters to increase the catching efficiency.
3. Mesh size regulations, size limits need to be implemented and monitored for sustainable exploitation.
4. Restrictions for using very small meshed nets like dragnets for preventing the capture of juvenile fishes.
5. Adequate stocking program with appropriate stocking measures and exploitation plans.
6. Application of electronic fish detection devices, bottom profiling and mapping, clearing of tree stumps/trunks, identification of potential fishing zones, exploratory fishing, acoustic surveys, and remote sensing for potential fishing area marking.
7. Minimizing harvest losses in gillnetting by decreasing the soaking duration.
8. Transfer of technologies and suitable training to fishers and stakeholders for adopting fishing gear developments.

14.5 CONCLUSION

Reservoir fisheries is a significant contributor to total inland fish production in the country and plays a major role in livelihood improvement of people displaced by the impoundment. The fish productivity of reservoir could be increased by using a number of approaches that combine better harvesting strategies, fertilization, stock enhancement, and extensive aquaculture (Kolding and van Zwieten, 2006). Among these strategies, exploitation methods have a greater role to play to augment fish production from reservoirs. Even though during the last few decades many technological advances occurred in fishing systems, there is a greater need for improvements/modifications in existing gears and application of new technologies for realizing optimum harvest from reservoirs.

KEYWORDS

- **fishing**
- **gillnets**
- **kernels**
- **monitoring**
- **reservoir**
- **stocking**
- **traps**

REFERENCES

Bhakta, D., Anand, M. W., Solanki, J. K., & Sah, R. K. (2016). Traditional fishing crafts and gears of Ukai reservoir, Gujarat, India. *International Journal of Fisheries and Aquatic Studies, 4*(4), 142–145.

George, V. C. (2002). Fishing techniques of riverine and reservoir systems, present status and future challenges. In M. R. Boopendranath, N. Meenakumari, J. Joseph, T. V. Sankar, P. Pravin & L. Edwin (Eds.), *Riverine and reservoir fisheries of India* (pp. 192–196). Society of Fisheries Technologists.

Kartha, K. N., & Rao, S. K. (1991). Selectivity of Gill Nets for *Catla catla* (Ham), *Cirrhinu smrigala* (Day) and *Labeo rohita*(Ham) in Gandhisagar Reservoir. *Fishery Technology, 28*, 5–10.

Khan, A. A., Kartha, K. N., Dawson, P., & George, V. C. (1991). Fish harvesting systems in Indian reservoirs. In *Proceedings of National Workshop on Low Energy Fishing* (pp. 152–155). SOFT(India).

Kolding, J., & Van Zwieten, P. A. M. (2006). Improving productivity in tropical lakes and reservoirs. Challenge Program on Water and Food. In *Aquatic ecosystems and fisheries review Series 1, Theme 3 of CPWF* (p. 139). World Fish Center.

Kuriyan, G. K. (1973). Fishing gears in the fresh water reservoirs of India. *Fishery Technology, 10*(1), 1–5.

NAAS. (2016). Reservoir Fisheries Development in India: Management and Policy Options. Policy Paper No. 78 (p. 20). National Academy of Agricultural Sciences, New Delhi.

Natarajan, A. V. (1976). Ecology and state of fishing development in some of the manmade reservoir in India. In *Proceedings of the Indo-Pacific Fisheries Council, 7 Session, Colombo, Sri Lanka.*

Pravin, P., Remesan, M. P., Thomas, S. N., Baiju, M., & Meenakumari, B. (2014). In *Fishing crafts and gears in reservoirs of Kerala* (p. 126). Central Institute of Fisheries Technology, Cochin.

Rajeswari, G., Raghu Prakash, R., Sreedhar, U., & Swamy, K. M. (2015). Studies on fishing Crafts and Gears in Tandava reservoir, Andhra Pradesh, India. *International Research Journal of Biological Sciences, 4*(11), 38–42.

Sugunan, V. V., & Suresh, V. R. (2022). Reservoir fisheries development and its trade-offs with conservation of natural fish genetic resources. *Indian Journal of Plant Genetic Resources, 35*(3), 308–311.

Sundaramoorthy, B., Parivallal, P., & Neethiselvan, N. (2013). Gillnet selectivity on *catla catla* (hamilton, 1822) in Aliyar reservoir, Tamil Nadu, South India. *Tamil Nadu Journal of Veterinary & Animal Sciences, 9*(5), 352–361.

Thomas, S. N. (2015). Review of fishing gear dimensions in the reservoirs of Himachal Pradesh. In *Report to the Department of Fisheries*, Government of Himachal Pradesh by CIFT, Kochi.

Thomas, S. N. (2019). Sustainable gillnet fishing. In ICAR-CIFT Winter school manual on *Responsible Fishing: Recent Advances in Resource & Energy Conservation*, 21 November–11 December 2019 (pp. 239–249).

Thomas, S. N., Edwin, L., Chinnadurai, S., Harsha, K., Salagrama, V., Prakash, R., Prajith, K. K., Diei-Ouadi, Y., He, P., & Ward, A. (2020). Food and gear loss from selected gillnet and trammel net fisheries of India. FAO Fisheries and Aquaculture Circular No. 1204, Rome, FAO.

Velmurugan, R., Neethiselvan, N., Xavier, P. R., Kalidoss, R., & Kalaiarasan, M. (2016). Gill net selectivity and fishing pressure on Indian Major Carp in Thirumoorthy Reservoir, Tamil Nadu. *Current World Environment, 11*(2), 554–559.

CHAPTER 15

Trends and Developments in Aquaponics Research: A Scientometric Analysis

K. ANANTHARAJA[1], P. KRISHNAN[2,*], A. K. VERMA[3], HEMA PRASANTH[1], SAROJ KUMAR SWAIN[4], G. KANTHARAJAN[5], S. ABUTHAGIR IBURAHIM[3], R. BHARATHI RATHINAM[3], and P. ROUTRAY[4]

[1]*RRC of ICAR-Central Institute of Freshwater Aquaculture, Bangalore, Karnataka, India*

[2]*Bay of Bengal Programme Inter-Governmental Organisation (BOBP-IGO), Abhiramapuram, Chennai, Tamil Nadu, India*

[3]*ICAR-Central Institute of Fisheries Education, Versova, Mumbai, Maharashtra, India*

[4]*ICAR-Central Institute of Freshwater Aquaculture, Kausalyaganga, Bhubaneswar, Odisha, India*

[5]*ICAR-National Bureau of Fish Genetic Resources, Lucknow, Uttar Pradesh, India*

*Corresponding author

ABSTRACT

Aquaponics is an evolving technology that integrates hydroponics with aquaculture to produce high-quality food sustainably. This study provides a comprehensive, up-to-date overview of the aquaponics research and adds value to its contents by identifying main research interests, journals,

Recent Advances in Sustainable Aquaculture and Fisheries. Shabir Ahmad Dar, Imtiyaz Qayoom, Gohar Bilal Wani, Adnan Abubakr, and Seemab Zehra (Eds.)

DOI: 10.1201/9781998511372-15

institutions, and countries, and how these are linked within currently available aquaponic studies using a scientometric approach based on metadata extracted from the Web of Science Core Collection. The results showed that 116 institutions from 52 countries were involved in aquaponics research. The United States had the most research publications (78) in aquaponics, followed by Germany, China, Netherlands, and Brazil. These top five countries contribute to 58% of the total research papers in aquaponics research. Tilapia and lettuce are the most studied species in aquaponics research. The research work of Rakocy was highly cited (417 citations) as per the author's cocitation analysis. The keyword analysis in CiteSpace and trend analysis in R revealed that the research on fish and plant growth performance, sludge management, water use, decoupled aquaponics, marine aquaponics, precision farming, integration of biofloc with aquaponics is emerging in recent times. In addition, by analyzing the document cocitations, we effectively clustered the popular topics and established future research directions to guide researchers. This study visualizes and provides extensive information on the trends of aquaponics research worldwide for researchers and policymakers.

15.1 INTRODUCTION

The United Nations (UN) Sustainable Development Goals (SDGs) 2030 promote a sustainable, prosperous, and equitable world for all (United Nations, 2015). As the population grows, we need to produce more food sustainably while minimizing food waste to meet food security. This has resulted in greater use of water and synthetic fertilizers in agriculture. Using chemical fertilizers excessively has resulted in environmental pollution, aggravating climate change issues. In 2016, worldwide nitrogen losses from aquaculture were estimated to be 7.52×10^6 t. A large portion of nitrogen loss from aquaculture is also accounted for by developing countries like India, Bangladesh, Nigeria, and so forth. Wastewater reuse through aquaponics is crucial for environmental and production sustainability. Therefore, global attention and policy formulation are needed (Adhikari et al., 2020). It is becoming increasingly important to ensure that food production systems are sustainable.

Moreover, recent coronavirus disease 2019 pandemic outbreaks have also highlighted the need for enhanced food security. The UN stated that the pandemic would "unleash a food security crisis not seen since the Great Recession" (Tiensin et al., 2020). To contend with the future threat of food security, countries cannot continue to increase production using conventional farming methods. According to Fan et al. (2012), it is likely to be challenging to continue to increase crop yields due to the rapid industrialization and urbanization process. Production systems that do not rely on soil have emerged as sustainable food production options. Aquaponics is one such nonsoil-based system, integrating aquaculture with hydroponics to produce vegetables in production systems, which can ensure future environmental and socioeconomic sustainability. When both food production systems are combined, plants receive essential nutrients for growth, while waste generated by fish production is managed (Shaalan et al., 2018). Plants grow rapidly in aquaponic systems using nutrients excreted by fish or generated by microbial activity in fish excretion (Love et al., 2014; Rakocy et al., 2006). There are many factors that drive the concentration of nutrients in aquaponics systems, including fish species, stocking density, growth stage, feeding rate, and the composition of the feed (Rakocy et al., 2006; Tyson et al., 2011; Yildiz et al., 2017). Aquaponics has many benefits from a sustainability perspective, including low water use, little to no pesticides or synthetic fertilizers, no chemical use, and waste recycling (Goddek et al., 2015).

Aquaponics has been popular on a small scale globally for decades and has established a strong public profile. In urban areas with limited space, small-scale aquaponic systems are particularly attractive for self-production. In a study comparing three small-scale aquaponic systems, the nutrient film technique outperformed the floating raft and vertical felt in terms of crop production and water use (Pérez-Urrestarazu et al., 2019). The popularity of aquaponics is apparent, as it ranks highly on Google, which has deemed aquaponics an emerging science and technology topic (Junge et al., 2017). However, commercial aquaponics is still in its early stages of development (Love et al., 2015; Villarroel et al., 2016). The key to the economic feasibility of aquaponics is the properly designed system and the appropriate management (Tyson et al., 2011). Determining the appropriate plant–fish biomass ratios plays a significant role in the design of aquaponic systems for improving biomass quality and production, saving water, and reducing greenhouse gas emissions (Paudel, 2020). Further, aeration is one of the most important components of operation cost that determine the feasibility of aquaponics during operation.

The findings of Fang et al. (2017) showed that semi-aeration is economically and environmentally beneficial and that it can be used to reduce aquaponic aeration. Few aquaponics literature reviews exist; however, they are generally focused on a specific concept. Many publications have focused on innovative system designs, hydroponic system components, plants and fish performance, and the role of microbe addition. Taking aquaponics in a broader sense, it is an increasingly complex system encompassing several disciplines, including engineering, aquaculture, horticulture, microbiology, ecology, and chemistry. Despite the vast amount of articles in this field, there are no literature reports on scientometric studies covering aquaponics research globally to the best of our knowledge. Scientistometric methods are commonly employed to study and measure research progress and effort from researchers, governments, research institutes, universities, professors, and scientific publishers and journals in a chosen field (Konur, 2012).

The Mapping Knowledge Domain is a visualization analysis in scientometrics that reveals knowledge structures and their development and displays research correlations of a specific field through the process of charting, mining, and analyzing (Dirk, 2004; Zou et al., 2018). Traditional reviews provide insight into a certain field in order to support further research (Hu et al., 2020). This approach has the disadvantage that it relies on the personal and intentional selection of limited publications so some bias may be imparted to the results (Zhu et al., 2020). By analyzing bibliometric data, we can explore the development of certain fields of science quantitatively (Secinaro et al., 2020). Among the benefits of this technique is that it allows for an objective and systematic investigation of research fields (Su et al., 2020). We can reveal knowledge structures and how they develop, as well as visualize research correlations, by mapping knowledge domains using scientometric techniques.

Research collaboration and networking among peers are considered the major drivers of research outcomes (Lee and Bozeman, 2005; Bhoomaiah et al., 2022); Advancement in aquaponics research demands collaboration and networking of peers from various disciplines across the globe. The data provided on the most productive organizations, cooperation, and coauthorship networks of researchers involved in aquaponics research would facilitate the global research community to find suitable collaborating partners to bring breakthroughs in research outcomes and thus contribute to the future development of the aquaculture industry. This scientometric review aims to provide researchers and practitioners with a thorough understanding of the current state of knowledge, quantitative research patterns, intellectual structure, and emerging topics in global aquaponics research.

15.2 MATERIALS AND METHODS

In the present study, we aimed to conduct a scientometric analysis of aquaponics research to assess the knowledge domains, quantified research patterns, intellectual structure, and emerging research topics to guide future research. Figure 15.1 shows a schematic of the research design.

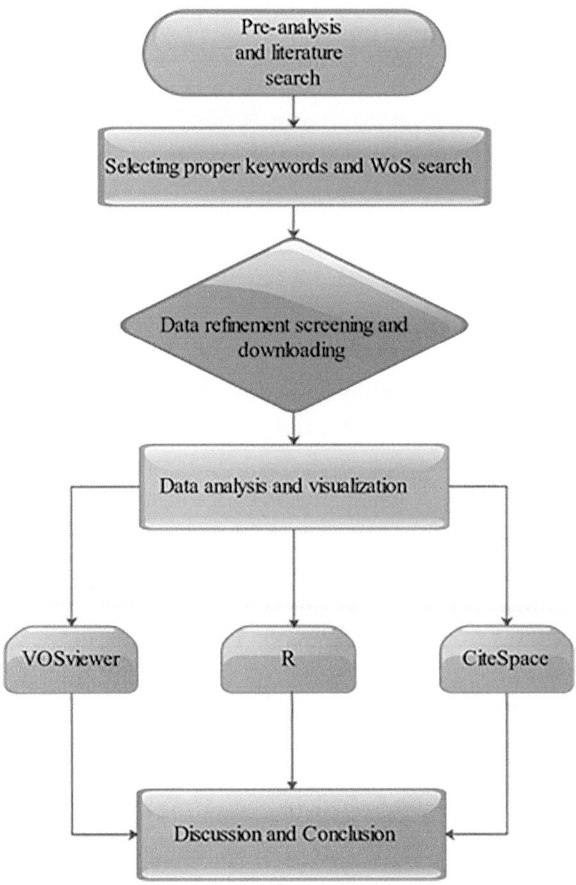

FIGURE 15.1 Schematic of the research design.

15.2.1 DATA SOURCES

The study utilized the Web of Science (WoS) database (Clarivate Analytics, Philadelphia, PA), which has a collection of research papers from over 21,100 peer-reviewed, high-quality scholarly journals published worldwide in >250 disciplines (https://clarivate.com/webofsciencegroup/solutions/web-of-science-core-collection/). We examined the literature and conducted several searches in the WoS in order to collect relevant aquaponics research publications using the following search terms: TITLE: ("Aquaponic*" OR "Gravel bed*" OR "Media bed*" OR "Ebb and flow" OR "Hydroponic*" OR "Deep Water Culture" OR "DWC" OR "Vertical farm*" OR "Nutrient Film Technique*" OR "NFT") AND TOPIC: (Aquaponic*) AND DOCUMENT TYPES: (Article OR Review) AND LANGUAGE: (English) Timespan: All years (i.e., 1990–2021). Indexes: SCI-EXPANDED. The search was performed on August 14, 2021, and retrieved bibliometric records of 324 articles from the Web of Science Core Collection, and all original papers published were selected for analysis.

15.2.2 DATA ANALYSIS

The bibliometrics records from the WoS database were exported, and analyses were carried out using Microsoft Excel 2019, VOSviewer, and Citespace. In this study, although the search was conducted from the years 1990 through 2021, we used data from 324 published articles between the years 2004 and 2021 since publications regarding aquaponics were indexed in the WoS database starting with the year 2004.

To assess the productivity of institutions and authors, scientometric indicators of publications such as total citation counts, average citation per paper (ACPP), and *h*-index were used. The *h*-index is a research output metric based on the most cited research articles and the number of citations received (Hirsch, 2005).

The VOSviewer software (Leiden University, Leiden, Netherlands) is used for constructing and visualizing bibliometric network maps, which are easy to interpret (van Eck and Waltman, 2010). In order to visualize leading organization, coauthorship analysis, and author and journal cocitation, we used the VOSviewer. The clusters were colored in different colors (Waltman et al., 2010).

CiteSpace is a progressive knowledge domain visualization tool for visualizing patterns and trends in the scientific literature (Chen, 2004). A network of country, keyword, document cocitation, and reference bursts were generated by using CiteSpace. On the cluster view and the timeline view of keyword analysis, and the Document cocitation analysis (DCA) network, we have applied the log-likelihood ratio (LLR) and latent semantic indexing (LSI) methods to determine cluster labeling. For assessing the structural properties of the networks, metrics such as modularity (*Q*) index and silhouette metrics, burstness, and betweenness were used. The modularity index (*Q*) indicates how a network can be divided into modules or clusters (Aryadoust et al., 2019). The *Q* value ranges from 0 to 1, and a value close to 1 indicates that the network is clearly divided into distinct groups. A silhouette score is used to figure out the homogeneity of these modules. It ranges between –1 and 1, depending on the module. In modules that have a high silhouette score, the nodes tend to be more consistent (Chen et al., 2010). Betweenness centrality measures the extent to which a single node connects two or more other nodes that would otherwise be separate. The centrality value ranges from 0 to 1, where higher values near 1 indicate a high level of influence (Aryadoust et al., 2020).

CiteSpace was used to analyze the keyword and DCA, as well as the detection of reference bursts (Chen, 2004). Betweenness centrality, modularity (*Q*), and silhouette are the metrics used in the cocitation cluster analysis (*S*). The ability of a node to connect to other nodes in the network is referred to as its betweenness centrality. The following equation describes the centrality of betweenness.

$$\text{centrality}(\text{node}_i) = \sum_{i \neq j \neq k} \frac{\rho_{jk}(i)}{\rho_{jk}} \tag{15.1}$$

where ρ_{jk} represents the number of shortest paths between node *j* and node *k* and ρ_{jk} (*i*) represents the number of paths that passed through node *i*. Modularity measures the cluster's interconnectivity of the networks. The following equation is used to calculate the modularity factor:

$$\text{Modularity}\, Q(G.C) = \frac{1}{2m} \sum_{i,j=0}^{n} \delta(c_i.c_j).\left(A_{ij} - \frac{\deg(n_i).\deg(n_j)}{2m} \right) \tag{15.2}$$

where *n* denotes the total number of nodes in the network *G* and the network *G* is restricted by a partition *C* and *m* (the total number of edges).

The term deg(n_i) denotes the degree of n_i. If nodes n_i and n_j belong to the same cluster, then δ ($c_i \cdot c_j$) is 1 or otherwise, it is 0. A silhouette metric determines the homogeneity of nodes in a cluster. The following equation describes the silhouette:

$$\text{Silhouette}\,(xi)\frac{b(x_i)-a(x_i)}{\max(a(x_i).b(x_i))} \tag{15.3}$$

where $a(x_i)$ represents the average intracluster path from node x_i to other nodes of the same cluster, and $b(x_i)$ represents the average intercluster paths of x_i to nodes in each cluster. The modularity and silhouette metrics reveal the "overall structural properties" of the nodes in the cocited network, ranging between 0 and 1, which are a crucial component of measuring the quality of the cocited node cluster. A larger Q value indicates better clustering of nodes, and a larger S value indicates high homogeneity among nodes in the cluster (Hamidi et al., 2021).

In the current study, the retrieved metadata from WoS was analyzed in R version 4.0.1 and Rstudio Version 1.3.959 using the bibliometrics R-package, where the Biblioshiny tool was used for analysis such as Trend topics, thematic evolution, and multiple-country publication (MCP) (Aria and Cuccurullo, 2017)/associations between nations, authors, and keywords were used to divide the data into multiple color-coded clusters (Van Eck et al., 2010; Waltman et al., 2010). The network visualization method was used to visualize the association between them, where the distance between two circles represents the similarity between them (Khalil and Gotway Crawford, 2015; Zhao et al., 2018).

15.3 RESULTS AND DISCUSSION

15.3.1 RESEARCH TREND IN AQUAPONICS RESEARCH

According to WOSCC's selection criteria, 324 publications were published in the field of aquaponics which includes 27 review articles. Aquaponics has been developed for several decades based on the development of aquaculture (particularly RAS) and hydroponics. During the period 2004–2014, there was an initial increase in aquaponic research publications, and publication numbers were below 10. Research achievements were preliminary and emerging, with fewer publications per year until 2014. Several publications have been published in this area since

2015, marking a significant breakthrough. It might be due to the increased urbanization, and this system has great potential to meet the needs of rapid urbanization. It is noteworthy that despite a stagnant trend between 2017 and 2018, overall publication volumes are still increasing. The peak in 2020 occurred with 76 publications, which accounted for 23.46% of the total. Figure 15.2 displays the distribution of aquaponics research publications and citations by year.

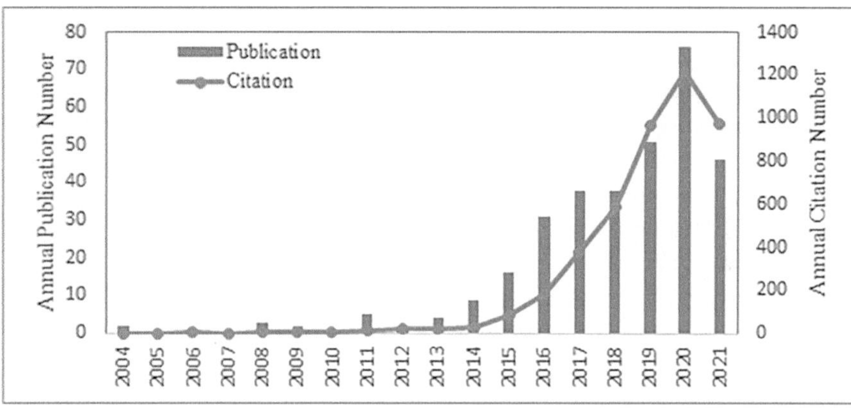

FIGURE 15.2 The distribution of annual publications and citations on aquaponics research.

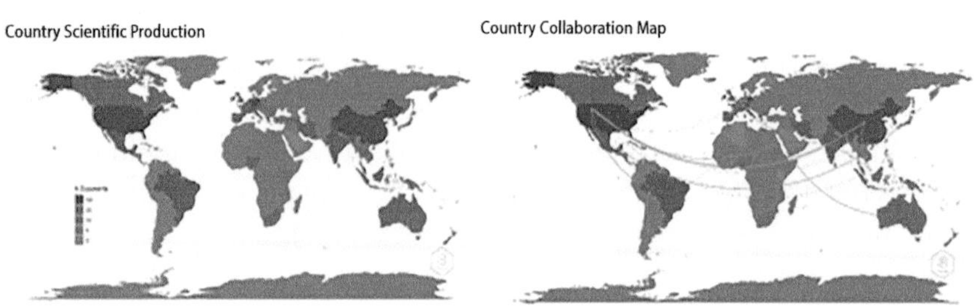

FIGURE 15.3 Geographic distribution of scientific publication and collaboration of the countries in the field of aquaponic research.

Note: The figure was generated based on WoS data in R Biblioshiny; Country boundaries are not to scale.

In terms of citations, the number of times cited was 4462 times, with an average of 13.77 citations per publication. As can be seen from the chart, citations have been steadily increasing during the past two decades, reaching 1214 times in 2020 indicating growth in aquaponic research in diverse fields of studies on hydroponic design and evaluation, fish and plant growth, water quality, and microbial community. Researchers and urban growers have focused on aquaponics in recent years, but the feasibility of its commercial scale still requires further study (Armanda et al., 2019).

15.3.2 LEADING COUNTRY IN AQUAPONIC RESEARCH

The publication volume reveals the strength and extent of the research activity in a country. Figure 15.3 shows the geographical distribution of scientific publications in aquaponics research. A total of 53 countries contributed to aquaponic research globally. The United States had the largest number of publications (78) in the field of aquaponics, followed by Germany (36), China (32), Netherlands (23), and Brazil (20). These top five countries contribute to 58% of the total research papers in aquaponics research.

The network of cooperation between the productive countries is displayed in Figure 15.4. From Figure 15.4, it was evident that the United States held a central position in the network and frequently cooperated with Italy, Israel, Spain, Netherlands, Brazil, South Korea, Australia, Thailand, Malaysia, Mexico, Iran, Japan, Vietnam, Ghana, and China. A purple ring outside each circle indicates centrality (Chen et al., 2014). Some countries do not communicate with each other well as evidenced by their lack of communication in the cooperation network. To promote development in this field, it is, therefore, recommended that some attention be paid to improve cooperation between countries. The analysis of MCP versus single-country publication in R shows that the trend of MCP was practiced mainly by the United States, Germany, and China, among others, in aquaponics research studies (Figure 15.5).

15.3.3 AUTHORS OF THE MOST CITED PAPERS

Literature with a high citation level is generally considered to be a high-quality document with vast innovation and influence in a field, as well as essential reading for researchers (Li et al., 2020; Zuanazzi et al., 2020). Table 15.1 presents a list of the 10 most-cited ("top 10") papers in aquaponic research. In aquaponics research, the most cited paper is "Aquaponic Systems: Nutrient recycling from fish wastewater by vegetable production"

TABLE 15.1 Top 10 Papers (The 10 Most-Cited Papers) in Aquaponic Research.

Rank	Title	Authors	Source Title	Publication Year	Total Citations	Average Per Year
1	Aquaponic systems: Nutrient recycling from fish wastewater by vegetable production	Graber, A. and Junge, R.	Desalination	2009	171	13.15
2	Commercial aquaponics production and profitability: Findings from an international survey	Love, D. C., Fry, J. P., Li, X., Hill, S., Genello, L., Semmens, K., and Thompson, R. E.	Aquaculture	2015	144	20.57
3	Challenges of sustainable and commercial aquaponics	Goddek, S., Delaide, B., Mankasingh, U., Ragnarsdottir, K. V., Jijakli, H., and Thorarinsdottir, R.	Sustainability	2015	132	18.86
4	Opportunities and challenges to sustainability in aquaponic systems	Tyson, R. V., Treadwell, D. D., and Simonne, E. H.	Hort Technology	2011	111	10.09
5	A comparison of three different hydroponic sub-systems (gravel bed, floating and nutrient film technique) in an aquaponic test system	Lennard, W. A. and Leonard, B. V.	Aquaculture International	2006	111	6.94
6	A study on the optimal hydraulic loading rate and plant ratios in recirculation aquaponic system	Endut, A., Jusoh, A., Ali, N., Nik, Wan, W. B., and Hassan, A.	Bioresource Technology	2010	102	8.5
7	Effect of plant species on nitrogen recovery in aquaponics	Hu, Z., Lee, J. W., Chandran, K., Kim, S., Brotto, A. C., and Khanal, S. K.	Bioresource Technology	2015	95	13.57
8	Nitrogen transformations in aquaponic systems: A review	Wongkiew, S., Hu, Z., Chandran, K., Lee, J. W., and Khanal, S. K.	Aquacultural Engineering	2017	84	16.8
9	An International Survey of Aquaponics Practitioners	Love, D. C., Fry, J. P., Genello, L., Hill, E. S., Frederick, J. A., Li, X., and Semmens, K.	PLoS One	2014	74	9.25
10	Effects of pH on nitrogen transformations in media-based aquaponics	Zou, Y., Hu, Z., Zhang, J., Xie, H., Guimbaud, C., and Fang, Y.	Bioresource Technology	2016	72	12

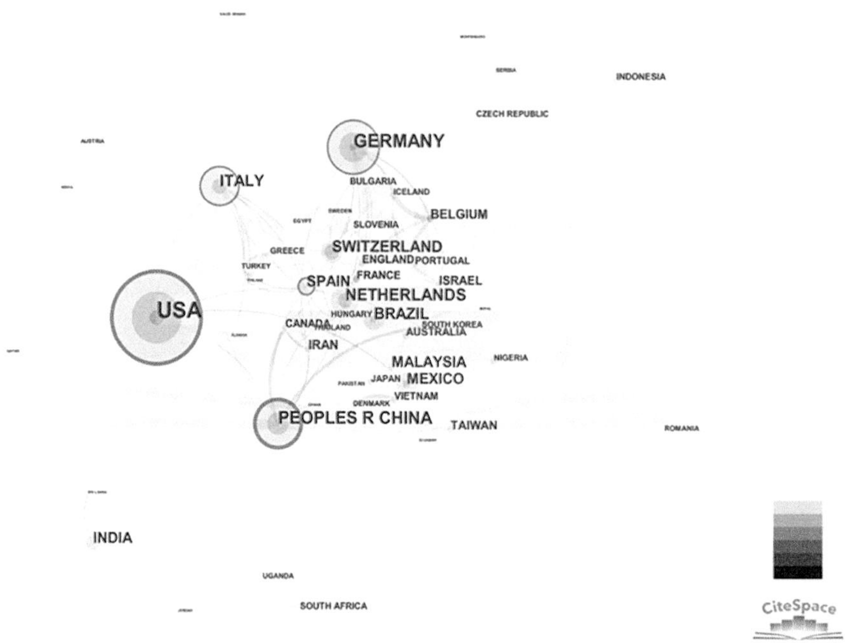

FIGURE 15.4 The cooperation network of the productive countries/regions.

Note: The figure was generated based on WoS data in Citespace.

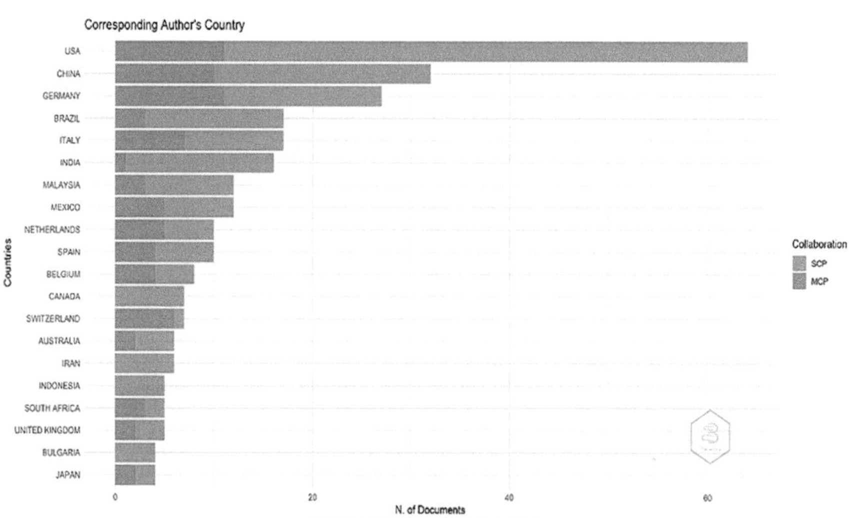

FIGURE 15.5 Collaboration between countries in authorship.

by Graber and Junge (2009), with 171 citations. However, the paper with the highest average citation per year is "Commercial aquaponics production and profitability: Findings from an international survey" by Love et al. (2015), demonstrating that this paper has received significant attention. In addition, older papers may be benefited from greater exposure over time as a result of citations that accrued over time.

Globally 1037 authors contributed to 324 papers in aquaponic research. The coauthorship network map constructed through VoSviewer indicates that there were about seven research groups that are predominant in this research field. It is also inferred from Figure 15.6 that the collaboration was not very close among the major research groups.

Among authors with publications in this field, Junge, R. authored 18 articles, followed by Goddek, S. (14), Keesman, K. J. (13), and Verma, A. K. (13). The majority of authors only contributed one or two articles, indicating that only a few researchers made substantial contributions to this field. The list of top authors working in a particular field will help research scholars to identify leading research communities for stronger academic and research collaborations.

Moreover, it would be beneficial for research communities to increase their use of standard citation practices (such as ORCID) so that researchers can be properly credited (Haak et al., 2012). Further, universities and research institutions undergo a variety of changes over time, so identifying each research unit with a unique identifier is necessary.

15.3.4 RESEARCH AREAS ANALYSIS

In the WoS database, bibliographic records are assigned subject area categories based on the scope of the journal corresponding to those records, and an article may be assigned to one or more subject areas (Olawumi and Chan, 2018). A total of 324 bibliographic records were analyzed in WoS results, yielding 53 research subject categories. The distribution of research areas shows that aquaponics involves a multidisciplinary approach which is evident from Figure 15.7.

The top 10 subject areas according to WoS category is fisheries (n = 102), environmental sciences (n = 91), water resources (n = 31), agricultural engineering (n = 30), green sustainable science technology (n = 29), engineering environmental (n = 26), marine freshwater biology (n = 26), biotechnology applied microbiology (n = 23), horticulture (n = 23), and environmental studies ($n = 20$). Of the top 53 subject categories, the most studied are fisheries, environmental sciences, and water resources, which together account for 31.481, 28.086, and 9.568% of the total literature.

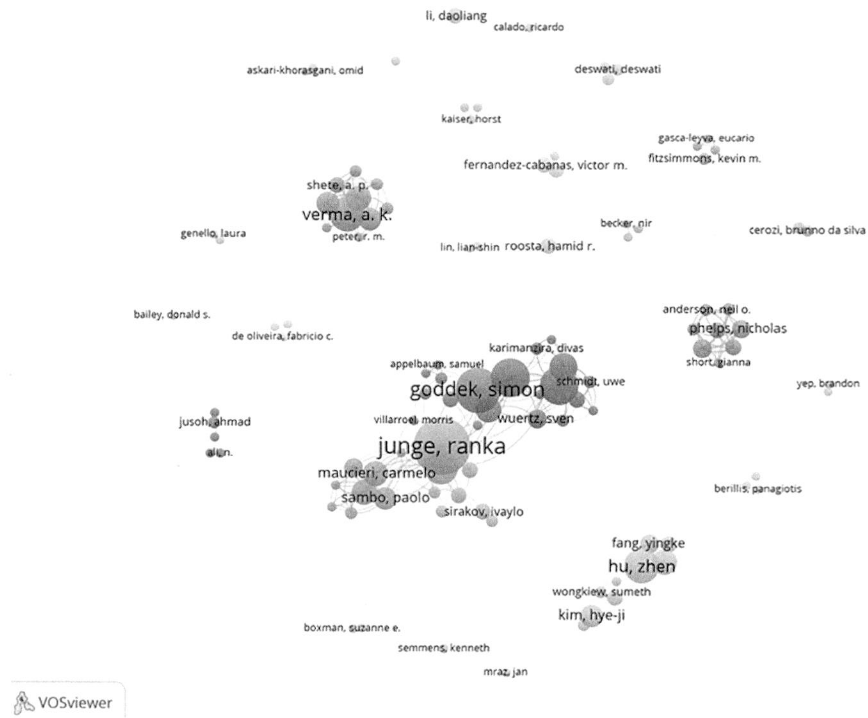

FIGURE 15.6 Coauthorship network of researchers involved in aquaponics research.

15.3.5 JOURNALS OF AQUAPONICS RESEARCH PUBLICATIONS

The number of publications in an academic journal over a given period is a measure of the journal's capacity to acquire and transmit knowledge. The research communications on aquaponic have been published in 101 diverse journals.

The top 10 highly cited journals are listed in Table 15.2, and the network visualization of journals of aquaponics research publications is illustrated in Figures 15.8 and 15.9. *Aquaculture International* has published the most papers (22), followed by *Aquacultural Engineering* (19), and *Aquaculture* (17). *Aquacultural Engineering* is the most-cited journal with 395 citations,

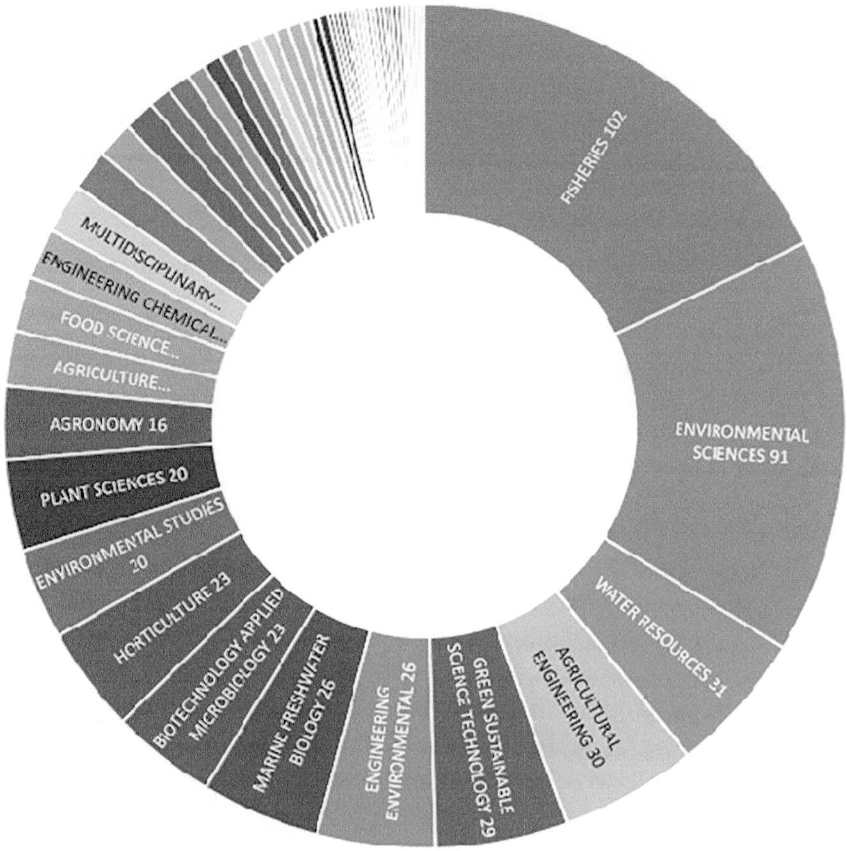

FIGURE 15.7 Subject area categories (WoS-based) of the aquaponics research.

followed by *Bioresource Technology* (376) and *Aquaculture International* (328). These top three journals demonstrate their importance in this field. In terms of ACPP, the desalination paper received 95 citations per paper. An evaluation of journal cocitations reveals the contribution of each journal to aquaponics research, and the journals that were highly cited will reflect the best sources in the field to some extent. *Aquacultural International* had the highest cocitation link strength of 376. Samuel C. Bradford developed Bradford's law in 1934 to study the distribution of scientific literature. Bradford's law states that a small number of journals account for a significant portion of total publications in a given subject area, whereas an increasing number of journals publish fewer articles in that area. Journals are ranked and classified into groups or categories based on the number of articles they contain. Bradford zones are the names given to these groups. *Aquaculture International*, *Aquaculture Engineering*, *Aquaculture*, *Water*, *Sustainability*, and *Journal of Cleaner Production* were in the core zone according to Bradford's law (Figure 15.10).

TABLE 15.2 Journal-wise Number of Publication on Aqquaponic Research.

Rank	Source	Documents	Citations	ACPP	Total Link Strength
1	*Aquaculture International*	22	328	14.91	376.00
2	*Aquacultural Engineering*	19	395	20.79	350.00
3	*Aquaculture*	17	250	14.71	280.00
4	*Water*	17	294	17.29	320
5	*Sustainability*	16	169	10.56	259
6	*Journal of Cleaner Production*	12	184	15.33	264.00
7	*Aquaculture Research*	11	84	7.64	119.00
8	*Environmental Science and Pollution Research*	11	65	5.91	108.00
9	*Horticulturae*	8	36	4.50	64.00
10	*Journal of the World Aquaculture Society*	8	159	19.88	133

15.3.6 LEADING ORGANIZATION

Understanding the distribution and variations in the contributing organizations allows us to identify important sites of knowledge production, dissemination, and evolution. Organizations can use this information to recruit talents and foster academic cooperation by assessing their strength and involvement. Table 15.3 list the most productive and influential organizations in aquaponics research. In total, 385 organizations published articles related to aquaponics. The Wageningen University Research was the most productive institute with 20 records (6.2%), followed by the Zurich University of Applied Sciences (19). The Humboldt University of Berlin, the Indian Council of Agricultural Research ICAR, University of Defense of the Czech Republic had 14 articles each.

TABLE 15.3 The 10 Most Productive and Influential Organizations in Aquaponics Research.

Rank	Organization-Enhanced	Country	Records	% of 324
1	Wageningen University Research	Netherlands	20	6.173
2	Zurich University of Applied Sciences	Switzerland	19	5.864
3	Humboldt University of Berlin	Germany	14	4.321
4	Indian Council of Agricultural Research ICAR	India	14	4.321
5	Leibniz Institut fur Gewasserokologie und Binnenfischerei IGB	Germany	14	4.321
6	ICAR Central Institute of Fisheries Education	India	13	4.012
7	University of Arizona	USA	12	3.704
8	Shandong University	China	11	3.395
9	State University System of Florida	USA	9	2.778
10	University of Liege	*Belgium*	9	2.778

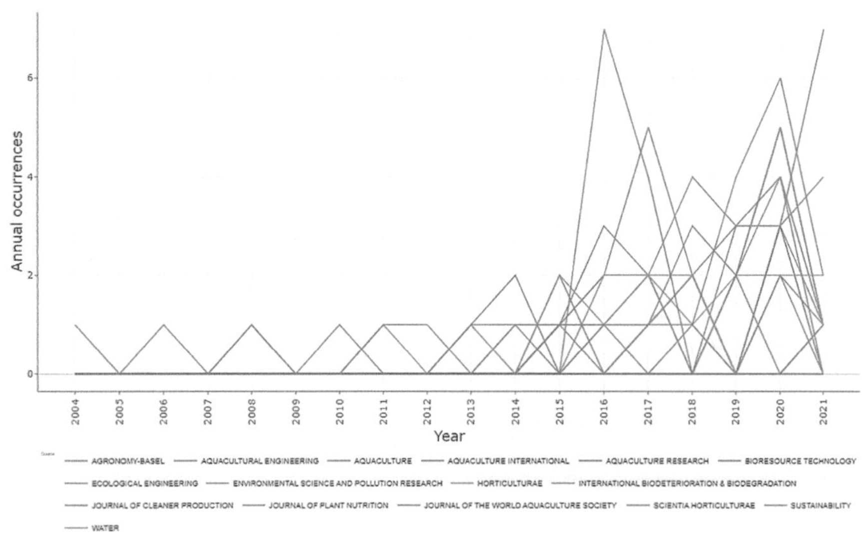

FIGURE 15.8 Source dynamics of aquaponics research worldwide.

Figure 15.8 shows the institute network visualization of aquaponics research publications. It is evident from the network that aquaponic research collaborations are not very close, typically linking five to seven institutions. There are many institutions that operate independently and rarely interact. Therefore, institutions involved in aquaponics research require an urgent improvement in communication.

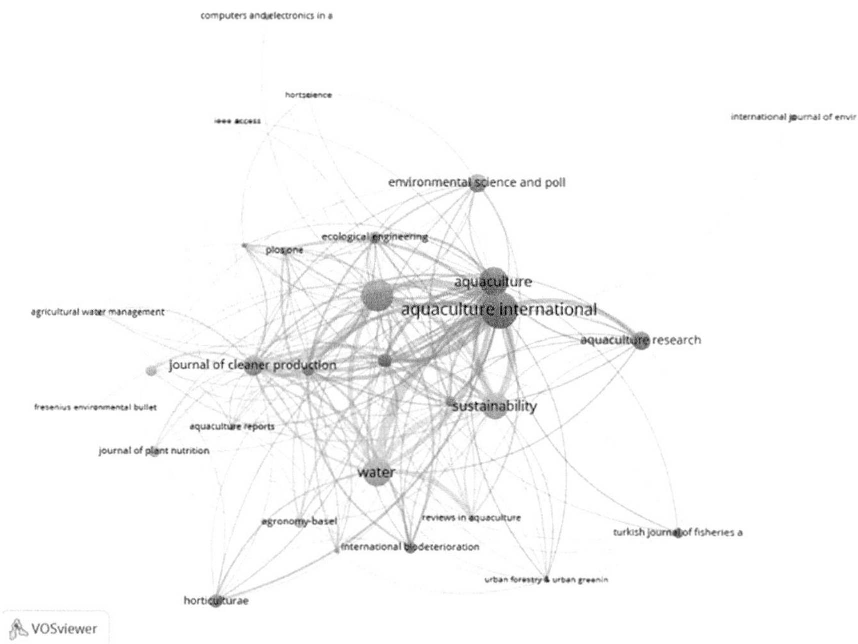

FIGURE 15.9 Network visualization of journals of aquaponics research publications.

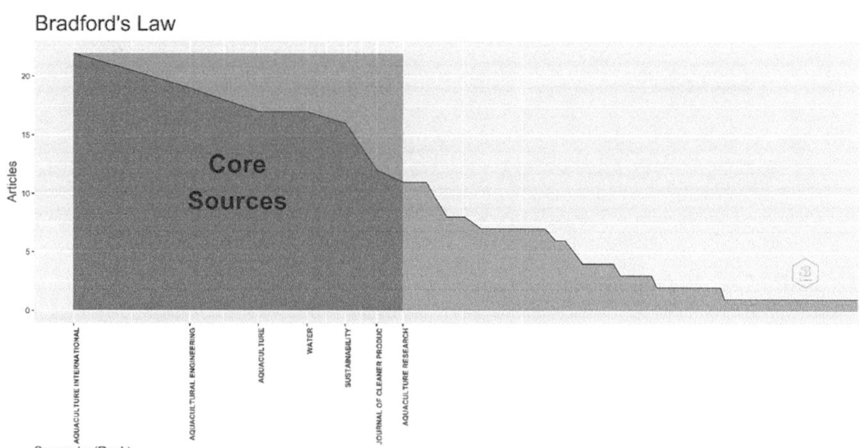

FIGURE 15.10 The top source for aquaponics research as per Bradford's law.

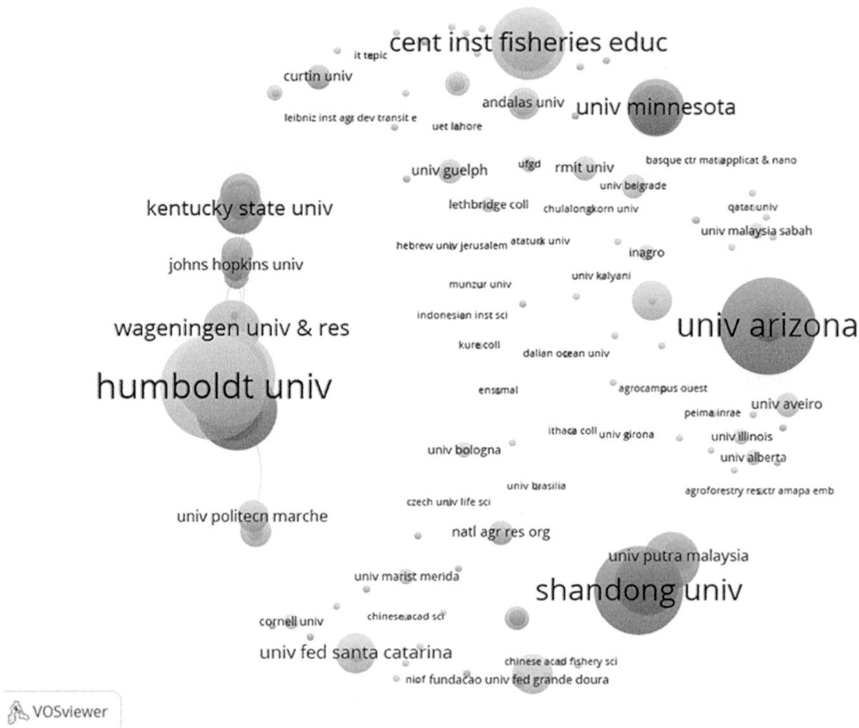

FIGURE 15.11 Institute network visualization of aquaponics research publications.

15.3.7 RESEARCH FOCUS IN AQUAPONICS RESEARCH: KEYWORDS BASED CO-OCCURRENCE ANALYSIS

The keywords in research articles serve as reference points that allow readers to locate articles and understand them better. Furthermore, it illustrates the evolution of the research field over time (Zhao, 2017). A network of co-occurring keywords is compiled in this study using the authors' keywords in CiteSpace.

The co-occurrence network of keywords is shown in Figure 15.12. Node size in a co-occurrence network indicates how frequently a keyword appears. Table 15.4 lists the top 20 effective keywords in aquaponics research ranked by frequency and centrality. The top five high-frequency keywords in this study include aquaponics (159), hydroponics (43), aquaculture (37), tilapia (28), and lettuce (25). This keyword describes the major aspects of aquaponic research in the past two decades. The betweenness centrality measures how close a node is to the shortest path connecting other nodes. An increasing value of betweenness centrality indicates a node with greater influence. Several keywords in our study indicate high betweenness centrality, including aquaponic (0.68), nitrate (0.39), murray cod (0.37), nutrient recycling (0.33), and hydroponics (0.33). A number of aquaponic research topics have been linked through these high-centrality keywords.

TABLE 15.4 Top 20 Effective Keywords in Aquaponics Research Ranked by Frequency and Centrality.

Ranking	Keyword	Frequency	Year	Keywords	Centrality	Year
1	Aquaponics	159	2004	Aquaponic	0.68	2006
2	Hydroponics	43	2004	Nitrate	0.39	2006
3	Aquaculture	37	2008	Murray cod	0.37	2004
4	Tilapia	28	2011	Nutrient recycling	0.33	2009
5	Lettuce	25	2004	Hydroponics	0.33	2004
6	Aquaponic	25	2006	Water spinach	0.28	2009
7	Sustainability	21	2008	Aquaculture	0.25	2008
8	Aquaponic system	15	2015	Aquaponics	0.25	2004
9	Hydroponic	13	2006	Foliar application	0.19	2011
10	Water quality	11	2011	Nitrifying bacteria	0.19	2004
11	Phosphorus	10	2016	Denitrification	0.18	2017
12	Nitrogen	9	2016	Aerobic digestion	0.18	2016
13	Basil	8	2014	Sludge	0.17	2016
14	Recirculating aquaculture system	8	2013	Recirculating system	0.16	2009
15	Nutrient	8	2011	Water use efficiency	0.15	2016
16	Nitrification	8	2008	Lactuca sativa	0.15	2014
17	Vegetable	7	2017	Nitrification	0.14	2008
18	Fish	7	2014	Wastewater	0.14	2006
19	Tomato	7	2011	Tilapia	0.13	2011
20	Water spinach	7	2009	African catfish	0.13	2010

In cluster analysis, bibliometric data are collected, and subgroups are created based on the structure of the research topics. Based on their co-occurrences, CiteSpace identifies and categorizes clusters of keywords. A label is generated from each cluster's keyword content. Accordingly, the derived cluster in CiteSpace is automatically labeled. Each cluster's top terms are the most frequently used keywords. CiteSpace's keyword analysis result showed the formation of 13 clusters (Table 15.5). Weighted mean silhouette and modularity value (*Q*) for the citation network are 0.9156 and 0.7777, respectively. For the 13 keyword clusters, the silhouette metric scores range between 0.804 and 1.000, which indicates that the cluster members are consistent. The largest cluster (#0) contains 21 members and has a silhouette value of 0.941. It is labeled as aquaculture and hydroponic by LLR and LSI, respectively. With 20 members, the second-largest cluster (#1) has a silhouette value of 0.972. Both LLR and LSI label it as hydroponics. The smallest cluster (#12) has eight members and a silhouette value of 0.986. LLR labels it as aquaculture, and LSI labels it as fish.

TABLE 15.5 Description of Keywords Co-occurrence Clusters.

Cluster ID	Size	Silhouette Mean	Citee Year	LSI	LLR
0	21	0.941	2011	Aquaculture	Hydroponic
1	20	0.972	2010	Hydroponics	Hydroponics
2	19	0.922	2017	Nile tilapia	Lactuca sativa
3	19	0.866	2014	Aquaculture	Nutrients
4	19	0.871	2017	Cyprinuscarpio	Cyprinuscarpio (koi)
5	17	0.952	2014	Hydroponics	Aquaponics
6	17	0.928	2012	Water spinach	Water spinach
7	12	0.902	2017	Digital twin	Anaerobic digestion
8	12	0.873	2017	N_2O emission	N_2O emission
9	12	0.946	2019	Litopenaeus vannamei	Marine aquaponics
10	11	0.929	2015	Nutrient recycling	Nutrient recycling
11	11	0.804	2018	Stocking density	Mint
12	8	0.986	2015	Aquaculture	Fish

A timeline view shows the changes in the keywords over time. This view sorts the keywords according to the mean year of occurrence. Accordingly, the size of the node is indicative of the frequency of its appearance; the larger it is, the more often it appears. The connections between nodes are represented by colored lines, which signal their co-occurrence. As shown in Figure 15.13, the emergence of keywords such as nutrient retention, biofloc, probiotics, recirculating aquaculture system, dryland, sustainable food production, biofloc technology, tomatoes, organic aquaponics, bacteria, marine aquaponics, urban agriculture, biochar, profitability analysis, life cycle assessment, decoupled aquaponics, precision farming, renewable energy source, and nutrient recovery in recent years is indicative of a growing interest in a wide range of topics within the aquaponics research domain. In recent years, aquaponics systems in association with biofloc, that is, hybrid biofloc technology-aquaponics getting popular and attention of researcher.

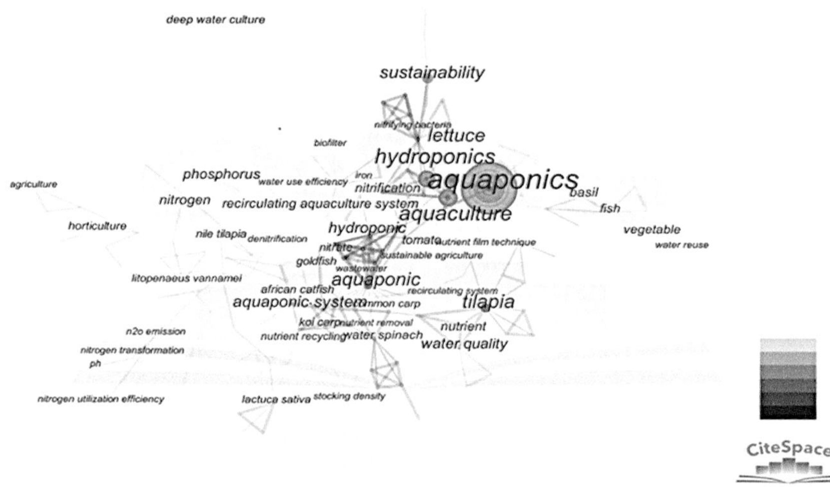

FIGURE 15.12 Keywords occurrence pattern. The size of nods reflects the frequency of each keyword.

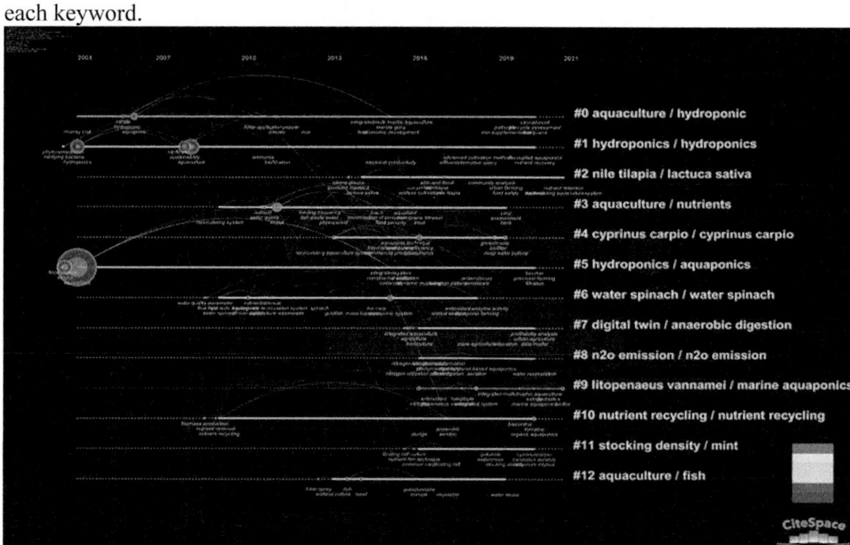

FIGURE 15.13 Timeline view of evolution trend of keywords. The size of nods reflects the frequency of each keyword.

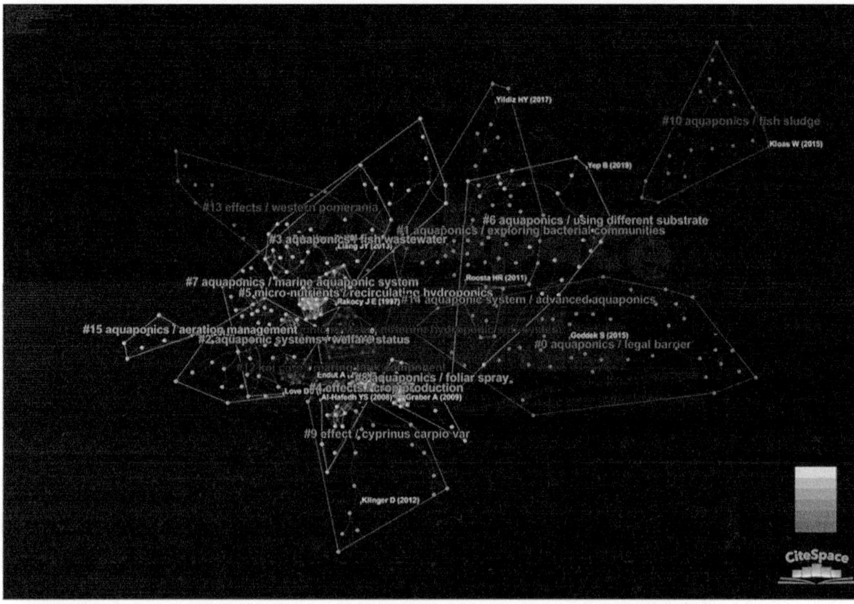

FIGURE 15.14 Cluster view of the document cocitation analysis (DCA).

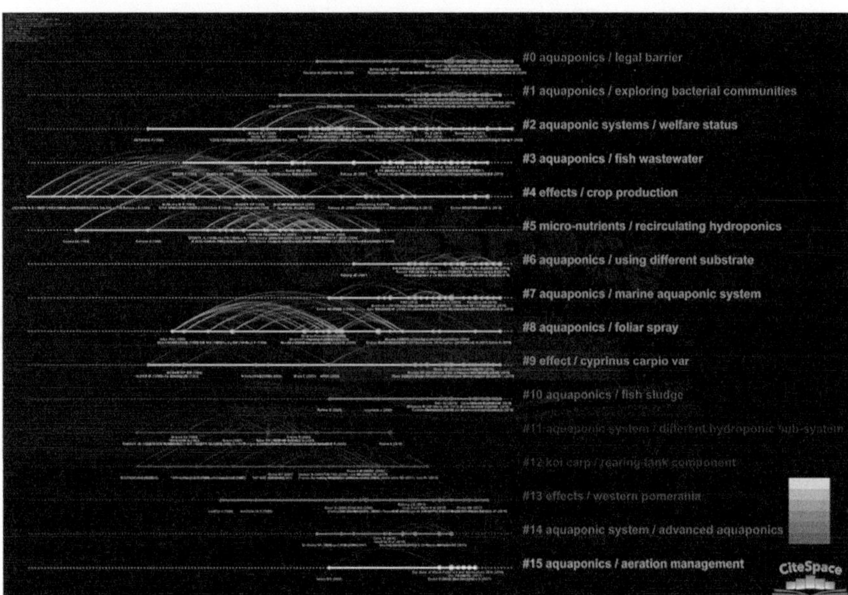

FIGURE 15.15 Timeline view of the document cocitation analysis (DCA) cluster.

15.3.8 DOCUMENT COCITATION ANALYSIS: MAIN RESEARCH TOPICS AND KNOWLEDGE BASES

Identifying a knowledge base in a particular subject can be done mainly through the study of cocitations since cocited documents have an impact on the research content in the field (Zou et al., 2020). The results of CiteSpace's document cocitation are shown in Figure 15.14, demonstrating the formation of 16 document cocitation clusters (Table 15.6).

TABLE 15.6 Description of 16 Document Cocitation (DCA) Clusters.

Cluster ID	Size	Silhouette Mean	Citee Year	Label (LSI)	Label (LLR)
0	49	0.936	2016	Aquaponics	Legal barrier
1	45	0.877	2014	Aquaponics	Exploring bacterial communities
2	42	0.834	2008	Aquaponic systems	Welfare status
3	31	0.936	2009	Aquaponics	Fish wastewater
4	28	0.963	1998	Effects	Crop production
5	27	0.954	2000	Micronutrients	Recirculating hydroponics
6	27	0.903	2015	Aquaponics	Using different substrate
7	25	0.952	2013	Aquaponics	Marine aquaponic system
8	25	0.986	2005	Aquaponics	Foliar spray
9	25	0.906	2009	Effect	Cyprinuscarpio var
10	23	1	2016	Aquaponics	Fish sludge
11	22	0.974	1998	Aquaponic system	Different hydroponic sub-system
12	22	0.982	2003	Koi carp	Rearing tank component
13	19	0.95	2010	Effects	Western pomerania
14	10	1	2009	Aquaponic system	Advanced aquaponics
15	8	1	2014	Aquaponics	Aeration management

The DCA network had an overall network modularity value (Q) of 0.8201. A silhouette score range of 0.834–1 indicates that the group of 16 salient document cocitation clusters are consistent. There are 49 members in the largest cluster (#0) with a silhouette value of 0.936. It is labeled as aquaponics by LLR, a legal barrier by TFIDF. With 45 members, the second largest cluster (#1) has a silhouette value of 0.877. It is labeled as aquaponics by LLR, exploring bacterial communities by TFIDF. The smallest cluster (#15) has eight members and a silhouette value of 1. It is labeled as aquaponics by LLR, and aeration management by TFIDF.

The timeline view (Figure 15.15) shows definite technological and scientific advances in this field. Initial efforts focused on the production of crops using fish wastewater. Afterward, the emphasis was on the use of micronutrients. Similarly, marine aquaponics is evolving. In recent years, research has focused on IOT, different substrate effects, and exploring different bacterial communities in aquaponic systems. The notion of recognizing aquaponics as organic is also gaining attention.

15.3.9 AUTHOR AND JOURNAL COCITATION

Cocitation analysis of the author and journal was performed using VOSviewer software. Cocitation networks could reveal which journals and authors have contributed most to the development of a study. In this study, we adopted the author and journal cocitation analysis that visualizes the similarities in the scholarship of authors and journals.

In VoSviewer, six citations were considered a threshold number, and a total of 312 sources met that criterion out of 7412 identified ones. In the cocitation analysis of authors, the authors are clustered in different colors based on their school of thought. Figure 15.16 shows that the authors of aquaponic research are grouped into seven clusters with different colors, each representing a school of thought. Nodes represent different authors in the cocitation map. Node size indicates the number of citations an author has received. The author Rakocy, J. E. was highly cited (417) as per cocitation analysis.

In Figure 15.17, the cocitations of sources are displayed more precisely. Three citations were chosen as the minimum threshold for a source, and out of 4231 sources found, 737 sources met this threshold. Sources are grouped into 10 clusters based on sources from aquaponics research. According to this study, *Aquaculture*, *Aquaculture Engineering*, *Bioresource Technology*, *Aquaculture International*, and *Acta Hortica* are the most commonly cited journals in aquaponics research. The 10 clusters in the aquaponics research demonstrate the interdisciplinarity of the field.

15.3.10 REFERENCE BURST-DETECTION ANALYSIS

References with the highest number of citations over given time periods are identified through citation burst detection analysis. Burst literature is distinguished by innovations, and all papers are often on the cutting edge of their subjects. Furthermore, the abrupt increase in citations is evidence that the paper or papers in question are being viewed as more valuable (Zou et al., 2020). By the analysis of burst detection using CiteSpace, Twenty-three (23) reference bursts were found and that is illustrated in Figure 15.18.

The paper "Integration of vegetable hydroponics with fish culture: a review" by Rakocy and Hargreaves (1993) had the highest burst strength with the longest duration from the year 1999–2018. The paper by Kloas et al. (2015) and Bosma et al. (2017) has shown a burst in recent years with a burst strength of 3.34 and 3.06, respectively.

15.3.11 THEMES OF AQUAPONICS RESEARCH

Topic trends are also part of this meta-analysis, where Figure 15.19 shows an overview of the development of the topic from time to time with the division per year in aquaponics research. The development of the topic began to experience a significant increase since 2015. Topics like

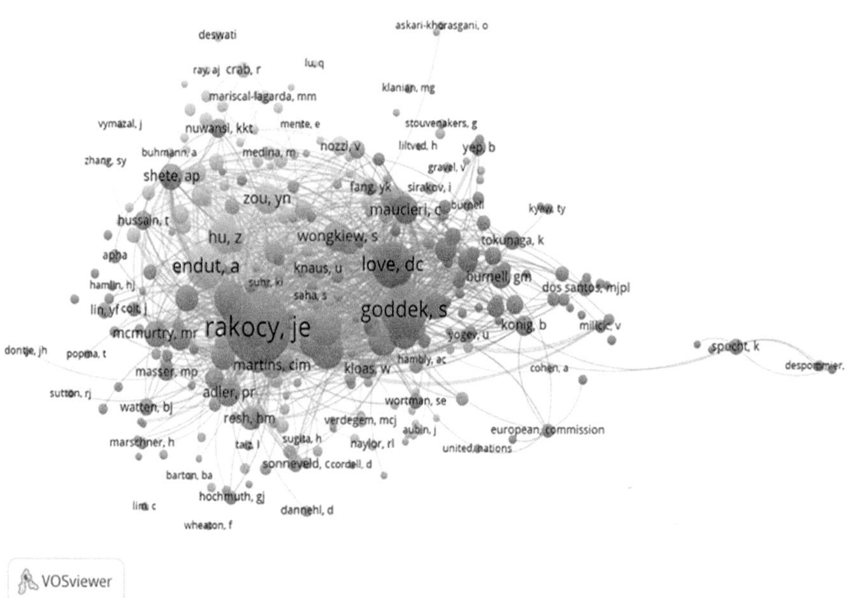

FIGURE 15.16 The network map shows author cocitation patterns of the 312 journals cited at least six times.

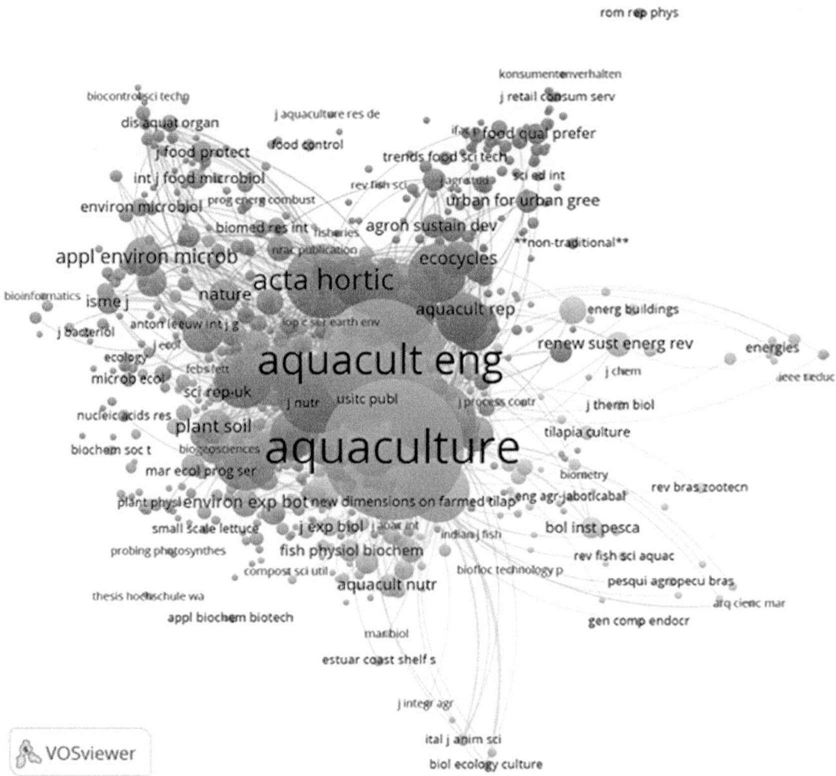

FIGURE 15.17 The network map shows journal cocitation patterns of the 737 journals cited at least three times.

growth performance, sludge management, water use, stocking density, and nitrogen cycle studies, and so forth are emerging related to aquaponics research in recent times (Figures 15.19 and 15.20). The clusters of keywords acquired using coword analysis (Callon et al., 1983) are referred to as research themes (Cobo et al., 2011) and are divided into four types depending on the density and centrality (Callon et al., 1991; He, 1999). Strong centrality and high-density themes (upper right quadrant) are known as motor themes (well developed and vital); weak centrality and high-density themes (upper left quadrant) are known as niche themes (highly developed and isolated themes); low-density and weak centrality themes (lower left quadrant) are known as marginal themes (emerging or declining themes); and strong centrality and low-density themes (lower

References	Year	Strength	Begin	End	2004 - 2021
RAKOCY JE, 1993, ASAE PUBL, V93, P112	1993	7.11	**2004**	2018	
Seawright DE, 1998, AQUACULTURE, V160, P215, DOI 10.1016/S0044-8486(97)00168-3, DOI	1998	5.67	**2004**	2018	
McMurtry MR, 1997, J WORLD AQUACULT SOC, V28, P420, DOI 10.1111/j.1749-7345.1997.tb00290.x, DOI	1997	4.06	**2004**	2018	
Adler PR, 2000, HORTSCIENCE, V35, P993, DOI 10.21273/HORTSCI.35.6.993, DOI	2000	3.17	**2004**	2015	
Adler P R, 1996, WORLD AQUACULTURE, V27, P24	1996	5.67	**2007**	2015	
Lennard WA, 2006, AQUACULT INT, V14, P539, DOI 10.1007/s10499-006-9053-2, DOI	2006	6.49	**2010**	2018	
Endut A, 2009, DESALIN WATER TREAT, V5, P19, DOI 10.5004/dwt.2009.559, DOI	2009	4.58	**2010**	2018	
Rakocy J E, 2006, SRAC Publication - Southern Regional Aquaculture Center, V0, P1	2006	4.49	**2010**	2018	
Rakocy J E, 1997, 4 INT S TIL AQ, V1, P357	1997	4.27	**2010**	2015	
Graber A, 2009, DESALINATION, V246, P147, DOI 10.1016/j.desal.2008.03.048, DOI	2009	4.14	**2010**	2018	
Ghaly AE, 2005, ENVIRON INT, V31, P1, DOI 10.1016/j.envint.2004.05.011, DOI	2005	3.45	**2010**	2015	
Schneider O, 2005, AQUACULT ENG, V32, P379, DOI 10.1016/j.aquaeng.2004.09.001, DOI	2005	3.38	**2010**	2015	
Piedrahita RH, 2003, AQUACULTURE, V226, P35, DOI 10.1016/S0044-8486(03)00465-4, DOI	2003	3.33	**2010**	2015	
Naylor RL, 2000, NATURE, V405, P1017, DOI 10.1038/35016500, DOI	2000	3.07	**2010**	2015	
APHA AWWA WEF, 2005, STANDARD METHODS EXA, Vfirst, P0	2005	5.22	**2013**	2018	
Hu Z, 2012, ENVIRON SCI TECHNOL, V46, P6470, DOI 10.1021/es300110x, DOI	2012	4.62	**2013**	2018	
Endut A, 2010, BIORESOURCE TECHNOL, V101, P1511, DOI 10.1016/j.biortech.2009.09.040, DOI	2010	4.52	**2013**	2018	
Rakocy JE, 2006, SRAC PUBLICATION, V454, P0	2006	4.08	**2013**	2015	
Zou YN, 2016, BIORESOURCE TECHNOL, V210, P81, DOI 10.1016/j.biortech.2015.12.079, DOI	2016	4.04	**2016**	2018	
Liang JY, 2013, INT BIODETER BIODEGR, V85, P693, DOI 10.1016/j.ibiod.2013.03.029, DOI	2013	3.94	**2016**	2018	
Kloas W, 2015, AQUACULT ENV INTERAC, V7, P179, DOI 10.3354/aei00146, DOI	2015	3.34	**2016**	2021	
Rakocy J, 2003, ACTA HORTIC, V648, P63	2003	3.18	**2016**	2018	
Bosma RH, 2017, AQUACULT ENG, V78, P146, DOI 10.1016/j.aquaeng.2017.07.002, DOI	2017	3.06	**2019**	2021	

FIGURE 15.18 References with strong citation burst.

right quadrant) are known as essential but not developed themes (basic and transversal themes). Thematic evolution on aquaponics research categorized the topics such as growth, water, nitrogen (one cluster), performance, plant, and lettuce (one cluster) as the basic themes (lower right); plant growth, stocking density, and Nile tilapia (one cluster) water-quality management, and growth performance (one cluster); plant growth, soil, and bacteria (one cluster) as the niche themes (upper left). However, fish, aquaculture, and wastewater were observed to be present in both basic and motor themes (Figure 15.20). Three field plots depicting the association between the elements such as major keywords (left), countries (middle), and journals (right) used in the publications on aquaponics research worldwide were shown in Figure 15.21. These three elements are connected by a gray plot that is related to one another. The size of the rectangle illustrates the large number of publications associated with each of these elements. The United States tops in country-wise publications with a diversity of keywords used in their publications, such as water, yield, growth, nitrification, and so forth, and they prefer the aquaculture engineering (JCR impact factor 3.281) and water (JCR impact factor 3.103) as their major publication source for their publications (Figure 15.21).

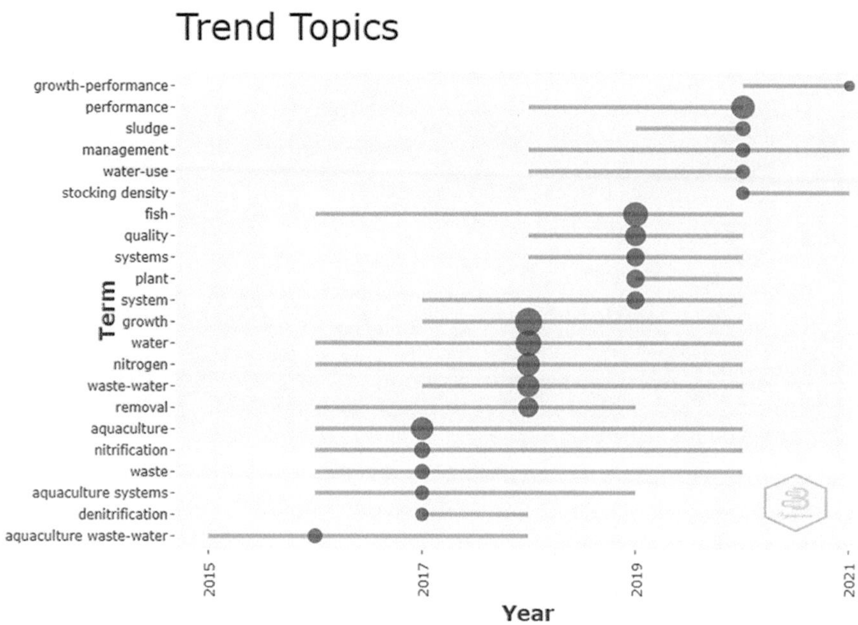

FIGURE 15.19 Major trend topics published on aquaponics research.

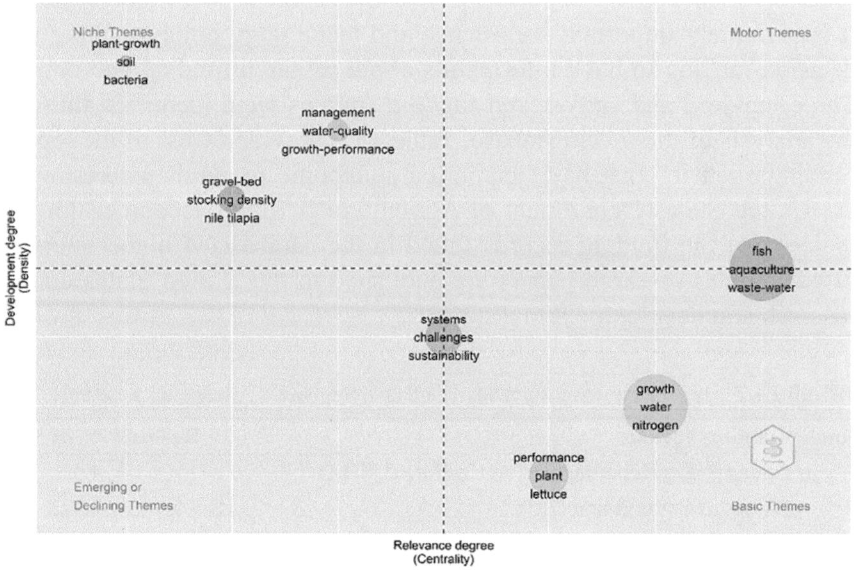

FIGURE 15.20 Thematic evolution of the publications on aquaponics research across the world.

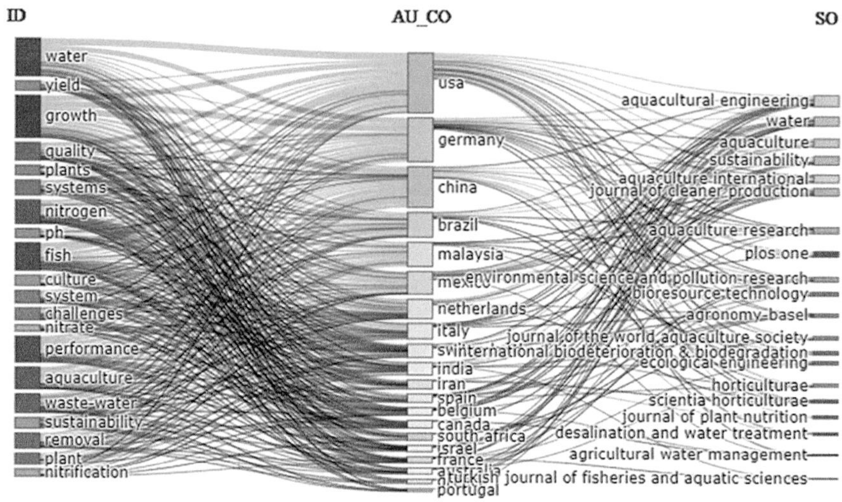

FIGURE 15.21 Three fields plot depicting the association between the major keywords (left), countries (middle), and journals (right) used in the publications on aquaponic research worldwide during the study period.

15.3.12 AQUAPONICS RESEARCH FUNDING

It is extremely important for science and technology policy-making and decision-making to have information about research funding and output. Three hundred and sixty-seven funding sources were identified through the analysis of the WoS database. Table 15.7 provides a list of the top 10 funding agencies that have supported aquaponic research. Interestingly, the United States Department of Agriculture (USDA) accounted for the majority of the funding records found in the database. A higher number of aquaponics research papers are published in the United States due to research and development investments.

TABLE 15.7 List of Funding Agencies That Have Supported Aquaponic Research.

Rank	Funding Agencies	Records	% of 324
1	United States Department of Agriculture (USDA)	15	4.63
2	European Commission (EU)	14	4.321
3	National Natural Science Foundation of China (NSFC)	11	3.395
4	Conselho Nacional De DesenvolvimentoCientifico E Tecnologico (CNPQ)	9	2.778

TABLE 15.7 *(Continued)*

Rank	Funding Agencies	Records	% of 324
5	Consejo Nacional De Ciencia Y Tecnologia (CONACYT)	8	2.469
6	European Cooperation in Science and Technology (ECOST)	7	2.16
7	Fundamental Research Funds of Shandong University	7	2.16
8	Indian Council of Agricultural Research (ICAR)	7	2.16
9	Coordenacao De Aperfeicoamento De Pessoal De Nivel Superior (CAPES)	6	1.852
10	Natural Sciences and Engineering Research Council of Canada (NSERC)	6	1.852

Researchers typically acknowledge sponsorship and outside funding they receive for their research, which constrains the results above. When national research labs use internal resources, the research undertaken is not explicitly described in the research papers and does not appear in the WoS dataset.

15.4 CONCLUSION

Aquaponics research has seen a significant increase in interest due to urbanization over the past few decades. Through bibliometric and visual analysis, we analyzed 324 publications from the Web of Science Core Collection regarding aquaponics research. According to the results, the research on aquaponics has been receiving a surge in interest. In spite of the fact that aquaponics research dates back to 1990, the number of relevant studies has increased significantly only since 2014, and from 2014 to 2020, the number of studies increased from 9 to 76. The trend indicates an increase in international interest in aquaponics research. The United States, Germany, and China made major contributions to aquaponics with the publication of productive information that became influential around the world. Wageningen University Research had the most records with 20 (6.2%) out of 324 publications. There are many institutions that are independent of each other and rarely interact. Thirteen (13) keyword clusters and sixteen (16) document cocitation clusters were identified based on the indexed research literature. The research hotspots in aquaponics research are Nile tilapia, *Cyprinus carpio*, *Litopenaeus vannamei*, *Lactuca sativa*, water spinach, mint N_2O emission, anaerobic digestion, stocking density, and nutrient recycling as per keyword cocitation analysis. DCA captures macro trends in the research, suggesting an ongoing conceptual and disciplinary shift in aquaponics. Based on the combination of keyword analysis and some influential references, some potential directions for future research have been identified. Research must first focus on how crops can be grown to accommodate organic production requirements by making suitable modifications. Since the analysis was based on a dataset extracted from WoS, the study was limited by the inherent limitations associated with its coverage of publications. A future study may attempt to overcome these limitations by using data from multiple sources and a variety of indicators for assessment. Overall, the present research provides an in-depth analysis of the trends and gaps in aquaponics research, which sets the stage for further investigation.

KEYWORDS

- **aquaculture**
- **aquaponics**
- **knowledge management**
- **scientometrics**
- **visualization**

REFERENCES

Adhikari, R., Rauniyar, S., Pokhrel, N., Wagle, A., Komai, T., & Paudel, S. R. (2020). Nitrogen recovery via aquaponics in Nepal: Current status, prospects, and challenges. *SN Applied Sciences*, *2*(7), 1–15.

Aria, M., & Cuccurullo, C. (2017). Bibliometrix: An R-tool for comprehensive science mapping analysis. *Journal of Informetrics*, *11*(4), 959–975. https://doi.org/10.1016/j.joi.2017.08.007

Armanda, D. T., Guinée, J. B., & Tukker, A. (2019). The second green revolution: Innovative urban agriculture's contribution to food security and sustainability – A review. *Global Food Security*, *22*, 13–24. https://doi.org/10.1016/j.gfs.2019.08.002

Aryadoust, V., Tan, H. A. H., & Ng, L. Y. (2019). A scientometric review of rasch measurement: The rise and progress of a specialty. *Frontiers in Psychology*, *10*, 2197. https://doi.org/10.3389/fpsyg.2019.02197

Aryadoust, V., Zakaria, A., Lim, M. H., & Chen, C. (2020). An extensive knowledge Mapping review of measurement and validity in language assessment and SLA research. *Frontiers in Psychology, 11*, 1–29. https://doi.org/10.3389/fpsyg.2020.01941

Bhoomaiah, D., Krishnan, P., Kantharajan, G., Agarwal, S., Hemalatha, M., Rajendran, M., & Rao, C. (2022). Mapping the research impact of collaboration and networking of ICAR fisheries research institutes in India: A scientometric study. *Indian Journal of Fisheries, 69*(1), 1–21.

Bosma, R. H., Lacambra, L., Landstra, Y., Perini, C., Poulie, J., Schwaner, M. J., & Yin, Y. (2017). The financial feasibility of producing fish and vegetables through aquaponics. *Aquacultural Engineering, 78*, 146–154. https://doi.org/10.1016/j.aquaeng.2017.07.002

Callon, M., Courtial, J. P., & Laville, F. (1991). Co-word analysis as a tool for describing the network of interactions between basic and technological research: The case of polymer chemsitry. *Scientometrics*, *22*(1), 155–205.

Callon, M., Courtial, J., & Turner, W., & Bauin, S. (1983). From translation to problematic network: An introduction to co-word analysis. *Social Science Information*, *23*.

Chen, C. (2004). Searching for intellectual turning points: Progressive knowledge domain visualization. In *Proceedings of the National Academy of Sciences of the United States of America* (Vol. 101, pp. 5303–5310). https://doi.org/10.1073/pnas.0307513100

Chen, C., Dubin, R., & Kim, M. C. (2014). Emerging trends and new developments in regenerative medicine: A scientometric update (2000–2014). *Expert Opinion on Biological Therapy*, *14*(9), 1295–1317. https://doi.org/10.1517/14712598.2014.920813

Chen, C., Ibekwe-SanJuan, F., & Hou, J. (2010). The structure and dynamics of co-citation clusters: A multiple-perspective co-citation analysis. *Journal of the American Society for Information Science and Technology, 61*(7), 1386–1409. https://doi.org/10.1002/asi.21309

Cobo, M. J., López-Herrera, A. G., Herrera-Viedma, E., & Herrera, F. (2011). An approach for detecting, quantifying, and visualizing the evolution of a research field: A practical application to the Fuzzy Sets Theory field. *Journal of Informetrics*, *5*(1), 146–166.

Dirk, L. (2004). Views Afield: Mapping knowledge domains. *Science Editor, 27*(6), 197.

Fan, M., Shen, J., Yuan, L., Jiang, R., Chen, X., Davies, W. J., & Zhang, F. (2012). Improving crop productivity and resource use efficiency to ensure food security and environmental quality in China. *Journal of Experimental Botany, 63*(1), 13–24. https://doi.org/10.1093/jxb/err248

Fang, Y., Hu, Z., Zou, Y., Fan, J., Wang, Q., & Zhu, Z. (2017). Increasing economic and environmental benefits of media-based aquaponics through optimizing aeration pattern. *Journal of Cleaner Production, 162*, 1111–1117. https://doi.org/10.1016/J.JCLEPRO.2017.06.158

Goddek, S., Delaide, B., Mankasingh, U., Ragnarsdottir, K. V., Jijakli, H., & Thorarinsdottir, R. (2015). Challenges of sustainable and commercial aquaponics. *Sustainability, 7*, 4199–4224. https://doi.org/10.3390/su7044199

Graber, A., & Junge, R. (2009). Aquaponic systems: Nutrient recycling from fish wastewater by vegetable production. *Desalination, 246*(1–3), 147–156. https://doi.org/10.1016/j.desal.2008.03.048

Haak, L. L., Fenner, M., Paglione, L., Pentz, E., & Ratner, H. (2012). ORCID: aA system to uniquely identify researchers. *Learned Publishing, 25*(4), 259–264. https://doi.org/10.1087/20120404

Hamidi, A., Ramavandi, B., & Sorial, G. A. (2021). Sponge city — An emerging concept in sustainable water resource management: A scientometric analysis. *Resources, Environment and Sustainability, 5*, 100028. https://doi.org/10.1016/J.RESENV.2021.100028

He, Q. (1999). Knowledge discovery through co-word analysis. *Library Trends*. https://doi.org/x

Hirsch, J. E. (2005). An index to quantify an individual's scientific research output. *Proceedings of the National Academy of Sciences, 102*(46), 16569–16572. https://doi.org/10.1073/PNAS.0507655102

Hu, C. Y., Huang, K., Fang, Y., Yang, X. J., Ding, K., Jiang, W., Hua, X. G., Huang, D. Y., Jiang, Z. X., & Zhang, X. J. (2020). Maternal air pollution exposure and congenital heart defects in offspring: A systematic review and meta-analysis. *Chemosphere, 253*, 126668. https://doi.org/10.1016/j.chemosphere.2020.126668

Junge, R., König, B., Villarroel, M., Komives, T., & Jijakli, M. H. (2017). Strategic points in aquaponics. *Water, 9*, 182. https://doi.org/10.3390/w9030182

Khalil, G. M., & Gotway Crawford, C. A. (2015). A bibliometric analysis of U.S.-based research on the Behavioral Risk Factor Surveillance System. *American Journal of Preventive Medicine, 48*(1), 50–57. https://doi.org/10.1016/J.AMEPRE.2014.08.021

Kloas, W., Groß, R., Baganz, D., Graupner, J., Monsees, H., Schmidt, U., Staaks, G., Suhl, J., Tschirner, M., Wittstock, B., Wuertz, S., Zikova, A., & Rennert, B. (2015). A new concept for aquaponic systems to improve sustainability, increase productivity, and reduce environmental impacts. *Aquaculture Environment Interactions, 7*(2), 179–192. https://doi.org/10.3354/aei00146

Konur, O. (2012). The evaluation of the global research on the education: A scientometric approach. *Procedia - Social and Behavioral Sciences, 47*, 1363–1367. https://doi.org/10.1016/j.sbspro.2012.06.827

Lee, S., & Bozeman, B. (2005). The impact of research collaboration on scientific productivity. *Social Studies of Science, 35*(5), 673–702. https://doi.org/10.1177/0306312705052359

Li, Q., Long, R., Chen, H., Chen, F., & Wang, J. (2020). Visualized analysis of global green buildings: Development, barriers and future directions. *Journal of Cleaner Production, 245*, 118775. https://doi.org/10.1016/j.jclepro.2019.118775

Love, D. C., Fry, J. P., Genello, L., Hill, E. S., Frederick, J. A., Li, X., & Semmens, K. (2014). An international survey of aquaponics practitioners. *PLoS One, 9*(7), 1–10. https://doi.org/10.1371/journal.pone.0102662

Love, D. C., Fry, J. P., Li, X., Hill, E. S., Genello, L., Semmens, K., & Thompson, R. E. (2015). Commercial aquaponics production and profitability: Findings from an international survey. *Aquaculture, 435*, 67–74. https://doi.org/10.1016/j.aquaculture.2014.09.023

Olawumi, T. O., & Chan, D. W. M. (2018). A scientometric review of global research on sustainability and sustainable development. *Journal of Cleaner Production, 183*, 231–250. https://doi.org/10.1016/j.jclepro.2018.02.162

Paudel, S. R. (2020). Nitrogen transformation in engineered aquaponics with water celery (*Oenanthe javanica*) and koi carp (*Cyprinus carpio*): Effects of plant to fish biomass ratio. *Aquaculture, 520*, 734971. https://doi.org/10.1016/J.Aquaculture.2020.734971

Pérez-Urrestarazu, L., Lobillo-Eguíbar, J., Fernández-Cañero, R., & Fernández-Cabanás, V. M. (2019). Suitability and optimization of FAO's small-scale aquaponics systems for joint production of lettuce (*Lactuca sativa*) and fish (*Carassius auratus*). *Aquacultural Engineering, 85*, 129–137.

Rakocy, J. E., & Hargreaves, J. A. (1993). Integration of vegetable hydroponics with fish culture: A review. *American Society of Agricultural Engineers*, 112–136. http://ci.nii.ac.jp/naid/10004852927/en/

Rakocy, J. E., Masser, M. P., & Losordo, T. M. (2006). Recirculating aquaculture tank production systems: Aquaponics- integrating fish and plant culture. *SRAC Publication - Southern Regional Aquaculture Center, 454*, 1–16.

Secinaro, S., Brescia, V., Calandra, D., & Biancone, P. (2020). Employing bibliometric analysis to identify suitable business models for electric cars. *Journal of Cleaner Production, 264*, 121503. https://doi.org/10.1016/j.jclepro.2020.121503

Shaalan, M., El-Mahdy, M., Saleh, M., & El-Matbouli, M. (2018). Aquaculture in Egypt: Insights on the current trends and future perspectives for sustainable development. *Reviews in Fisheries Science and Aquaculture, 26*(1), 99–110. https://doi.org/10.1080/23308249.2017.1358696

Su, Y., Yu, Y., & Zhang, N. (2020). Carbon emissions and environmental management based on big data and streaming data: A bibliometric analysis. *Science of the Total Environment, 733*, 138984. https://doi.org/10.1016/j.scitotenv.2020.138984

Tiensin, H., Kalibata, A., & Cole, M. (2020). Ensuring Food Security in the Era of COVID-19 – United Nations Sustainable Development. United Nations. https://www.un.org/sustainabledevelopment/blog/2020/04/ensuring-food-security-covid-19/

Tyson, R. V., Treadwel, D. D., & Simonne, E. H. (2011). Opportunities and challenges to sustainability in aquaponic systems. *HortTechnology, 21*(1), 6–13. https://doi.org/10.21273/horttech.21.1.6

United Nations. (2015). The 17 goals of sustainable development goals. https://sdgs.un.org/goals

Van Eck, N. J., & Waltman, L. (2010). Software survey: VOSviewer, a computer program for bibliometric mapping. *Scientometrics, 84*(2), 523–538. https://doi.org/10.1007/s11192-009-0146-3

Van Eck, N. J., Waltman, L., Dekker, R., & Van Den Berg, J. (2010). A comparison of two techniques for bibliometric mapping: Multidimensional scaling and VOS. *Journal of the American Society for Information Science and Technology, 61*(12), 2405–2416. https://doi.org/10.1002/ASI.21421

Villarroel, M., Junge, R., Komives, T., König, B., Plaza, I., Bittsánszky, A., & Joly, A. (2016). Survey of aquaponics in Europe. *Water, 8*, 468. https://doi.org/10.3390/w8100468

Waltman, L., van Eck, N. J., & Noyons, E. C. M. (2010). A unified approach to mapping and clustering of bibliometric networks. *Journal of Informetrics, 4*(4), 629–635. https://doi.org/10.1016/j.joi.2010.07.002

Yildiz, H. Y., Robaina, L., Pirhonen, J., Mente, E., Domínguez, D., & Parisi, G. (2017). Fish welfare in aquaponic systems: Its relation to water quality with an emphasis on feed and faeces-A review. *Water, 9*, 13. https://doi.org/10.3390/w9010013

Zhao, X. (2017). A scientometric review of global BIM research: Analysis and visualization. *Automation in Construction, 80*, 37–47. https://doi.org/10.1016/j.autcon.2017.04.002

Zhao, S. X., Lou, W., Tan, A. M., & Yu, S. (2018). Do funded papers attract more usage? *Scientometrics, 115*(1), 153–168. https://doi.org/10.1007/S11192-018-2662-5

Zhu, Q., Yang, Y., Zhong, Y., Lao, Z., O'Neill, P., Hong, D., Zhang, K., & Zhao, S. (2020). Synthesis, insecticidal activity, resistance, photodegradation and toxicity of pyrethroids (A review). *Chemosphere, 254*, 126779. https://doi.org/10.1016/j.chemosphere.2020.126779

Zou, X., Vu, H. L., & Huang, H. (2020). Fifty years of accident analysis & prevention: A bibliometric and scientometric overview. *Accident Analysis and Prevention, 144*, 105568. https://doi.org/10.1016/j.aap.2020.105568

Zou, X., Yue, W. L., & Vu, H. L. (2018). Visualization and analysis of mapping knowledge domain of road safety studies. *Accident Analysis and Prevention, 118*, 131–145. https://doi.org/10.1016/j.aap.2018.06.010

Zuanazzi, N. R., Ghisi, N. de C., & Oliveira, E. C. (2020). Analysis of global trends and gaps for studies about 2,4-D herbicide toxicity: A scientometric review. *Chemosphere, 241*, 125016. https://doi.org/10.1016/j.chemosphere.2019.125016

CHAPTER 16

Energy Use and Carbon Footprint of Fishing Systems

PARAS NATH JHA[1,2,*] and LEELA EDWIN[1]

[1] *Fishing Technology Division, ICAR-Central Institute of Fisheries Technology, Cochin, India*

[2] *Faculty of Marine Sciences, Cochin University of Science and Technology, Cochin, India*

*Corresponding author

ABSTRACT

Fish harvest (fishing) is an external energy-intensive operation which produces greenhouse emissions mainly due to the burning of non-renewable fossil fuels, which releases high quantities of carbon dioxide and other greenhouse gases (GHGs). Global warming potential (GWP) is one of the environmental impacts due to GHGs which quantitatively relate to the heat produced by GHG (CO_2, CH_4, N_2O, chloroflouro carbons, and so forth) and traps in the atmosphere. Fossil fuel is the predominant source of energy in different fishing methods, such as longlining, gillnetting, trawling, ring seining, and purse seining. In 2011, the world fishing fleet consumed about 40 billion liters of fossil fuel which resulted in the emission of 179 million tons of CO_2 eq. (CO_2equivalent) and other GHGs to the environment. In 2016, the total CO_2 emission of the industrial fishing sector was 159 million tons compared with 39 million tons in 1950. Due to increased levels of fuel consumption in fishing operations, processing, and storage of fish catch, emission of GHGs also increased, leading to climate change. Fuel use varies widely with fishing type, level of fishing effort, and profit margins. India contributes 134 million tons (MT) (2.7%) of CO_2

Recent Advances in Sustainable Aquaculture and Fisheries. Shabir Ahmad Dar, Imtiyaz Qayoom, Gohar Bilal Wani, Adnan Abubakr, and Seemab Zehra (Eds.)

DOI: 10.1201/9781998511372-16

emission due to total marine capture fisheries, against 90 MT (3.9% of global production) of fish production. The harvest phase contributes the maximum equivalent carbon emission due to the burning of fossil fuel. In this context, analysis of energy and its use in different fishing systems is important and relevant.

16.1 INTRODUCTION

Fishing is a livelihood activity which nourishes the mushrooming anthropogenic population. The fisheries sector provides income generation to a large section of the society. Globally, 59.6 million people are involved in fisheries and aquaculture directly or indirectly (FAO, 2018) from which 37% of people are directly dependent on fishing and the rest are in processing and marketing activities (Matthews et al., 2012). Fishing is one of the ancient practices which provides about 10% of the protein need in humans.

Fish harvest/fishing is an external energy-intensive operational process which ends up with huge greenhouse emissions due to the burning of non-renewable fossil fuels (Parker et al., 2018; Drisscoll and Tyedmers, 2010). Fishing encompasses the dissipation of energy to complete the harvest of fishery resources. The indirect cost of the fishing process is less studied and also paid less attention than the direct impact on fishery and ecosystems. The availability of fossil fuel for fishing is one of the reasons for the decline of the stock health of fisheries resources. Compared with the inland fishing sector, the marine fishing sector consumes more fossil fuel, so higher carbon footprint for landed catch. Also, a significant presence of non-mechanized fishing systems in many inland waters reduces the demand for fuel. (Consumed by boats using active fishing gear in major marine fisheries).

Fossil fuel is the only source of energy for fish landing by different fishing systems viz., long lining, gillnetting, trawling, ring seining, and purse seining. Greenhouse gases (GHGs), such as carbon dioxide (CO_2), methane (CH_4), nitrous oxide (N_2O), chloroflourocarbons (CFCs), HFCs, and PFCs are the main reasons for the rising global temperature. Anthropogenic activities would result in emissions of these GHGs which affect the environment severely. It affects several ecological parameters, its composition, distribution areas, pattern of migration, seasonal profile, and growth rate. The fast depletion of fossil fuel is a major concern to global warming potential (GWP). GWP is the ability to trap extra heat in the atmosphere over a period of time relative to CO_2. GWP is expressed in kilograms of CO_2 equivalents. CO_2 equivalent is a measure to compare the emissions of different GHGs with their GWP. It is calculated by converting levels of other gases to the equivalent amount of CO_2 with the same GWP. The fuel consumption of the world fishing fleet was about 40 billion liters which is equivalent to 179 million tons of CO_2 equivalent and other GHGs in the atmosphere (Parker et al., 2018). India contributes 134 MT (2.7%) of CO_2 emission due to total marine capture fisheries, compared with 3.9% of total global fish production. While the marine landings have stagnant over the year and reached the plateau, equivalent carbon emissions due to fuel consumption constantly rising over the year and reached 3.6 MT in the year 2010 from 0.3 MT in the year 1960 (Vivekanandan et al., 2013).

16.2 ENERGY INPUTS TO FISHERIES

There are about 1.6 million fisheries-related vessels operating at the marine fisheries sector of India; about 70% of them are fossil fuel-driven (CMFRI, 2017–2018). During the past decade, the fuel price tremendously increased. According to one estimate, in 2001, value-wise fuel accounted for 21% of the total catch landed, this percentage had increased to about 50% in 2008. Fuel use depends upon the fishing type and its effort level, but it is one of the key factors which control the price of catch and over which the fishing sector has no direct control. Profit of the catch is highly dependent upon fuel/energy costs (FAO, 2015). Some work has been carried out along the Indian coast to establish the relationship between fuel/energy use with different fishing systems and operations (Edwin and Hridayanathan, 1997; Boopendranath, 2000; Boopendranath and Hameed, 2009; Boopendranath and Hameed, 2010; Vivekanandan et al., 2013; Jha and Edwin, 2018). The enhanced effort and higher efficiency in the past decades are also one of the reasons for the hike in fuel consumption. There is also a significant change in CO_2 emission depending upon craft types and their service life. According to a report (Vivekanandan et al., 2013), the mechanized fishing vessel emitted 1.18 t CO_2/t of fish landed, and the smaller motorized boats emitted 0.59 t CO_2/t of fish landed. Generally, in comparison to motorized fishing vessels, mechanized fishing vessels have more CO_2 emission. Due to more drag and fuel consumption, trawlers have a higher footprint.

In the context of the same study, the author also reported that trawlers release more CO_2 (1.43 t CO_2/t of fish) than the gillnetters, bagnetters, seiners, liners, and dolnetters (0.56–1.07 t CO_2/t of fish) in atmosphere. The post-publication of Gerald Leach's handbook *"Energy and Food Production"* in 1976, researchers from across the world have started working on energy and fisheries from different angles. Overall, one can say that fishing is a set of different processes which requires energy in different forms in order to capture target species. However, as of now, a limited type of fisheries has been taken for specific energy analyses in order to account for energy use. Harvest requires energy to accomplish propelling the fishing boat and operate fishing gears and accessories. Generally, three common forms of energy used in the area of harvest sector which include animate, wind, and fossil fuel energy. Animate energy is the simplest form of energy for all fisheries irrespective of their technological complexity. In the traditional artisanal fisheries sector, manpower which uses human muscle power as a source of energy and which is used for propulsion scouting, deploying/hauling the gears, and catch handling. For as long as people have sailed, wind energy has been used to support fishing activities. However, in the case of single-day small-scale fisheries, still it prevails, mostly along East Coast due to rough weather conditions. The sail/wind energy not only helps in fishing vessels for propulsion but also facilitates other auxiliary activities too, which is related to harvest. Out of all the different sources of energy, fossil fuel energy remains the dominant form and source of energy used in fish harvest and post-harvest also. During the early 1900s, fossil fuel-based engines were first adapted for use on fishing vessels. Global fishing fleet size and number along with engine power increased post-2nd World War. on the other hand, relatively, small engines in the form of outboard motors are also used in small-scale fisheries. These phenomena of increasing dominance and size of engines resulted in tremendously enhanced fossil fuel consumption for the global fish harvest sector. Fossil fuel is an essential component of motorized and mechanized fishing, which is non-renewable and scarce.

16.2.1 DIRECT AND INDIRECT ENERGY INPUTS

Energy inputs in the form of direct and indirect can be scientifically analyzed with the help of process analysis and methods based on input and output. Apart from the method of fishing, target species, or gear used, the major chunk of the requirement is primarily used by the propulsion of the vessel, and on average, about 80–90% of the total energy is used for the same. The remaining quantity depends upon routine maintenance of the vessel, provision of labor, fishing gear, bait, and ice, which would depend on the type of fishery and the scope of the improvement. The activities related to primary energy consumption and required fuel burned are higher in quantum than primary activity viz., on-board processing, storage, and so forth. However, squid jigging is an example where a large proportion of energy chunks are required to be used for activities other than sailing, which are mainly battery-operated high-intensity lamps, automated jigging machines, and on-board storage facilities. The energy requirement is fulfilled by diesel-operated generators which serve the purpose of luring and preserving the catch while fishing. In this case, energy demands on average for non-propulsion account for 38–40% of the total fuel required. In the case of fisheries sector, mostly, the direct fuel energy is considered to ease the comparison, and it is expressed as liters of fuel burned per live weight ton of aquatic resource landed. The sector's fuel needs and its requirements in the sector are important instruments for learning its vulnerability to changing prices. During the last decade, the relationship between energy and fossil fuels in food production has been more prominent. These detailed the analysis of the energy and carbon footprints of various food production systems. Energy use analysis is one of the important tools to assess the health of the stock, exploitation of aquatic resource-use, trade trends and its potential, and carbon and related GHG impacts in climate change and its mitigation (FAO, 2012). Direct or indirect energy used in the fisheries sector is commonly transformed to work as propulsive hauling and other supplementary work.

16.3 ENERGY SPENT IN DIFFERENT FISHING OPERATIONS

There are several different types of fishing methods based on the behavior and habitat of target species, and based on their operation the amount of fuel/energy required would be also different. About 14–19 million t and 1–2.5 million t of fuel oil for globally large-scale industrial fishing sector and small-scale fishing sector consumed, respectively.

16.3.1 TRAWLING

Trawling is a highly intensive fishing method (Endal, 1989). It required more fuel (about five times) than other passive fishing methods (longlining and gillnetting) and over 11 times to purse seining for landing each kg of fish and aquatic organism (Gulbrandson, 1986). According to a report suggested from the southwest coast of India, trawl operation requires six to seven times more fossil fuel compared with purse seining and about two times more fuel than gill net operation, to land unit kg of fish to land unit kg of fish or aquatic organisms. For large trawlers, 90% of fuel is required for active trawling consumption accounts during active trawling operations. About 45–75% is the average estimate of fuel cost in the total operational expenditure of trawlers, which varies depending on the power of the engine and the duration of the voyage. Boopendranath (2000) studied 674 trawlers of length class LOA13.1–14.0 m, equipped with 99–106 hp engines, in which 20% had steel hulls and the remaining wooden hulls. The gross energy requirement was estimated as 31.40 GJ/t fish for wooden trawlers to 36.97 GJ/t fish for steel trawlers.

16.3.2 GILLNETTING/LONGLINING

Gillnetting and longlining are the passive types of fishing for which the energy requirement is lower than other active fishing. Passive gears can be operated by both fixed or drifting types in which the energy requirement is minimal energy for the operation process whereas hauling is performed by a mechanical method which requires comparatively higher energy. According to a study on 210 numbers of wooden gillnetter-cum-liners operating off Cochin with length class of 9–11 m L_{OA} and installed engine power with 89–99 hp inboard engines had the average fuel consumption was 0.36 kg fuel/kg of fish landed. Out of different operational costs, fuel contributed the most, that is, 95% of the energy requirement (Boopendranath, 2000). A study conducted by Vivekanandan et al. (2013) clearly revealed that mechanized fishing vessels emitted 1.18 t CO_2/t of live fish landed, and the smaller motorized fishing vessels (with outboard motor) emitted 0.59 t CO_2/t of live fish landed. Among the mechanized fishing vessels, the trawlers emitted more CO_2 (1.43 t CO_2/t of live fish landed) than the gillnetters, bagnetters, seiners, liners, and dolnetters (0.56–1.07 t CO_2/t of live fish landed).

16.3.3 PURSE SEINING

Purse seining is an encircling fishing technique used to target pelagic schooling resources. Operation of this gear includes shoal detection prior to actual operation which is highly energy-intensive, after detecting the shoal, the encircling, capture, and hauling process would be followed. The operation of purse seine is relatively energy efficient and lower in carbon footprint compared with other method like trawling. Small-scale mechanized purse seine is less intensive than other large-scale operations and is low compared with trawling, gillnetting, and lining operations. The energy requirement/t of fish were 5.54, 5.93, and 6.4 for 156 hp wooden seiner, 156 hp steel seiner, and 235 hp wooden seiner, respectively (Boopendranath, 2000). There are several other energy-conserving fishing practices viz., large-scale purse seining became possible after the introduction of synthetic netting material.

16.3.4 TRAPS AND POTS

Traps or pots are designed in a way that the entrance itself becomes a non-return device, allowing the fish to enter the trap but making it impossible to leave the chamber. Generally, passive fishing gears require little energy for fishing and in some cases, absolutely nil mechanical energy. However, to go to the fishing ground and come back to harbors, setting and hauling of gear may use energy; target species are attracted by bait or are carried to the gear or encounter it by chance and are trapped. Tyedmers (2001) studied and collected data over 20-years (1980s–1990s) and concluded that 330 L of fuel was used to catch per ton of catch in a crab trap.

16.3.5 ENERGY-INTENSIVE ILLEGAL FISHING

Light fishing is one of the fish harvest practices which requires more energy. This practice has been popular since ancient times because of the behavior of fish that fishes are attracted by the light source. Chinese dipnet is an excellent example of last 200 years. Chinese dip net uses lights (hurricane lamp) to attract fish to the net. It is a sustainable fishing operation which mostly is based on animate energy. On an average, In India, more than 50% of the purse seiners and squid jiggers use artificial light (as of now is banned). As per report of the ICES-FAO Working Group on Fishing Technology and Fish Behavior (WGFTFB), 2012, suggests that marine catches using lights is about 1.6% of global catches (1.09 million tons).

16.3.6 SMALL-SCALE FISHERIES

Small-scale fisheries comprise a range of practices, but mainly artisanal activities which involve simple gear and less capital, commonly with small fishing vessels, making smaller fishing trips. Worldwide 4.5 million vessels are operating in the marine fisheries sector, out of which 2.1 million vessels are less than 12 m overall. Due to the small size of the vessel, the area of operation is limited and performs. Fishing operations are on daily basis, which for an average of 1–3 tons of live fish per person per year. It requires less energy for operation. Among different fish harvesting systems, traditional gillnetting is the most energy efficient one having the lowest gross energy requirement. Stake net is a type of non-motorized system that is comparatively high energy-intensive type. Ring seines type of motorized operations need a lower gross energy per ton of fish landed. The gross energy equirement of mechanized seine was found more compared with non-mechanized seine and this may be due to the lower fuel efficiency of outboard motors compared with inboard high-capacity diesel engines. Scouting-based fishing operations at fishing grounds which may be distantly located need relatively high gross energy per ton of live fish landed.

16.4 CURRENT ESTIMATES OF FUEL USE AND COST

Fisheries contribute about 50 million m^3 of oil consumption per year (1.2% of total global) Tyedmers et al. (2005). With 80.4 million tons of landings of marine fish, the global value of fuel-use intensity approximately was 620 L (527 kg) per live weight ton. With burning of this much fuel by fishing vessels it released a total of about 134 million kg of CO_2 into the environment with 1.7 kg of CO_2 per ton of live-weight landings on an average. Compared with other food production systems of the world, fisheries require 12.5 times more the amount of fuel energy for the produce. It has been noticed that during the last five decades due to the increased fishing effort of Indian marine capture fisheries, the substantial increase in fossil fuel consumption resulted in increased CO_2 emission from 0.30 to 3.60 MT in the year 1961–2010. In general, over a period of 50 years, CO_2 emission has increased from 0.50 to 1.02 t.

16.5 CONSERVATION OF FUEL AS A PART OF RESPONSIBLE WAY OF FISHING

As far as responsible fishing is concerned, the energy security has a direct significance in terms of meeting the requirement gap of fossil fuel. So, the limitation of fossil fuel because of its non-renewable nature and the effects of its burning on the environment should be taken completely. In fisheries, various harvesting processes and their systems, and different environmental conditions greatly influence the fuel consumption. Also fuel consumption can be controlled by different approaches in order to conserve fuel and energy as well as natural assets and environmental safety. According to Gulbrandson (1986) 0.8, 0.15, and 0.25 kg of fuel is required for trawling, longlining, and gillnetting, whereas purse seining requires 0.07 kg of fuel per kg of live weight of catch. So, there is a huge scope to conserve fuel in trawl operations. In addition, the time consumed in trawling is mostly contributed by towing. While towing, the resistance produced due to fishing vessels is insignificant compared with fishing gear. So the overall fuel consumption and its economy depend upon gear resistance. The total cost of fuel is about 50–60% of the total operational cost. Fuel costs can be over 50% of the total expenses on a fishing trip. There are different factors that contribute different on drag, and so, fuel consumption viz., floats, sweeps, warp, otter boards, foot rope, and webbing are contributing 3%, 4%, 5%, 20%, 10%, and 58%, respectively. There are several preventive measures that can help in enhancing the efficiency of fuel intensity in trawling as well as other fishing practices; some of them are knotless netting, thinner twine, large meshes, cambered otter boards, optimal angle of attack of otter boards, slotted otter boards, multi-rig trawling, pair trawling, and so forth.

The fuel consumption is directly proportional to the speed and RPM of the vessel which increases at a maximum speed of the vessel, this is because an increase in wave-breaking resistance requires more power thus fuel/energy. Studies have proven that a 10–20% reduction in speed can lead to saving fuel by 35–61%. Among two- and four-stroke engines, it is shown that two-stroke outboard engines have less fuel efficiency than four-stroke petrol outboard engines (Gulbrandson, 1986). Turbo-charged diesel engines are always better than other engines and are about 15% more fuel efficient than normally aspirated engines. They have much better fuel economy and emission standards and are also being introduced in small-scale fisheries. As per Gulbrandsen (2012), a 10% and 20% reduction in engine RPM resulted in the reduction of 20% and 40% in fuel consumption, respectively. Compared with two-stroke outboard petrol engines, inboard diesel engine consumes 62% less fuel at the same speed and hence is considered more fuel efficient. Boopendranath (2009, 2010) suggested higher fuel efficiency of four-stroke fuel-efficient diesel engines compared with two-stroke engines. He further added energy requirement drastically increases at its maximum speed and 35% of savings in fuel noticed at a reduction of 10% in speed. A study revealed the technical efficiency of the different types of ring seines using lesser horsepower engines performed well with respect to the catch per adjusted horsepower (Edwin and Hridayanathan, 1997). Energy requirement is an important tool to evaluate the efficiency of food production systems in terms of energy input and carbon footprint.

16.6 CONCLUSION

Harvest technique is highly variable worldwide. The main types of harvest and energy input required for it are extensive and exhaustive. The effective and long-term solutions to the problem of fuel conservation due to fishing are enhancement of coastal fishery resources andtheir restoration with the help of responsible fishing practices, monitoring of conservation resources and implementation of its rules and reducing overcapacity and along with other enhancement strategies. Indirect energy or directly fuel price and its relation to the fisheries sector and its dynamism toward demand and supply are very important as sector's energy and fuel needs and interactions and their future trends are concerned. This required considering different harvesting conditions and parts of the sector, economic impacts, and the mitigation measures of the effects of increased fuel requirement and fuel costs. Implications of intensification of agricultural and allied sectors have developed vigorous analyses of energy in support of the ecosystem with the help of carbon footprints inputs of various food commodities. With growing global trade, the outlook has also changed from the input of fuel/ energy for food production/harvest to complete supply chain demand, produce, and related products from original sources to consumers and up to their waste scenario. The analysis and estimation of energy use is important in relation to resource-use analysis and GHG impacts in order to take a mitigation approach for climate change mitigation (Poseidon, 2011). It has got emphasis due to the high instability in fossil fuel cost which has potentially lasting impacts on the economic performance of various fishing systems. These effects and their implications for fish harvest are likely to have profound aftermaths for resource impacts and food security across the globe.

KEYWORDS

- **carbon footprint**
- **energy use**
- **fuel consumption**
- **fishing systems**
- **global warming potential**

REFERENCES

Boopendranath, M. R. (2000). Studies on energy requirement and conservation of selected fish harvesting systems. Ph.D. Thesis, Cochin University of Science and Technology, Cochin, India.

Boopendranath, M. R., & Hameed, M. S. (2009). Energy analysis of traditional non-motorised gill net operations in Vembanad lake, Kerala, India. *Fishery Technology, 46*, 15–20.

Boopendranath, M. R., & Hameed, M. S. (2010). Energy analysis of stake net operations in Vembanad lake, Kerala, India. *Fishery Technology, 47*, 35–40.

Central Marine Fisheries Research Institute. (2017–2018). Annual Report ICAR (p. 320).

Drisscoll, J., & Tyedmers, P. (2010). Fuel use and greenhouse gas emission implications of fisheries management: The case of the New England Atlantic herring fishery. *Marine Policy, 34*, 353–359.

Endal, A. (1989). Future outlook-vessels, Keynote paper-Session 3, In *Proceedings of International Fisheries Energy Optimisation Working Group Meeting, University of British Columbia, Vancouver, B.C. Canada.*

Edwin, L., & Hridayanathan, C. (1997). Energy efficiency in the ring seine fishery of south Kerala coast. *Indian Journal of Fisheries, 44*(4), 387–392.

FAO. (2012). *Report of the Expert Workshop on Greenhouse Gas Emissions Strategies and Methods in Seafood*; FAO Fisheries and Aquaculture Report No. 1011: Rome, 23–25 January 2012, p. 117. (www.fao.org/docrep/017/i3062e/i3062e00.htm).

Gulbrandsen, O. (1986). Reducing the fuel costs of small fishing boats. Bay of Bengal Programme, Working Paper 27 (p. 29). Chennai

Gulbrandsen, O. (2012). *Fuel savings for small fishing vessels - a manual*. FAO, Rome.

Jha, P. N., & Edwin, L. (2019). Energy use in fishing. In L. Edwin, S. N. Thomas, M. P. Remesan, P. Muhamed Ashraf, M. V. Baiju, N. Manju Lekshmi & V. R. Madhu (Eds.), *ICAR winter school training manual: Responsible fishing: Recent advances in resource and energy conservation* (p. 424). ICAR-CIFT, Kochi.

Matthews, E., Bechtel, J., Britton, E., Morrison, K., & Mc Clennen, C. (2012). A gender perspective on securing livelihoods and nutrition in fish dependent coastal communities, Report to The Rockefeller Foundation from Wildlife Conservation Society, Bronx, NY.

Parker, R. W. R., & Blanchard, J. L., Gardner, C., Green, B. S., Hartmann, K., Tyedmers, P. H., & Watson, R. A. (2018). Fuel use and greenhouse gas emissions of world fisheries. *Nature Climate Change, 8*(1), 333–337.

Poseidon. (2011) Lifecycle analysis and greenhouse gas emissions methods in seafood production systems. Background paper for FAO Expert Workshop on GHG and LCA methodologies in the fisheries sector. FAO, Rome.

Tyedmers, P. (2001). Energy consumed by North Atlantic fisheries. In D. Zeller, R. Watson & D. Pauly (Eds.), *Fisheries impacts on North Atlantic ecosystems: Catch, effort and national/regional datasets* (Vol. 9(3), pp. 12–34). Fisheries Center Research Report, Vancouver, British Columbia.

Vivekanandan, E., Singh, V. V., & Kizhakudan, J. K. (2013). Carbon footprint by marine fishing boats of India. *Current Science, 3*, 361–366.

CHAPTER 17

Gamete Quality Assessment: An Important Tool for Aquaculture

ISHRAT MOHD[1,*], FAROOZ AHMAD BHAT[2], TASADUQ HUSSAIN SHAH[1], ASIFA WALI[1], and SHABIR A. DAR[1]

[1]*Division of Fisheries Resource Management, Faculty of Fisheries-Rangil, Sher-e-Kashmir University of Agricultural Sciences and Technology, Rangli, Jammu and Kashmir, India*

[2]*Division of Fisheries Resource Management, Faculty of Fisheries, Skuast-K, Rangil, Ganderbal, Jammu and Kashmir, India*

*Corresponding author

ABSTRACT

A key necessity for the successful development of aquaculture is the availability of large quantities of high-quality seeds "on demand." The quality of gametes depends on having a thorough understanding of the diverse environmental and dietary requirements of captive fish. However, such knowledge is scarce or not yet available. As enhanced qualities in stock are chosen for farming environments, the degree of domestication is also crucial. Importantly, there are presently no reliable measures of egg quality. Studies on sperm and egg physiology may provide us with the basics of managing fertilization. In order to determine the reproductive potential of farmed fish and assure controlled and successful production, a full study of the physical and chemical features of the milt is required to assess the reproductive capacity of farmed fish in order to ensure controlled and successful production in aquaculture systems. The potential

Recent Advances in Sustainable Aquaculture and Fisheries. Shabir Ahmad Dar, Imtiyaz Qayoom, Gohar Bilal Wani, Adnan Abubakr, and Seemab Zehra (Eds.)

DOI: 10.1201/9781998511372-17

hatchery-related factors influencing gamete quality are briefly described. Overview of techniques for evaluating the quality of eggs and sperm as well as many of the critical elements influencing the quantity and quality of gametes, such as nutrition for broodstock, environmental conditions, spawning induction procedures, genetic management, gamete preservation, and innovative reproductive techniques.

17.1 INTRODUCTION

Eggs and sperm of high quality are essential for aquaculture to succeed. The main obstacles to the expansion of fish aquaculture are poor egg and sperm quality. According to Kjorsvik et al. (1990), a key limiting factor in the production of fry, and subsequently, the growth of the aquaculture industry, is the quality of eggs and larvae produced in hatcheries (Bromage, 1995). The ability of eggs to generate viable fry is referred to as egg quality (Brooks et al., 1997). Importantly, there are presently no reliable measures of egg quality. Equally crucial is sperm quality, which can influence fertilization success and egg viability (Bromage, 1995). An essential criterion for determining a finfish species' prospective fertility is the examination of milt quality during reproduction. A population's ability to reproduce depends on how the brooders are chosen for production (Najar and Qadri, 1999; Najar et al., 2000). Sperm quality is a significant management factor in aquaculture since it can guarantee the proportion of fertilized eggs, and consequently, the production of viable larvae from a given brood stock. A measurement of sperm quality is any quantifiable physical characteristic that is directly related to fertilization.

The possibility of having viable offspring carrying the paternal genes will be reduced if poor-quality sperm cannot be produced. Low mortalities at fertilization, hatching, and the first feeding have been recognized as good quality eggs in the fish farming sector (Bromage et al., 1992). While Coban et al. (2011) define egg quality as the potential that an egg will hatch into a healthy larva, Bobe and Labbe (2010) defined fish egg quality as the capacity of the egg to be fertilized and subsequently grow into a normal embryo. According to Bobe and Labbe (2010), sperm quality is also the capacity of sperm to effectively fertilize an egg and consequently permit the development of a healthy embryo. The production of high-quality fish larvae and the efficient use of hatcheries depend on egg and sperm quality

(Coban et al., 2011). Good quality eggs and sperm increase fish production and year-round availability of fish, thereby increasing market demand for fish and fish protein. It also increases the revenue of the nation and those directly involved in fish farming. Eggs and sperm generally play an important role in the quality of fish farming.

17.2 SPERM PHYSIOLOGY IN OPTIMIZING FERTILIZATION SUCCESS IN AQUACULTURE

For fish farms to successfully fertilize fish, gamete management is essential (Billard et al., 1995). The fundamentals of controlling fertilization can be learned from studies on sperm physiology (Alavi and Cosson, 2005). A thorough understanding of the physical and chemical properties of the milt is required to assess the reproductive capacity of farmed fish in order to ensure controlled and successful production in aquaculture systems. Fish fertilization potential can be determined by the ionic and organic components of the seminal fluid (Ciereszko et al., 2000). Fertilization is a crucial step in achieving better yield in aquaculture. Due to the difficulty and length of time required to test fertilizing capacity, it is simpler for both fish farmers and researchers to employ other metrics that enable a quicker evaluation. As several authors have examined, some scientists have attempted to associate fertilization with other criteria that are more or less simple to measure (Bobe and Labbe, 2010; Cabrita et al., 2008; Rurangwa et al., 2004).

Fertilization success is determined by the intrinsic gamete quality and quantity as well as the environment of gamete fusion. Therefore, fertilization is an integrative reaction to a variety of circumstances that may conceal changes in sperm intrinsic quality. In factorial crossings of trout (Nagler et al., 2000) and seabass (Saillant et al., 2001) when the egg quality effect was large, several studies found no sperm quality influence on fertilization. According to the most recent reviews by Alavi et al. (2008) and Cabrita et al. (2009), the majority of sperm characteristics contribute to the overall quality of sperm, but none of them are sufficiently integrative to fully characterize the sperm's ability to fertilize eggs. Since studies describing milt characteristics, which can influence fertilization ability, have shown large individual variations in the different parameters investigated (Dreanno et al., 1998), this rendered difficult the use of a single sperm characteristic to define good sperm quality. Other works show that non-motile spermatozoa can fertilize eggs (Truscott and Idler, 1969).

All these instances demonstrate the limitations of defining the sperm's quality based on a single feature. However, there is no one predictive parameter because quality does not depend on a certain milt quality. To forecast a sample's ability to fertilize, we could gain a consistent sense of the sample quality by evaluating a few criteria together. Furthermore, sperm quality varies greatly between species, as do the criteria that must be utilized to assess the quality in each species. When determining a male's degree of maturity, identifying males with greater reproductive performance, determining when *in vitro* fertilization is required, or determining the ideal conditions for brood stock rearing in two separate settings, the evaluation of sperm samples is crucial.

17.2.1 PARAMETERS FOR ASSESSING SPERM QUALITY

The components of sperm, such as spermatozoa and seminal plasma can be used to evaluate its quality. Seminal plasma constituents and their variation, such as enzymes (lytic, oxidative, metabolic), metabolites, sugars, vitamins, amino acids, fatty acids, and other inorganic compounds can provide very useful information on sperm condition. Standard analysis may include parameters, such as spermatozoa concentration, motility, sperm volume, seminal plasma osmolarity, and pH (Rurangwa et al., 2004; Cabrita et al., 2008). So far, spermatocrit, sperm density, osmolarity and pH of seminal plasma, chemical composition of seminal plasma, enzymatic activity, adenosine triphosphate (ATP) concentration, motility, morphology and ultrastructure, fertilizing capacity, and a number of other factors have been identified as biomarkers of sperm quality (Billard and Cosson, 1992; Ciereszko and Dabrowski, 1993; Lahnsteiner et al., 1998; Geffen and Evans, 2000).

The spermatozoa are supported by some components of the seminal plasma, while other components show how the reproductive system and the spermatozoa function (Ciereszko et al., 2000). The primary function of seminal plasma is to provide the best conditions for the storage of spermatozoa, frequently for an extended period (Ciereszko, 2008). Sperm fertility, motility, and viability must all be preserved while being stored. Failure of fertilization is likely to come from damage to any of these processes. In contrast to the situation in reptiles or mammals, the majority of fish species now under study have immotile spermatozoa in the testis and seminal plasma (Stoss, 1983; Billard, 1986; Krasznai et al., 1995). Sperm motility in the fish sperm duct is typically inhibited by the seminal plasma's osmolality and chemical composition (Billard, 1986). The spermatozoa are both immobilized and safeguarded by the seminal fluid (Cosson et al., 1997). When sperm is diluted in a medium with a high osmotic pressure, activation typically occurs in marine species (Morisawa, 1985). As a result, fish sperm motility duration and seminal plasma composition and osmolality have a definite link; sperm are often active only after being released into the marine environment.

Fish sperm preservation for a brief period is generally advantageous from a commercial standpoint and makes a variety of hatchery procedures easier. The short-term storage of sperm at a low temperature (4°C) is mostly used in the short-distance transport of gametes collected in various locations, in synchronizing the timing of collecting high-quality gametes from males and females during artificial insemination, in preventing the aging of sperm, in facilitating hatchery operations, as well as in experimental programs for genetic studies. There has been evidence of sex-specific asynchrony in gamete maturation and spawning times in various animals. When trying to create hybrids, this problem can get really severe. Researchers have looked at methods like milt cryopreservation and short-term refrigeration storage to address these issues. For a number of teleost species, including walleye (Moore, 1987), red drum, Atlantic sturgeon (*Acipenser oryrinchus*) (Dilauro et al., 1994), Mozambique tilapia (*Tilapia mossambica*) (Harvey and Kelley, 1984), and paddlefish (*Polyodon spathula*), methods for the (Brown and Mims, 1995).

17.3 SHORT-TERM SPERM PRESERVATION TECHNIQUES IN AQUACULTURE

By keeping the fish sperm in both undiluted and diluted form, it might be stored. It has been found that undiluted sperm held at low temperatures reduces the ability to fertilize (Lahnsteiner et al., 1997). Compared with undiluted storage, sperm storage with an extender that is diluted offers better control (Harvey and Kelley, 1984). Many fish species benefit from the sperm being diluted in a suitable medium because it keeps the physicochemical qualities steady and allows for short-term storage, such as during transportation from the location of collection to the location of employment (Gwo, 1994; Ohta and Izawa, 1996). However, these treatments must temporarily inhibit motility in order to prevent the spermatozoa's energy reserves from being depleted (Gwo, 1993).

17.4 ENVIRONMENTAL INFLUENCES AND DOMESTICATION EFFECTS ON FISH REPRODUCTION IN CAPTIVITY

Although the environment in captivity will never completely replicate the environment in the wild, holding conditions should mimic as closely as possible. Environmental adjustments, such as photoperiod, water temperature, or spawning substrate, can be used to regulate fish reproduction in captivity. However, some fish's ecobiology is poorly understood, or simulating the necessary environmental factors for natural reproductive success is difficult or impossible (i.e., spawning migration, depth, riverine hydraulics, and so forth). Fish kept in aquariums are susceptible to stress, which can harm their ability to reproduce and the quality of their gametes (Schreck et al., 2010). Fish overstocking and confinement in aquariums can both lower the quality of eggs. The endocrinology that supports the normal growth and development of the ovaries in trout has been demonstrated to be disrupted by both chronic confinements experienced during the latter stages of reproductive development and times of acute stress and may lead to noticeably decreased progeny survival rates (Campbell et al., 1994). In Atlantic cod, stress can result in irregular spawning intervals, low fertilization rates, and an increase in the prevalence of malformed embryos (Kjesbu, 1989). It is unknown how stressing brood fish affects the quality of the eggs that are laid. Some fish that are raised in captivity rarely release their eggs, which causes the eggs to age or over-ripen inside the body cavity. Over-ripening of eggs is perhaps the most common reason for poor egg quality in captive brood fish. After fertilization, dying, or dead eggs become colonized with bacteria/fungus, and if these eggs are not removed quickly viable eggs may also be colonized.

Concerns regarding the impact of domestication on fish have been raised by the rapid rise in worldwide aquaculture production (Naylor et al., 2005). When compared with the wild, the aquaculture setting offers fish a radically different environment, changing the selective forces that can result in fundamental genetic alterations at the population level (Skaala et al., 2004; Jonsson and Jonsson, 2006). Due to genetic bottlenecks and new selective pressures brought on by domestication, farming practices frequently cause a decrease in genetic diversity and the divergence of farmed stocks from wild populations (Einum and Fleming 1997; Norris et al., 1999; Skaala et al., 2004). The lower allelic diversity of farmed salmon populations when compared with wild salmon populations has been used to demonstrate the loss of genetic diversity (Norris et al., 1999; Skaala et al., 2004). This can happen when a large number of offspring are produced from a small number of breeding individuals. Small breeding populations result in smaller effective populations, which can increase the likelihood of inbreeding (Bentsen and Olesen, 2002).

Adult fish raised in captivity and spontaneously generated (wild) fish have quite distinct environmental regimes. Fish raised in captivity are fed diets that have been specially developed, and they do not make long migrations. Wild fish and fish raised in ponds have different morphologies, and the degree of divergence seems to be positively correlated with the length of confinement. The common carp has been subjected to mass selection after several centuries of regulated and cultured spawning, resulting in several stocks that are not correctly identifiable in fish culture units in Europe and Asia. Ever since fish for aquaculture are taken from the wild, the characteristics of domestic races have changed progressively because of both broodstock selection and rearing methods (association with other species, type of fertilization, preferred harvest size).

Age, management, feeding, chemical and physical elements, water quality, and other variables all have an impact on the quality of fish gametes, both male and female, which in turn has an effect on the short- and long-term viability of eggs, larvae, and/or fry. Gamete aging is one of the most crucial elements, particularly for animals that cannot reproduce normally outside of hatcheries. In general, males are more resilient to stress conditions brought on by insufficient feeding, bad management, and/or poor water conditions than females are to stress conditions caused by inadequate feeding, management, and/or poor water conditions. The stress produced on broodstock by inadequate conditions in hatcheries can produce adverse effects on gamete quality, survival rates, and the embryonic eggs after hatching.

17.5 CONCLUSION

The growth of the aquaculture industry depends to a large extent on the ability of hatcheries to supply good quality seed. Eggs and sperm of high quality are essential for aquaculture to succeed. The main obstacles to the expansion of fish aquaculture are poor egg and sperm quality, and measuring all these parameters in fish sperm provides complex knowledge regarding male fertility and helps to improve broodstock maintenance protocols as well as gamete handling and fertilization processes.

KEYWORDS

- **aquaculture**
- **egg**
- **fertilization**
- **gamete quality**
- **milt**

REFERENCES

Alavi, S. M. H., & Cosson, J. (2005). Sperm motility and fertilizing ability in the Persian sturgeon (*Acipenser persicus*). *Aquaculture Research, 36*, 841–850.

Alavi, A. M. H., Rodina, M., Policar, T., & Linhart, O. (2008). Relationship between semen characteristics and body size in *Barbus barbus* and effect of ions and osmolality on sperm motility. *Comparative Biochemistry and Physiology, 153*, 430–437.

Bentsen, H. B., & Olsen, I. (2002). Designing aquaculture mass selection programs to avoid high inbreeding rates. *Aquaculture, 204*, 349–359.

Billard, R. (1986). Spermatogenesis and spermatology of some teleost fish species. *Reproduction Nutrition Development, 26*, 877–920.

Billard, R., & Cosson, M. P. (1992). Some problems related to the assessment of sperm motility in freshwater fish. *Journal of Experimental Zoology, 261*, 122–131.

Billard, R., Cosson, J., Crim, L. W., & Suquet, M. (1995). Biology of sperm and artificial reproduction in carp. *Aquaculture, 124*, 95–112.

Bobe, J., & Labbe, C. (2010). Egg and sperm quality in fish. *General and Comparative Endocrinology, 165*(3), 535–548.

Bromage, N. R. (1995). Broodstock management and seed quality-general considerations. In N. R. Bromage & R. J. Roberts (Eds), *Brood stock management and egg and larval quality* (pp. 1–24). University Press, Cambridge.

Bromage, N., Jones, J., Randal, C., Thrush, M., Davies, B., Springate, J., Duston, J., & Baker, G. (1992). Brood stock management, fecundity, egg quality and the timing of egg production in the rainbow trout (*Oncorhynchus mykiss*). *Aquaculture, 100*, 141–166.

Brooks, S., Tyler, C. R., & Sumpter, J. P. (1997). Department of Biology and Biochemistry, Brunel University, Uxbridge, Middlesex Ub8 3PH, United Kingdom. *Reviews in Fish Biology and Fisheries, 7*, 387–416.

Brown, G. G., & Mims. S. D. (1995). Storage, transportation and fertility of undiluted and diluted paddlefish milt. *Progressive Fish-Culturist, 57*, 64–69.

Cabrita, E., Robles, V., & Herraez M. P. (2008). Sperm quality assessment. In E. Cabrita, V. Robles & M. P. Herráez (Eds.), *Methods in reproductive aquaculture: Marine and freshwater species. Biology series* (Vol. 9, pp. 31–47). CRC Press (Taylor and Francis group), London, UK.

Cabrita, E., Robles, V., & Herraez, P. (2009). Sperm quality assessment. Methods in reproductive aquaculture. In E. Cabrita, V. Robles & P. Herraez (Eds.), *Marine and freshwater species* . CRC Press Taylor & Francis Group, Boca Raton, USA.

Campbell, P. M., Pottinger, T. G., & Sumpter, J. P. (1994). Preliminary evidence that chronic confinement stress reduces the quality of gametes produced by brown and rainbow trout. *Aquaculture, 120*, 151–169.

Ciereszko, A. (2008). Chemical composition of seminal plasma and its physiological relationship with sperm motility, fertilizing capacity and cryopreservation success in fish. In S. M. H. Alavi, J. Cosson, K. Coward & G. R. Rafiee (Eds.), *Fish spermatology* (pp. 215–240).Alpha Science, Oxford.

Ciereszko, A., & Dabrowski, K. (1993). Estimation of sperm density of rainbow trout, whitefish and yellow perch using a sperctrophotometric technique. *Aquaculture, 109*, 367–373.

Ciereszko, A., Glogowski, J., & Dabrowski, K. (2000). Biochemical characteristics of seminal plasma and spermatozoa of freshwater fishes. In T. R. Tiersch & P. M. Mazik (Eds.), *Cryopreservation in aquatic species* (pp. 20–48).) World Aquatic Society Baton Rouge, LA, USA.

Coban, D., Kamaci, H. O., Cuneyt, S., Yildirim, S., Arda, G., Korkut, A. Y., Saka, S., & Firat, K. (2011). Effect of some morphometric characteristics on egg quality in Common Dentex, *Dentex dentex* (Linnaeus, 1758*). Turkish Journal of Fisheries and Aquatic Sciences, 11*, 425–431.

Cosson, J., Billard, R., Cibert, C., Dreanno, C., Linhart, O., & Suquet, M. (1997). Movements of fish sperm flagella studied by high speed video microscopy coupled to computer assisted image analysis. *Poliskie Archiwum Hydrobiologii, 10*, 518–527.

Dilauro, M. N., Krise, W. F., Hendrix, M. A., & Baker. S. E. (1994). Short-term cold storage of Atlantic sturgeon sperm. *Progressive Fish-Culturist, 56*, 143–144.

Dreanno, C., Suquet, M., Desbruyeres, E., Cosson, J., Delliou, H., & Billard, R. (1998). Effect of urine on semen quality in turbot (*Psetta maxima*). *Aquaculture, 169*, 247–262.

Einum, S., & Fleming, I. A. (1997). Genetic divergence and interactions in the wild among native, farmed and hybrid Atlantic salmon. *Journal of Fish Biology, 50*, 634–651.

Geffen, A. J., & Evans, J. (2000). Sperm traits and fertilization success of male and sex reversed female rainbow trout. *Aquaculture, 182*, 61–72.

Gwo, J. C. (1993). Cryopreservation of black grouper (*Epinephelus malabaricus*) spermatozoa. *Theriogenology, 39*, 1331–1342.

Gwo, J. C. (1994). Cryopreservation of yellowfin seabream (*Acanthopagrus latus*) spermatozoa (teleost, perciformes, sparides). *Theriogenology, 41*, 989–1004.

Harvey, B., & Kelley R. N. (1984). Chilled storage of *Sarotherodon mossambicus* semen. *Aquaculture, 36*, 85–95.

Jonsson, B., & Jonsson, N. (2006). Cultured Atlantic salmon in nature: A review of their ecology and interaction with wild fish. *ICES Journal of Marine Science, 63*, 1162–1181.

Kjesbu, O. S. (1989). The spawning activity of cod, *Gadus morhua* L. *Journal of Fish Biology, 34*, 195–206.

Krasznai, Z., Marian, T., Balkay, L., Gasper, R. J., & Tron, L. (1995). Potassium channels regulate hypoosmotic shock induced motility of Common Carp, *Cyprinus carpio*, sperm. *Aquaculture, 129*, 123–128.

Lahnsteiner, F., Berger, R., Weismand, T., & Patzner, R. (1997). Sperm motility and seminal fluid composition in the burbot, *Lota lota*. *Journal of Applied Ichthyology, 13*, 113–119.

Lahnsteiner, F., Berger, B., Weismann, T., & Patzner, R. A. (1998). Determination of semen quality of the rainbow trout, *Oncorhynchus mykiss*, by sperm motility, seminal plasma parametersand spermatozoa metabolism. *Aquaculture, 16*, 163–181.

Moore, A. A. (1987). Short-term storage and cryopreservation of walleye semen *Progressive Fish-Culturist, 49*, 40–43.

Morisawa, M. (1985). Initiation mechanism of sperm motility at spawning in teleosts. *Zoological Science, 2*, 605–615.

Nagler, J. J., Parsons, J. E., & Cloud, J. G. (2000). Single pair mating indicates maternal effects on embryo survival in rainbow trout, *Oncorhynchus mykiss*. *Aquaculture, 184*(1-2), 177–183.

Najar, A. M., & Qadri, M. Y. (1999). Progesterone levels and ovarian development in *Schizothorax niger* (Heckel*). Oriented Science, 4*(1), 31–43.

Najar, A. M., Qadri, M. Y., Wani, G. M., & Zargar, M. A. (2000). Estradoil 17-β levels in relationship with various events of ovarian development in *Schizothorax niger* (Heckel). *Applied Biological Research, 2*(1), 7–13.

Naylor, R., Hindar, K., Fleming, I. A., Goldburg, R., Williams, S., Volpe, J., Whoriskey, F., Eagle, J., Kelso, D., & Mangel, M. (2005). Fugitive salmon: Assessing the risks of escaped fish from net-pen aquaculture. *Bioscience, 55*, 427–437.

Norris, A. T., Bradley, D. G., & Cunningham, E. P. (1999). Microsatellite genetic variation between and within farmed and wild Atlantic salmon (*Salmo salar*) populations. *Aquaculture, 180*, 247–264.

Ohta, H., & Ilzawa, T. (1996). Diluent for cold storage of the Japanese eel (*Anguilla japonica*) spermatozoa. *Aquaculture, 142*, 107–118.

Rurangwa, E., Kime, D. E., Ollevier, F., & Nash, J. P. (2004). The measurement of sperm motility and factors affecting sperm quality in cultured fish. *Aquaculture, 234*, 1–28.

Schreck, C. B. (2010). Stress and fish reproduction, the roles of allostasis and hormesis. *General and Comparative Endocrinology, 165*, 549–556.

Skaala, O., Hoyheim, B., Glover, K., & Dahle, G. (2004). Microsatellite analysis in domesticated and wild Atlantic salmon (*Salmo salar* L.): Allelic diversity and identification of individuals. *Aquaculture, 240*, 131–143.

Stoss, J. (1983). Fish gamete preservation and spermatozoan physiology. In W. S. Hoar, D. J. Randall & E. M. Donaldson (Eds.), *Fish physiology* (pp. 305–350). Academic Press.

Truscott, B., & Idler, D. R. (1969). An improved extender for freezing Atlantic salmon spermatozoa. *Journal of the Fisheries Research Board of Canada, 26*, 3254–3258.

CHAPTER 18

Vitamin Nutrition for Sustainable Aquaculture of Cultivable Fish Species

SEEMAB ZEHRA[1], JOSEPH LEOPOLDO Q. LARANJA[1], EDOARDO PANTANELLA[1], ASAAD H. MOHAMED[1], Y. M. KHAN[2], and A. AL SHAIKHI[3]

[1]*Beacon Development, King Abdullah University of Science and Technology, Thuwal, Jeddah, Saudi Arabia*

[2]*Cluster University of Srinagar, Srinagar, Jammu and Kashmir, India*

[3]*Ministry of Environment, Water and Agriculture, Riyadh, Saudi Arabia*

ABSTRACT

Vitamins are organic chemical substances that must be obtained in trace amounts from an external source in order to support normal development, reproduction, and health. The vitamins do not have structural purposes like other nutrition classes do, and their catabolism does not produce a lot of energy either. Rather, each of their distinct applications is typically somewhat specialized, which means that only trace amounts of the vitamins are needed in the diet. Every vitamin serves a distinct purpose in the body, and one vitamin cannot take the place of another. If they are not consumed or are absorbed or used poorly, they result in a particular deficiency illness. Vitamins are among the priciest elements used in full food formulation for aquatic animals. The research on the vitamin nutrition of several cultivable fish species is compiled in this chapter.

Recent Advances in Sustainable Aquaculture and Fisheries. Shabir Ahmad Dar, Imtiyaz Qayoom, Gohar Bilal Wani, Adnan Abubakr, and Seemab Zehra (Eds.)

DOI: 10.1201/9781998511372-18

18.1 INTRODUCTION

Aquaculture is expected to be the most dependable source of seafood in the future, as seen by the rise in the production of seafood worldwide in recent years. Nonetheless, there are a lot of contentious concerns in aquaculture when it comes to food safety, nutrition, and sustainability, and a lot of them have to do with the feeds and nutrition that farmed fish receive. In order to achieve sustainability in aquaculture and balance in the production of safe and nutritious food, several nutrition-related factors need to be taken into account. Among these factors, vitamins are organic chemical substances that, unlike lipids, carbohydrates, and amino acids, are needed in minute quantities through diet for overall development, health, and reproduction (Jobling, 2012). Like other nutrition classes, vitamins neither carry out the structural role, nor they serve as energy-generating compounds. Rather, they play a specialized and distinct role in the body and one vitamin cannot switch for another. If they are not consumed or are absorbed or used poorly, they result in a particular deficiency illness. Vitamins are among the most expensive elements utilized in the preparation of balanced diets for aquatic animals (Gaylord, 1998). Among their many biochemical roles, vitamins primarily serve as cofactors for a number of enzymes. Reduced enzyme activity from inadequate vitamin consumption can lead to both more specific deficiency signs and symptoms as well as general symptoms like stunted growth, low survival, and increased susceptibility to infections and illnesses (Halver, 2003; Yossa et al., 2015).

The difference between water-soluble and fat-soluble vitamins is that the former are found in every living tissue, whereas the latter are entirely absent from certain tissues. Based on their solubility, vitamins are categorized as either fat-soluble or water-soluble. Eight vitamins of the vitamin B complex family are required in relatively trace quantities and have primarily coenzyme functions. Choline, inositol, and vitamin C being the macrovitamins perform other than coenzymes functions and hence needed in higher quantities. The inclusion of one or more polar or ionizable groups (carboxyl, keto, hydroxyl, amino, or phosphate) is what distinguishes these vitamins. The fat-soluble vitamins—A, D, E, and K—achieve their effects without the aid of enzymes. The properties that make up fat-soluble vitamins are primarily aromatic and aliphatic (Combs Jr and McClung, 2016; Yossa et al., 2015). Some of them function as hormones and others perform structural roles in certain cells since they are a component of cell membranes.

Eight notable members of the vitamin B complex family having distinctive coenzyme functions *in vivo* include pantothenic acid, pyridoxine, biotin, folic acid, thiamine, riboflavin, niacin, and cyanocobalamin. According to conventional definitions, all currently studied species of farmed fish appear to depend on these water-soluble vitamins. Historical fish requirement trials are not well-researched, and the functions of individual B vitamins in fish bodies remain incompletely understood (Halver, 2003; Waagbø, 2010; Webster and Lim, 2002) and investigations with other species undertaken under modern intensive production conditions are not as well-researched as they should be. Fish require B vitamins based on their growth and metabolic rate; therefore, the amount needed during the larval stage is significantly larger than during the juvenile and adult phases (Waagbø, 2010). Since fish do not store water-soluble vitamins in their bodies, they must always have access to them in order to avoid showing symptoms of shortage.

18.1.1 WATER-SOLUBLE VITAMINS

18.1.1.1 THIAMINE

Among all the water-soluble vitamins thiamin or vitamin B_1 plays a much-diversified role in the growth and metabolism of all animals including plants and microorganisms (Oser, 1979). Thiamin serves as the coenzyme thiamin pyrophosphate, required for a number of metabolic transketolation and decarboxylation processes, in all cells. It involves in the oxidative decarboxylation of α-keto acids, such as pyruvate to α-ketoglutarate, and in the transketolase reactions in the pentose phosphate pathway that are essential to provide pentose phosphate for nucleotide synthesis and dihydronicotinamide adenine dinucleotide (NAD) phosphate for fatty acid synthesis (Jobling, 2012). Lack of adequate amount of thiamin in diet results in deficiency signs such as anorexia, reduced weight gain, nervous symptoms, convulsions, loss of equilibrium, muscle atrophy, abnormal coloration, fin congestion, subcutaneous hemorrhages, and high mortality (Jobling, 2012; Xiang et al., 2016; Zehra and Khan, 2017).

18.1.1.2 RIBOFLAVIN

Riboflavin or vitamin B_2 is a member of the group of vitamins referred to as vitamin B complex, which is a water-soluble vitamin. Riboflavin is involved in the metabolism of macronutrients as well as the production of some other B complex vitamins. It is known to participate in redox reactions in the metabolic pathways through cofactors flavin adenine dinucleotide and flavin mononucleotide, derived from riboflavin, by acting as electron carriers (Jobling, 2012). Inadequate riboflavin intake would be expected to lead to a disturbance in the intermediate steps of metabolism, with specific functional implications. Riboflavin has been documented to increase fish's capacity for antioxidants by enhancing the activities of antioxidant enzymes (Huang et al., 2010; Zehra and Khan, 2018). Deficiency signs reported in different fish species include eye disorders, high mortality, uncoordinated swimming, photophobia, abnormal skin coloration, anemia, shortening of the operculum and fin erosions (Halver, 2003; Jobling, 2012; Zehra and Khan, 2017). Lipid accumulation and degeneration of the pancreas with melanin deposition are recorded in riboflavin-deficient seabream, *Sparus aurata* (Morris et al., 1995), while hemorrhages in the hepatopancreas have been found in riboflavin-deficient carp, *Cyprinus carpio* (Aoe et al., 1967).

18.1.1.3 NIACIN

One of the most important water-soluble vitamins is niacin because of its major impact on fish body growth and health. It is one of the most stable vitamins and exists in two forms, nicotinic acid and nicotinamide. Both forms are readily absorbed from the stomach and the small intestine. The production of the pyridine nucleotides NAD and nicotinamide adenine dinucleotide phosphate (NADP), which are both necessary for redox reactions that take place within cells, is one of the many metabolic processes in which niacin plays a crucial part (Jobling, 2012). NAD is specific for hydrogenases involved in passing electrons to oxygen in the electron transport chain, whereas NADP is specific in hydrogenases for fatty acid synthesis and glucose metabolism. Furthermore, during DNA repair NAD also supplies essential ADP-ribose moiety involved in the ADP-ribosylation of proteins (Carson et al., 1988) and in calcium signaling by releasing calcium from intercellular spaces by cyclic ADP-ribose (Lee and Aarhus, 1991). Signs of niacin deficiency such as loss of appetite, poor-feed utilization, high-mortality, skin hemorrhage, erratic behavior, and retarded growth were observed in fish-fed niacin-deficient diets (Halver, 1982; Jobling, 2012; Li et al., 2016a; Morris et al., 1998).

18.1.1.4 PANTOTHENIC ACID

A required water-soluble vitamin, pantothenic acid is a component of coenzyme A (CoA), which is involved in many activities related to the metabolism of proteins, carbohydrates, and fats. CoA transfers acetyl (two-carbon) units. The presence of CoA within a cell is fundamental and has manifold functions, both anabolic and catabolic, involving the release of energy by way of tricarboxylic acid cycle. In biological acetylation reactions, such as the transformation of choline into acetylcholine and oxaloacetic acid into citric acid, CoA is also involved. Being an essential vitamin and precursor of CoA, pantothenic acid also donates a 4-phosphopantetheine moiety to acyl carrier protein a cofactor involved in the synthesis of fatty acids, delivering precursors for the condensation reactions and transporting the expanding acyl chain between different enzymes. Fish require pantothenic acid in their diets; a deficit causes dietary gill illness, which includes severe hyperplasia of the gill lamellae's epithelial cells, fused gill filaments, poor growth, bleeding, sluggishness, high mortality, and anorexia (Jobling, 2012; Soliman and Wilson, 1992; Wen et al., 2009). The dietary deficiency of pantothenic acid can also lead to the hepatic fat deposition and distorted lipid metabolism in fish (Jobling, 2012; Raggi et al., 2016). Furthermore, lower survival, short breath, behavior, anorexia, abnormal swimming, anemia, gill lesions, and death were also documented (Halver, 2003; Jobling, 2012; Karges and Woodward, 1984; Lin et al., 2012; Murai and Andrews, 1979; Qian et al., 2015; Soliman and Wilson, 1992; Wen et al., 2009; Wilson et al., 1983).

18.1.1.5 PYRIDOXINE

Pyridoxine influences growth, development, immunological response, and the activity of steroid hormones, among other aspects of metabolism (Combs Jr and McClung, 2016). Pyridoxal phosphate, the metabolically active form of pyridoxine, is a coenzyme for many other enzymes, the majority of which are involved in the metabolism of amino acids and the generation of red blood cells. Additionally, it participates in antioxidant processes and controls the expression of the glucocorticoid hormone and albumin genes (Natori and Oka, 1997) and the production of DNA and RNA (Trakatellis et al., 1997). Enzymes dependent on pyridoxal phosphate are involved in the manufacture of neurotransmitters, which has an antistress impact (Gerster, 1996). Symptoms of pyridoxine deficiency include tetany, convulsions, pale skin, anorexia, and hemorrhages in the intestine and kidney were documented in stinging catfish at 10 weeks of the experiment (Mohamed, 2001). Huang et al. (2005) have reported these deficiency signs after three weeks of the experimental period. Various histopathological changes have been observed in the liver, intestinal tissue, and kidney of rainbow trout (Jurss and Jonas, 1981), gilthead sea bream (Kissil et al., 1981), and Indian catfish (Mohamed, 2001) due to pyridoxine deficiency.

18.1.1.6 BIOTIN

Biotin plays a multifaceted role in several metabolic processes, including cellular respiration, gluconeogenesis, lipogenesis, and amino acid metabolism (Christodoulou et al., 2006; McMahon, 2002; Tomlinson et al., 2004). Biotin plays biochemical roles in carboxylation reactions. The biotin-dependent enzyme acetyl CoA carboxylase catalyzes the carboxylation of acetyl CoA to malonyl CoA, which is a crucial step in the extramitochondrial biosynthesis pathway for fatty acids. In Kreb's cycle, pyruvate is converted to oxalate by the biotin-dependent enzyme pyruvate carboxylase (Mohamed, 2001). In addition, biotin plays a role in three noncoenzyme processes, including the control of gene expression, the cell cycle, and reproduction (Bender, 2003; Combs Jr and McClung, 2016; Dakshinamurti, 2005; Landenberger et al., 2004; Yossa et al., 2015). Biotin deficiency resulted in poor growth, dark skin color, anorexia, convulsions, and loss of appetite in several fish species, such as *Salmo gairdneri* (Woodward and Frigg, 1989), *Clarias batrachus* (Mohamed et al., 2000), *Heteropneustes fossilis* (Shaik Mohamed, 2001), and *Lateolabrax japonicus* (Li et al., 2010a).

18.1.1.7 FOLIC ACID

Vitamin B_9 or folic acid is a well-known member of the vitamin B complex family and is indispensable nutrient for fish, as it is for other vertebrates. Folic acid is transformed to its active coenzyme form, tetrahydrofolic acid in the tissues. This coenzyme is functional in the transfer of single-carbon units in a number of complex enzymatic systems such as interconversion of serine and glycine, synthesis of histidine, methionine–homocysteine synthesis, and pyrimidine synthesis for nucleotide bases of DNA and RNA (Jobling, 2012). Folic acid is involved in the metabolism of several amino acids and is crucial for the development and maturation of red blood cells (Miao et al., 2013). Folic acid enhances blood glucose regulation, enhances cell membrane function, and promotes egg hatchability. Adequate supplementation of folic acid has been found to improve normal growth and health status in most of the aquatic animals (Ying et al., 2013). Deficiency of folic acid influences fish health and development. Typical symptoms of folic acid deficiency in fish and crustacean species include poor growth, poor feed conversion, anorexia, general anemia, macrocytic normochromic anemia, lethargy, fragility of the caudal fins, dark skin pigmentation, dark coloring, and splenic infarction (Asaikkutti et al., 2016; Halver, 1982; Sesay et al., 2017; Shiau and Huang, 2001).

18.1.1.8 VITAMIN B_{12}

Cyanocobalamin, also called Vitamin B_{12}, is a water-soluble vitamin and is involved in the preservation of the brain's and the central nervous system's regular operation, as well as in the metabolism of fatty and amino acids (Jobling, 2012; Lin et al., 2010). Additionally, cyanocobalamin is necessary for the regular recycling of tetrahydrofolic acid and for the methylation of homocysteine to methionine (Jobling, 2012). Cyanocobalamin is involved with folic acid in hemopoiesis. It is also involved in cholesterol metabolism, in purine and pyrimidine biosynthesis, and in the metabolism of glycols. Because the synthesis of cyanocobalamin is limited almost exclusively to bacteria, neither fungi, plants, nor animals are capable of producing cyanocobalamin (Jobling, 2012). Cyanocobalamin is found generally in animal-derived products and is seldom found in foods derived from plants (Combs Jr and McClung, 2016). Because of the limited supply and high cost of ingredients of animal proteins, plant protein ingredients are considered as alternative in commercial feed used for intensive aquaculture which may cause cyanocobalamin deficiency symptoms such as anorexia, megaloblastic anemia, growth retardation, poor feed conversion, and high mortality in cultured animals (Hansen et al., 2017; Li et al., 2016a; Lin et al., 2010).

18.1.1.9 INOSITOL

Inositol is a water-soluble vitamin and myo-inositol (MI) is the most prevalent naturally occurring biologically active isomer of it. MI is a structural component of biological membranes in the phospholipid form (Khosravi et al., 2015b; Shirmohammad et al., 2016). As part of the phospholipid form, phosphatidylinositol participates in transmembrane signal transfer (Aukema and Holub, 1994). *De novo* synthesis of inositol from glucose by liver, brain, and other tissues have been documented in several fish species such as common carp *C. carpio* (Aoe and Masuda, 1967), channel catfish *Ictalurus punctatus* (Burtle and Lovell, 1989), sea bass *Lates calearifer* (Boonyaratpalin and Wanakowat, 1993), and sunshine bass *Morone chrysops* × *Morone saxatilis* (Deng et al., 2002). However, in some fishes, *de novo* synthesis of MI is not enough to satisfy their physiological demands and have the need of an exogenous dietary inclusion of this vitamin in various studies (Aoe and Masuda, 1967; Jiang et al., 2009; Khosravi et al., 2015b; Shiau and Su, 2004; Shirmohammad et al., 2016). Inositol deficiency has been reported to be linked to a range of lipid metabolic disorders that cause fat buildup in liver, decreased hepatic lipoprotein output, and intestinal lipodystrophy. Cell damage (medium lactate dehydrogenase activity) was induced by MI deficiency in carp, but with increasing levels of MI supplementation, the cell damage gradually disappeared. However, lipid peroxidation and protein oxidation were all gradually inhibited by higher levels of MI supplementation (Jiang et al., 2013). Protein oxidation influences amino acid metabolism (Dabrowski and Guderley, 2003).

18.1.1.10 CHOLINE

Choline is a vitamin-like nutrient that was formerly known as vitamin B_4. Choline, like other vitamins, is important for nutrition, however, compared to other vitamins, choline appears to have a much higher requirement (Mai et al., 2009). Being a constituent part of phosphatidylcholine and acetylcholine, it aids in lipid transport, inhibits fatty livers, and acts as a metabolic methyl donor (Griffin et al., 1994; Halver, 2003). Choline can be synthesized by most of the animals from methyl donors like methionine present in the diet (Shiau and Lo, 2000). Although being widely available in almost all feedstuffs, animals do have a dietary choline requirement because not all naturally occurring choline is bio-available and hence could not meet the requirement. Therefore, dietary supplementation of extra choline chloride becomes a prerequisite. Griffin et al. (1994) reported that it appears that choline chloride is a superior dietary choline source. Choline is a lipotropic factor preventing the development of fatty liver (Halver, 2003). Fatty liver has been reported as the most characteristic sign of choline deficiency in lake trout (Ketola, 1976) and common carp (Ogino et al., 1970b). Choline-deficient rainbow trout (Hung, 1989), channel catfish (Wilson and Poe, 1988), and white sturgeon (Rumsey, 1991) did not develop fatty liver. Thus, there appears to be species specificity in the manifestations of choline deficiency among fishes (Hosokawa et al., 2001). Several choline-deficiency signs involve poor growth and feed conversion, with impaired fat metabolism. Hemorrhagic kidneys and intestines have been reported in trout (Mclaren et al., 1947) and increased gastric emptying time has been observed in salmon (Halver, 1957). Similar deficiency signs were observed in catfish (Dupree, 1966), carp (Ogino et al., 1970a), eel (Arai, 1972), and lake trout (Ketola, 1976). However, in the present study, when fish were fed choline-free and choline-deficient diets, there were no deficiencies seen except poor growth and anorexia. Hosokawa et al. (2001) were also found poor growth and anorexia in yellowtail.

18.1.1.11 ASCORBIC ACID (VITAMIN C)

Ascorbic acid is an essential micronutrient that plays various roles in physiological processes. It acts as an antioxidant to inactivate the free radicals such as reactive oxygen and reactive nitrogen species and avoid cellular components from radical damage (El-Gawad et al., 2014; Jiménez-Fernández et al., 2015). It also takes part in collagen and steroid hormone synthesis and enhances tolerance and immune response to toxicants and environmental stressors (Dawood and Koshio, 2018; Roosta et al., 2014; Shahkar et al., 2015; Waagbø, 2010). It facilitates the absorption of iron (Corti et al., 2010; Harrison et al., 2010; Jiménez-Fernández et al., 2015). Due to several reported vitamin C deficiency signs such as spinal deformity, impaired collagen production, internal hemorrhaging, anemia, retarded growth, and depressed immunity in various fish species

(Chen et al., 2015b; Ebi et al., 2018; Jobling, 2012) and also the lack of L-gulonolactone oxidase in fish which is required for the biosynthesis of vitamin C (Wilson and Poe, 1973), its dietary supplementation is crucial in fishes.

18.1.2 FAT-SOLUBLE VITAMINS

18.1.2.1 VITAMIN A

Vitamin A, a fat-soluble vitamin, is involved in a number of physiological processes, including hematopoiesis, differentiation, and maintenance of epithelial cells, bone development, reproduction, and vision (Duester, 2008; Semba et al., 1998; Stephensen, 2001). Thus, it is considered an essential nutrient for normal growth and development of an organism. Quantitative requirements of vitamin A for several fish species have been determined but the values vary among species (Council, 1993; Jobling, 2012). Hemorrhages, delayed growth, poor feed efficiency, anorexia, abnormal coloration, bone deformity, fin erosion, exophthalmia, lesions and hemorrhages in the eyes, increased erythrocyte osmotic fragility, anemia, lethargy, and high mortality are among the symptoms of vitamin A deficiency that have been reported for various fish species (Blazer, 1992; Council, 1993, 2011; Mohamed et al., 2003). In addition to these indicated functions, vitamin A has been identified to play an important role in the immune system (both innate and adaptive immunity) therefore helping to protect against infections in a number of ways (Guimarães et al., 2014; Semba, 1998; Stephensen, 2001; Tomkins and Hussey, 1989).

18.1.2.2 VITAMIN D

Fat-soluble vitamin D is essential for maintaining calcium and phosphate homeostasis as well as for safeguarding the integrity of the skeleton. Furthermore, osteoblasts—the native cells in the skeleton that make bones—are directly affected by vitamin D, which inhibits their proliferation, modifies their differentiation, and controls the extracellular matrix's mineralization (Darias et al., 2011). Fish are an essential dietary source of vitamin D because they store a lot of the vitamin in their liver and fat cells, particularly the fat that is connected with muscle. Fish must get all of their vitamin D from food because they are unable to produce it. In the aquatic food chain, planktonic vitamin D naturally accumulates. In aquaculture, vitamin D intake can be controlled in the formulated feeds. For a number of fish species, the minimal need for vitamin D in the diet has been determined. It is still unclear how vitamin D functions in fish physiology. It was widely accepted until the 1970s that fish do not metabolize vitamin D; instead, they accumulate it. Four decades later, there is strong evidence that fish have vitamin D (Lock et al., 2010).

18.1.2.3 VITAMIN E

Tocopherols and tocotrienols are a class of lipid-soluble compounds that are collectively known as vitamin E. They protect organisms from lipid oxidation and may have additional, more specialized biological roles. Similar to other vertebrates, fish also preferentially retain α-tocopherol (TOH) over the other tocopherols. This is likely due to the liver's tocopherol transfer protein (TTP), which binds tocopherols with varying affinities and releases them back into the bloodstream. Bile contains a higher amount of tocopherols that are expelled than those that bind poorly to TTP. Because α-TOH interacts with other nutrients, the amount needed changes depending on the makeup of the food. The need is increased by high concentrations of polyunsaturated fatty acids and insufficient intakes of vitamin C, selenium, and astaxanthin. The reason for this is that damaged vitamin E is recycled by other antioxidants in the antioxidant defense system, which is a dynamic process that also involves lipid oxidation (Hamre, 2011).

18.1.2.4 VITAMIN K

Phylloquinone (vitamin K1) and menaquinone (vitamin K2) are naturally occurring forms of vitamin K, which is a member of the lipid-soluble vitamin family. Menadione, often known as vitamin K3, is a synthetic provitamin that is mostly utilized in animal feed as a source of vitamin K. Menadione can become dangerously low in the diet due to its instability during feed preparation and storage. The presence of menadione in salmonid feed has also been questioned in recent papers. Fish blood coagulation and bone mineralization depend heavily on vitamin K, yet different vitamin K sources have different recommended minimum requirements. The symptoms of vitamin K insufficiency include anemia, increased blood clotting time, death, and histological alterations in the gills and liver. But it is important to evaluate both natural and supplemental vitamins. For precise estimates, however, it is important to evaluate both naturally occurring and augmented forms of vitamin K in feeds, since pertinent new feed components derived from plants would be adequate to fulfill the recommended daily intake of vitamin K (Krossøy et al., 2011).

18.1.2.5 VITAMIN INTERACTIONS

Certain vitamins can be easily produced from other metabolic elements, making them metabolic essentials but not nutritional essentials for certain animals. For instance, channel catfish appear to synthesize choline provided sufficient methyl donors, such as methionine, are included in the diet; yet, a choline need can be shown if the amount of dietary methionine is limited (Wilson and Poe, 1988). Numerous animals have been shown to convert tryptophan to niacin through metabolism (Council, 2011; Xia et al., 2015). A number of fish species, including hybrid tilapia (Shiau and Suen, 1992), channel catfish (Ng et al., 1997), trout, and salmon (Poston and Combs Jr, 1980) have a limited capacity to convert tryptophan to niacin effectively.

18.1.2.6 VITAMIN REQUIREMENT OF INDIAN MAJOR CARPS

The large ranges described in the literature highlight uncertainties in vitamin need estimates. It is commonly known that warm-water fishes require certain vitamins in their diet. Natural food sources are few in intensive production, so vitamin supplementation is necessary to ensure appropriate fish growth. Therefore, the natural content of live feed, feed ingredients, and further dietary enrichment or supplementation through vitamin premixes constitute the exogenous supply of vitamins. Premix is the phrase used to describe the vitamin supplement that is added to the meal. A premix is made to make up for vitamins that are partially or fully absent from feed ingredients or to augment the vitamins found in the dietary ingredients. Similar to those for essential amino acids, the requirements for vitamins are both qualitative and quantitative ascertained by feeding purified diets lacking in a particular vitamin and then comparing the growth and other symptoms of deficiency in the vitamin-deprived fish to control fish that have been fortified with all the vitamins. Many researchers were able to ascertain the water-soluble vitamin requirements for salmon and trout, both qualitatively and quantitatively by developing H-440 diet (Halver, 1957). A number of cultivable fish species that have been studied thus far have proven their quantitative vitamin requirements (Council, 2011). As a result of their widespread distribution and global cultivation, cyprinids are plentiful in a wide range of environments, constitute the largest group of freshwater teleost fishes numerically, and are hence highly valuable to humankind. Among these are the Indian major carps (IMCs), which grow quickly, reaching a marketable size of 800–1000 g in less than a year for catla, *Catla catla*, rohu, *Labeo rohita*, and mrigal, *Cirrhinus mrigala* (Jhingran and Pullin, 1988). Compared to other fish species, IMCs have a number of advantages. Compared to other fish species, these carps have cheaper production costs, less environmental issues, and a decreased chance of disease outbreaks (Abidi and Khan, 2008). The information exist on vitamin requirements of different cultivable fish species is presented in Table 18.1.

It has been proposed that variations in fish species, sizes, lab settings, feed quality, methodology, and evaluation standards could account for the vast range of variation and consistency in fish vitamin requirements (Jiang et al., 2013; Mohamed and Ibrahim, 2001; Morris et al., 1998; Zehra and Khan, 2018). The large differences in vitamin requirements among the different fish species may also be caused by the test diets' dietary energy levels and the availability of dietary vitamins. Different estimations of fish requirements may also arise from the use of different statistical models for fitting the response and estimating requirements. The different species' requirements for vitamins may also be affected by their nutrient and energy digestibility (Council, 2011; Xia et al., 2015). The differing factors used, such as weight increase and enzyme assay, may potentially contribute to variations in vitamin requirements. As a result, requirements for growth in the same species have been observed by several studies to vary very widely. Given that many vitamins have coenzymatic properties, Jauncey (1982) proposed that the best way to determine the ideal levels of vitamins in the diet is to define the vitamin requirement as the amount of a vitamin in the diet that allows for the maximum activity of all the enzymes for which the vitamin functions as a coenzyme. Given that many vitamins have coenzyme functions, it has been proposed that the best way to determine the appropriate amounts of vitamins in the diet is to define the vitamin requirement as the amount of a vitamin in the diet that allows

TABLE 18.1 Vitamin Requirements of Cultivable Fish Species.

Common Name	Scientific Name	Source	Requirement	Response Criteria	References
			Thiamin		
Channel catfish	*Ictalurus punctatus*		1 mg	WG, ADS	Murai and Andrews (1979)
Common carp	*Cyprinus carpio*		0.5 mg	WG, ADS	Aoe et al. (1969)
Grass carp	*Ctenopharyngodon idella*		1.3 and 5 mg	WG, MLS	Jiang et al. (2014b)
Jian carp	*Cyprinus carpio* var. Jian		1.02 mg	PWG	Huang et al. (2011)
Mrigal carp	*Cirrhinus mrigala*		1.79–3.34 mg	WG, MLS	Zehra and Khan (2017)
Nile tilapia	*Oreochromis niloticus*		3.5 mg	WG	Lim et al. (2011)
Pacific salmon	*Oncorhynchus* spp.		10–15 mg	MLS	Halver (1982)
Rainbow trout	*Oncorhynchus mykiss*		1–10 mg	WG, ADS	Mclaren et al. (1947)
			1 mg	WG, ADS	Morito et al. (1986)
Turbot	*Scophthalmus maximus*		1 mg	WG	Cowey et al. (1975)
Catla	*Catla catla*		0.74–0.79 mg	WG, MLS	Khan and Khan (2022)
Oriental river prawn	*Macrobrachium nipponense*		66.80–67.57 mg	WG	Sun et al. (2022)
			Riboflavin		
Blue tilapia	*Oreochromis aureus*		6 mg	WG, ADS	Soliman and Wilson (1992)
Common carp	*Cyprinus carpio*		4 mg	WG, ADS	Aoe et al. (1967)
Channel catfish	*Ictalurus punctatus*		9 mg	WG, ADS	Murai and Andrews (1979)
Grass carp	*Ctenopharyngodon idella*		6.65 mg	ED	Chen et al. (2015a)
Hybrid striped bass	*M. chrysops female × M. saxatilis*		4.1 - 5.0 mg	WG, MLS	Deng and Wilson (2003)
Jian carp	*Cyprinus. carpio* var. Jian		5.0 mg	WG	Li et al. (2010b)
Pacific salmon	*Oncorhynchus* spp.		20–25 mg	MLS	Halver (1982)
Rainbow trout	*Oncorhynchus mykiss*		5–15 mg	WG, ADS	Mclaren et al. (1947)

TABLE 18.1 *(Continued)*

Common Name	Scientific Name	Source	Requirement	Response Criteria	References
			3 mg	ED	Hughes et al. (1981)
			2.7 mg	MLS, ED	Amezaga and Knox (1990)
Spotted snakehead	*Channa punctatus*		5.7–7.7 mg	WG, MLS	Zehra and Khan (2017)
			Niacin		
African catfish	*Clarias gariepinus*		33.1 mg	FE	Morris et al. (1998)
Blunt snout bream	*Megalobrama amblycephala*		30.62–31.25 mg	MLS, WG	Li et al. (2017)
Brook trout	*Salvelinus fontinalis*		95.0 mg	MLS	Phillips Jr and Brockway (1949)
Common carp	*Cyprinus carpio*		28 mg	WG, ADS	Aoe et al. (1967)
Channel catfish	*Ictalurus punctatus*		14 mg	WG, ADS	Andrews and Murai (1978)
			7.4 mg	WG, ED	Ng et al. (1997)
Gilthead seabream	*Sparus aurata*		63.0–83.0 mg	WG	Morris et al. (1995)
Grass carp	*Ctenopharyngodon idella*		34.01, 42.08 mg	WG, ED	Feng et al. (2016)
Common carp	*Cyprinus carpio*		98 mg	WG	Choopani et al. (2022)
Hybrid tilapia	*Oreochromis niloticus* × *O. aureus*		26 mg	WG	Shiau and Suen (1992)
Mrigal carp	*Cirrhinus mrigala*		30 mg	WG, FE	Ahmed (2011)
Rohu	*Labeo rohitta*		33.0	WG, FE	Ahmed (2011)
Pacific salmon	*Oncorhynchus spp.*		150–200 mg	MLS	Halver (1982)
Rainbow trout	*Oncorhynchus mykiss*		1–5 mg	WG, ADS	Mclaren et al. (1947)
			10 mg	WG, ADS	Poston and Wolfe (1985)
Stinging catfish	*Heteropneustes fossilis*		25.0 mg	FE	Mohamed and Ibrahim (2001)
Tilapia	*Oreochromis niloticus*		20.4, 84.6 mg	WG, MLS	Jiang et al. (2014a)

TABLE 18.1 *(Continued)*

Common Name	Scientific Name	Source	Requirement	Response Criteria	References
Catla	*Catla catla*		31.7 mg	WG, MLS	Khan and Khan (2021a)
			Pantothenic acid		
Blue tilapia	*Oreochromis aureus*		10 mg	WG, ADS	Soliman and Wilson (1992)
Blunt Snout Bream	*Megalobrama amblycephala*		24.08 mg	WG	Qian et al. (2015)
Channel catfish	*Ictalurus punctatus*		10 mg	WG, ADS	Murai and Andrews (1979)
			15 mg	WG, ADS	Wilson et al. (1983)
Juvenile black carp	*Mylopharyngodon piceus*		10–20 mg	WG	Jia et al. (2022)
Hybrid striped bass	*Morone chrysops* × *M. saxatilis*		18.8 mg	WG	Raggi et al. (2016)
Juvenile blunt snout bream	*Megalobrama amblycephala*		240 mg	WG	Qian et al. (2022)
Grouper	*Epinephelus malabaricus*		11 mg	WG	Lin et al. (2012)
Hybrid striped bass	*Morone chrysops* × *M. saxatilis*		18.8 mg	WG	Raggi et al. (2016)
Jian carp	*Cyprinus. carpio* var. Jian		23 mg	WG	Wen et al. (2009)
Catla	*Catla catla*		28.70–38.46 mg	WG, MLS	Khan and Khan (2022)
Pacific salmon	*Oncorhynchus* spp.		40–50 mg	MLS	Halver (1982)
Rainbow trout	*Oncorhynchus mykiss*		10–20 mg	WG, ADS	Mclaren et al. (1947)
			20 mg	WG, ADS	Cho and Woodward (1990)
			Pyridoxine		
Atlantic salmon	*Salmo salar*		5 mg	WG, ADS	Lall and Weerakoon (1990)
Channel catfish	*Ictalurus punctatus*		3 mg	WG, ADS	Murai and Andrews (1979)
Yellow catfish	*Pelteobagrus fulvidraco*		6.70 –7.42 mg	WG	Li et al. (2022)
Catla	*Catla catla*		3.20–6.87 mg	WG, MLS	Khan and Khan (2021b)

TABLE 18.1 *(Continued)*

Common Name	Scientific Name	Source	Requirement	Response Criteria	References
Grass carp	*Ctenopharyngodon idella*		1.13–3.73 mg	SGR, ED	Yang et al. (2017a)
Hybrid tilapia	*Oreochromis niloticus* × *Oreochromis aureus*		15–16.5 mg	WG, ED	Shiau and Hsieh (1997)
Indian catfish	*Heteropneustes fossils*		3.21 mg	WG	Shaik Mohamed (2001)
Jian carp	*Cyprinus. carpio* var. Jian		6.07 mg	WG	He et al. (2009)
Mrigal carp	*Cirrhinus mrigala*		5.63–8.61 mg	WG, MLS	Zehra and Khan (2017)
Snakehead	*Channa punctatus*		7.6 and 10.4 mg	WG, MLS	Zehra and Khan (2017), Zehra and Khan (2018)
Nile tilapia	*O. niloticus*		10 mg	MLS	Teixeira et al. (2012)
Pacific salmon	*Oncorhynchus* spp.		10–20 mg	WG	Halver (1982)
Rainbow trout	*Oncorhynchus mykiss*		1–10 mg	WG, ADS	Mclaren et al. (1947)
			2 mg	WG, ADS	Woodward and Frigg (1989)
			3–6 mg	ED	Woodward and Frigg (1989)
Red hybrid tilapia	*Oreochromis mossambicus* × *O. niloticus*		3 mg	WG	Lim et al. (1995)
Grass carp	*Ctenopharyngodon idella*		4.62 mg	skin hemorrhage and lesions morbidity	Zheng et al. (2022)
			Biotin		
Asian catfish	*Clarias batrachus*		2.49 mg	WG	Shaik Mohamed (2001)
Blunt snout bream	*Megalobrama amblycephala*		0.063–0.075 mg	WG, ED	Qian et al. (2014)
Common carp	*Cyprinus carpio*		1 mg	WG, ADS	Ogino et al. (1970b)
Channel catfish	*Ictalurus punctatus*		R		Robinson and Lovell (1978)

TABLE 18.1 *(Continued)*

Common Name	Scientific Name	Source	Requirement	Response Criteria	References
Hybrid tilapia	*Oreochromis niloticus* × *Oreochromis aureus*		0.06 mg	WG, ED	Shiau and Chin (1999)
Japanese sea bass	*Lateolabrax japonicas*		0.046 mg	WG	Li et al. (2010a)
Jian carp	*Cyprinus. carpio* var. Jian		0.15 mg	SGR	Zhao et al. (2012)
Lake trout	*Salvelinus namaycush*		0.1 mg	WG, ADS	Poston (1976)
Pacific salmon	*Oncorhynchus* spp.		1–1.5 mg	MLS	Halver (1982)
Rainbow trout	*Oncorhynchus mykiss*		0.05–0.25 mg	WG, ADS	Mclaren et al. (1947)
			0.08 mg	WG, ADS	Woodward and Frigg (1989)
Yellow tail zebrafish	*Seriola lalandi*		0.67 mg	MLS, WG	Yossa et al. (2014)
	Danio rerio		0.51 mg		
Catla	*Catla catla*		0.41–0.87 mg	WG, MLS	Khan and Khan (2019)
Grass carp	*Ctenopharyngodon idella*		0.245–0.354 mg	ROS content, *caspase 3* and *claudin-3c* expression levels	He et al. (2022)
Vitamin B_{12}					
Blunt snout bream	*Megalobrama amblycephala*		0.06–0.12 mg	WG, ED	Li et al. (2016b)
Common carp	*Cyprinus carpio*		NR		Kashiwada et al. (1970)
Channel catfish	*Ictalurus punctatus*		R		Limsuwan and Lovell (1981)
Hybrid tilapia	*Oreochromis niloticus* × *Oreochromis aureus*		NR		Shiau and Lung (1993)
Nile tilapia	*O. niloticus*		NR		Lovell and Limsuwan (1982)
Pacific salmon	*Oncorhynchus* spp.		0.015–0.02 mg	MLS	Halver (1982)
Rainbow trout	*Oncorhynchus mykiss*		R		Phillips et al. (1964)

TABLE 18.1 *(Continued)*

Common Name	Scientific Name	Source	Requirement	Response Criteria	References
Catla	*Catla catla*		0.034 and 0.037 mg	WG, MLS	Mohd Khan and Khan (2021)
Snakehead	*Channa punctatus*		0.068–0.15 mg	WG, MLS	Zehra and Khan (2019a)
Folic acid					
Blunt snout bream	*Megalobrama amblycephala*		0.68–0.82	WG, SGR, HSI	Sesay et al. (2016)
Common carp	*Cyprinus carpio*		NR		Aoe et al. (1967)
Channel catfish	*Ictalurus punctatus*		1–1.5 mg	WG, NHV	Duncan and Lovell (1994)
Grass carp	*Ctenopharyngodon idella*		3.6–4.3	WG, STG, STC, NHV	Zhao et al. (2008)
Green cat fish	*Mystus nemerus*		1 mg	WG	Hien and Doolgindachbaporn (2011)
Hybrid tilapia	*Oreochromis niloticus* × *Oreochromis aureus*		0.82 mg	WG, MLS, HIS	Shiau and Huang (2001)
Juvenile abalone	*Haliotis discus hannai*		2.62–5.29 mg	WG, VFAC	Miao et al. (2013)
Pacific salmon	*Oncorhynchus* spp.		6–10 mg	MLS	Halver (1982)
Rainbow trout	*Oncorhynchus mykiss*		1.0 mg	WG, ADS	Cowey and Woodward (1993)
Snakehead	*Channa punctatus*		0.74–1.15 mg	WG, MLS	Zehra and Khan (2020)
Catla	*Catla catla*		0.22–0.56 mg	WG, MLS	Khan and Khan (2020)
Largemouth bass	*Micropterus salmoides*		1.42–1.46 mg	WG, MLS	Hang et al. (2022)
Choline					
Common carp	*Cyprinus carpio*		1500 mg	WG, LLC	Ogino et al. (1970a)
Channel catfish	*Ictalurus punctatus*		400 mg	WG, LLC	Wilson and Poe (1988)
Grass carp	*Ctenopharyngodon idella*		3000 mg	WG, LLC, FE	Wang et al. (1995)

TABLE 18.1 *(Continued)*

Common Name	Scientific Name	Source	Requirement	Response Criteria	References
Grouper	*Epinephelus coioides*		1093.7 and 1579.7 mg	SGR, LLC	Qin et al. (2017)
Hybrid tilapia	*Oreochromis niloticus* × *Oreochromis aureus*		1000 mg	WG	Shiau and Lo (2000)
Hybrid striped bass	*Morone chrysops* × *M. saxatilis*		500 mg	WG	Griffin et al. (1994)
Lake trout	*Salvelinus namaycush*		1000 mg	WG	Ketola (1976)
Pacific salmon	*Oncorhynchus* spp.		600–800 mg	MLS	Halver (1982)
Parrot fish	*Oplegnathus fasciatus*		230 mg	WG	Khosravi et al. (2015a)
Rainbow trout	*Oncorhynchus mykiss*		714–813 mg	WG, LLC	Rumsey (1991)
Red drum	*Sciaenops ocellatus*		588 mg	WG	Craig and Gatlin III (1996)
Siberian sturgeon	*Acipenser baerii*		1500 mg	WG	Yazdani Sadati et al. (2014)
Yellow perch	*Perca flavescens*		598–634 mg	WG	Twibell and Brown (2000)
Yellow tail	*Seriola lalandi*		2920 mg	MLS	Hosokawa et al. (2001)
Myo-inositol					
Common carp	*Cyprinus carpio*		440 mg	WG, ADS	Aoe and Masuda (1967)
Channel catfish	*Ictalurus punctatus*		NR		Burtle and Lovell (1989)
Grouper	*E.* spp.		335–365 mg	WG, MLS	Su and Shiau (2004)
Grass carp	*Ctenopharyngodon idella*		166–214 mg	WG, FE	Wen et al. (2007)
Gibel carp	*Carassius auratus gibelio*		165.3 mg	WG	Gong et al. (2014)
Hybrid tilapia	*Oreochromis niloticus* × *Oreochromis aureus*		400 mg	WG, MLS	Shiau and Su (2005)
Hybrid striped bass	*Morone chrysops* × *M. saxatilis*		NR		Deng et al. (2002)

TABLE 18.1 *(Continued)*

Common Name	Scientific Name	Source	Requirement	Response Criteria	References
Nile tilapia	*Oreochromis niloticus*		NR	WG	Peres et al. (2004)
Parrot fish	*Oplegnathus fasciatus*		100 mg	WG	Khosravi et al. (2015b)
Pacific salmon	*Oncorhynchus* spp.		300–400 mg	MLS	Halver (1982)
Rainbow trout	*Oncorhynchus mykiss*		250–500 mg	WG, ADS	Mclaren et al. (1947)
Snakehead	*Channa punctatus*		374.1–504.9 mg	WG, MLS	Zehra and Khan (2019b)
Hybrid grouper	Brown-marbled Grouper *Epinephelus fuscoguttatus* ♀ × Giant Grouper *E. lanceolatus* ♂		181.47 mg	WG	Zhu et al. (2022)
Jian carp	*Cyprinus carpio*		800 mg/kg C-MI and 400 mg/kg E-MI	WG	Luo et al. (2022)
Hybrid Grouper	♀ *Epinephelus fuscoguttatus* × ♂ *E. lanceolatu*		1.23–1.67 g	WG	Pan et al. (2022)
Vitamin C					
African catfish	*Clarias gariepinus*	ECAA	45 mg	WG	Eya (1996)
Atlantic salmon	*Salmo salar*	AMP	10–20 mg	WG	Sandnes et al. (1992)
Blue tilapia	*Oreochromis aureus*	AA	50 mg	WG	Stickney et al. (1984)
Channel catfish	*Ictalurus punctatus*	AA	25–50 mg	WG	Andrews and Murai (1975)
Coho salmon	*Oncorhynchus kisutch*	AA	60 mg	WG	Lim and Lovell (1978)
		LAPP	11 mg	WG	El Naggar and Lovell (1991)
Chinese sucker	*Myxocyprinus asiticus*	LAPP	84.6–126.1 mg	WG, MLS	Huang et al. (2017)
Cobia	*Rachycentron canadum*	LAPP	53.9–104 mg	WG, MLS	Xiao et al. (2010)
Common carp	*Cyprinus carpio*	AA	45 mg	WG	Gouillou-Coustans et al. (1998)

TABLE 18.1 *(Continued)*

Common Name	Scientific Name	Source	Requirement	Response Criteria	References
Mrigal carp	*Cirrhinus mrigala*	LAPP	35.65–41.99 mg	WG, MLS	Zehra and Khan (2012)
Snakehead	*Channa punctatus*	LAPP	38.2–47.9 mg	WG, MLS	Zehra and Khan (2021)
European sea bass	*Scophthalmus maximus*	LAPP	20 mg	WG	Merchie et al. (1996)
Gilthead seabream	*Sparus aurata*	AA	R		Alexis et al. (1997)
Jian carp	*Cyprinus carpio* var. Jian	AMP	40.9 mg	WG	Liu et al. (2011)
Loach	*Misgurnus anguillicaudatus*	AA	200 mg	WG	Zhao et al. (2017)
Mexican cichlid	*Cichlasoma urophthalmus*	AA	40 mg	WG	Chavez de Martinez (1990)
Nile tilapia	*Oreochromis niloticus*	AA	50 mg	WG	Abdelghany (1996)
Parrot fish	*Oplegnathus fasciatus*	AP	118 mg	WG	Wang et al. (2003)
Rainbow trout	*Oncorhynchus mykiss*	AA	50–100 mg	WG	Halver et al. (1969)
Red drum	*Scianops ocellatus*	AMP	15 mg	WG	Aguirre and Gatlin III (1999)
Red sea bream	*Pagus major*	LAPP	400 mg	WG	Gao et al. (2013)
Rohu	*Labeo rohitta*	AA	200 mg		Misra et al. (2007)
Sunshine bass	*Morone chrysops* × *M. saxatilis*	LAPP	22 mg	MLS	Sealey and Gatlin III (1999)
Spotted rose snapper	*Lutjanus guttatus*	LAPP	29 mg	WG	Chávez-Sánchez et al. (2014)
Striped catfish	*Pangasianodon hypophthalmus*	LAPP	46–76 mg	WG	Daniel et al. (2018)
Tilapia	*Oreochromis niloticus*	LAPP	45.0–114.9 mg	WG	Huang et al. (2016)
	Oreochromis karongae	LAPP	53.05–180.53 mg	WG	Wu et al. (2015)
Turbot	*Scophthalmus maximus*	LAPP	51 mg	WG, ED	Nsonga et al. (2009)
Tiger puffer	*Takifugu rubripes*	AA	29 mg	WG	Eo and Lee (2008)
Vundu	*Heterobranchus longifilis*	AA	82.2 mg	WG	Ibiyo et al. (2007)
White sturgeon	*Acipenser transmontanus*		NR		Hung (2017)

TABLE 18.1 *(Continued)*

Common Name	Scientific Name	Source	Requirement	Response Criteria	References
			Vitamin A		
Amur sturgeon	*Acipenser schrenckii*		0.27–0.59 mg	WG, MLS	Wen et al. (2008)
Atlantic halibut	*Hipoglossus hipoglossus*		2.5 mg	WG	Moren et al. (2004)
European sea bass	*Dicentrarchus labrax*		31 mg	WG	Villeneuve et al. (2006)
Grouper	*Epinephelus* spp.		0.93 mg	WG	Mohamed et al. (2003)
Gibel carp	*Carassius auratus*		0.53 mg	WG	Yang et al. (2017b)
			0.80 mg	WG	Shao et al. (2016)
Hybrid striped bass	*Morone chrysops* × *M. saxatilis*		0.51–40.52 mg	WG	Hemre et al. (2004)
Hybrid tilapia	*Oreochromis niloticus* ×*Oreochromis aureus*		1.76–2.09 mg	WG, MLS	Hu et al. (2006)
Japanese flounder	*Paralichthys olivaceus*		2.7 mg	WG	Hernandez et al. (2005)
			Vitamin E		
Atlantic salmon	*Salmo salar*		35 mg	WG	Hardy and Kaushik (2021)
Blue tilapia	*Oreochromis aureus*		25 mg	WG, ED	Roem et al. (1990)
Cobia	*Rachycentron canadum*		78–111 mg	WG, ADS	Zhou et al. (2013)
Channel catfish	*Ictalurus punctatus*		25 mg	WG, ADS	Murai and Andrews (1974)
Common carp	*Cyprinus carpio*		50 mg	AASLP, WG, ADS	Wilson et al. (1984)
Grouper	*Epinephelus malabaricus*		61–68 and 104–115 mg	WG and TBARS	Huang et al. (2005)
Grass carp	*Ctenopharyngodon idella*		200 mg	MLS	Takeuchi et al. (1992)
			62.9 mg	WG, FE	Wu et al. (2017)
Hybrid tilapia	*Oreochromis niloticus* × *O. aureus*		60–67 mg	WG, AASLP	Shiau and Shiau (2001)

TABLE 18.1 *(Continued)*

Common Name	Scientific Name	Source	Requirement	Response Criteria	References
Hybrid striped bass	*Morone chrysops* ♀ × *M. saxatilis*		28 mg	WG	Kocabas and Gatlin Iii (1999)
Juvenile grass carp	*Ctenopharyngodon idellus*		100.36 mg	WG	Li et al. (2014)
Mrigal (fry)	*Cirrhinus mrigala*		99 mg	WG	Paul et al. (2004)
Korean rock fish	*Sebastes schlegeli*		45 mg	WG	Bai and Lee (1998)
Murrel	*Channa punctatus*		140–169 mg	WG	Abdel-Hameid et al. (2012)
Pacific salmon	*Oncorhynchus* spp.		30 mg	WG, ADS	Halver (2003)
			40–50 mg	WG, MLS	Halver (2003)
Rohu	*Labeo rohita*		131.91 mg	WG	Sau et al. (2004)
Red drum	*Sciaenops ocellatus*		31 mg	WG	Peng and Gatlin Iii (2009)
Rainbow trout	*Oncorhynchus mykiss*		25 mg	WG, ADS	Hung et al. (1980)
			50 mg	AASLP	Cowey et al. (1983)
			Vitamin D		
Channel catfish	*Ictalurus punctatus*		12.5 µg	WG	Lovell and Li (1978)
			25 µg	WG	Andrews et al. (1980)
			6.25 µg	WG	Brown (1987)
Hybrid tilapia	*Oreochromis niloticus* × *O. aureus*		9.35 µg	WG	Shiau and Hwang (1993), Halver (1982)
Pacific salmon	*Oncorhynchus* spp.		NR		
Rainbow trout	*Oncorhynchus mykiss*		40–60 µg	WG, FE	Barnett et al. (1982)
			Vitamin K		
Atlantic salmon	*Salmo salar*		< 10 mg	WG	Krossøy et al. (2011)
Atlantic cod	*Gadus morhua*		0.2 mg	WG	de los Angeles Pereira et al. (2016)

TABLE 18.1 *(Continued)*

Common Name	Scientific Name	Source	Requirement	Response Criteria	References
Channel catfish	*Ictalurus punctatus*		R		Dupree (1966)
Jain carp	*Ctenopharyngodon idella*		3.13 mg	WG	Yuan et al. (2016)
Lake trout	*Salvelinus namaycush*		0.5–1 mg		Halver (1982)
Pacific salmon	*Oncorhynchus* spp.		R		Halver (1982)
Largemouth bass	*Micropterus salmoides*		9.93–15.22 mg	WG, antioxidant activity	Wei et al. (2023)

for the best possible activity of all the enzymes for which the vitamin acts as a coenzyme. Professional judgment must be used to determine which requirement value best suits the user's needs because, in addition, requirement values based on weight gain and the absence of deficiency signs are frequently much lower than requirement values determined by maximum liver storage or based on specific enzyme data. Therefore, further research is required to clarify what different species need in order to grow normally, stay healthy, and strengthen their defense systems.

18.2 CONCLUSION

Inadequate availability of quality seed of fish primarily because of unavailability of nutritionally balanced starter feeds due to fragmented information on nutrients requirements is one of the important factors limiting their production. In this chapter, available information on vitamin requirements of fish has been compiled. This chapter of the book is a sound database that would be helpful in preparing vitamin-balanced feeds for boosting the aquaculture production of various cultivable fish species.

ACKNOWLEDGMENTS

The authors are thankful to the Ministry of Environment Water and Agriculture, Saudi Arabia (Applied Research Support for Enhancing Fisheries Production, Initiative No. 368) awarded to Beacon Development Department, King Abdullah University of Science and Technology, Thuwal, Jeddah, Saudi Arabia.

KEYWORDS

- **aquaculture**
- **fish**
- **requirement**
- **vitamin**

REFERENCES

Abdelghany, A. E. (1996). Growth response of Nile tilapia *Oreochromis niloticus* to dietary L-ascorbic acid, L-ascorbyl-2-sulfate, and L-ascorbyl-2-polyphosphate. *Journal of the World Aquaculture Society, 27*(4), 449–455.

Abdel-Hameid, N. A. H., Abidi, S. F., & Khan, M. A. (2012). Dietary vitamin E requirement for maximizing the growth, conversion efficiency, biochemical composition and haematological status of fingerling *Channa punctatus*. *Aquaculture Research, 43*(2), 226–238.

Aguirre, P., & Gatlin III, D. (1999). Dietary vitamin C requirement of red drum *Sciaenops ocellatus*. *Aquaculture Nutrition, 5*(4), 247–249.

Ahmed, I. (2011). Effect of dietary niacin on growth and body composition of two Indian major carps rohu, *Labeo rohita*, and mrigal, *Cirrhinus mrigala* (Hamilton), fingerlings based on dose–response study. *Aquaculture International, 19*(3), 567–584.

Alexis, M., Karanikolas, K., & Richards, R. (1997). Pathological findings owing to the lack of ascorbic acid in cultured gilthead bream (*Sparus aurata* L.). *Aquaculture (Amsterdam, Netherlands), 151*(1–4), 209–218.

Amezaga, M., & Knox, D. (1990). Riboflavin requirements in on-growing rainbow trout, *Oncorhynchus mykiss*. *Aquaculture, 88*(1), 87–98.

Andrews, J. W., & Murai, T. (1975). Studies on the vitamin C requirements of channel catfish (*Ictalurus punctatus*). *The Journal of Nutrition, 105*(5), 557–561.

Andrews, J. W., & Murai, T. (1978). Dietary niacin requirements for channel catfish. *The Journal of Nutrition, 108*(9), 1508–1511.

Andrews, J. W., Murai, T., & Page, J. W. (1980). Effects of dietary cholecalciferol and ergocalciferol on catfish. *Aquaculture (Amsterdam, Netherlands), 19*(1), 49–54.

Aoe, H., & Masuda, I. (1967). Water soluble vitamin requirements of carp—II. Requirements for para-aminobenzoic acid and inositol. *Bulletin of the Japanese Society of Scientific Fisheries, 33*(7), 674–680.

Aoe, H., Masuda, I., Mimura, T., Saito, T., Komo, A., & Kitamura, S. (1969). Water-soluble vitamin requirements of carp—6: Requirement for thiamine and effects of antithiamines. *Bulletin of the Japanese Society of Scientific Fisheries, 35*, 459–465.

Aoe, H., Masuda, I., Saito, T., & Takada, T. (1967). Water-soluble vitamin requirements of carp. 5. Requirement for folic acid. *Bulletin of the Japanese Society of Scientific Fisheries, 33*, 1068–1071.

Arai, S. (1972). Qualitative requirements of young eels, Anguilla japonika, for water-soluble vitamins and their dificiency symptoms. *Bulletin of the Freshwater Fisheries Research Laboratory, Tokyo, 22*, 69–83.

Asaikkutti, A., Bhavan, P. S., & Vimala, K. (2016). Effects of different levels of dietary folic acid on the growth performance, muscle composition, immune response and antioxidant capacity of freshwater prawn, Macrobrachium rosenbergii. *Aquaculture, 464*, 136–144.

Aukema, H., & Holub, B. (1994). Inositol and pyrroloquinoline quinone. In *Modern nutrition in health and disease* (8th ed., pp. 466–472). Lea & Febiger.

Bai, S. C., & Lee, K.-J. (1998). Different levels of dietary DL-α-tocopheryl acetate affect the vitamin E status of juvenile Korean rockfish, Sebastes schlegeli. *Aquaculture (Amsterdam, Netherlands), 161*(1–4), 405–414.

Barnett, B. J., Cho, C. Y., & Slinger, S. J. (1982). Relative biopotency of dietary ergocalciferol and cholecalciferol and the role of and requirement for vitamin D in rainbow trout (*Salmo gairdneri*). *The Journal of Nutrition, 112*(11), 2011–2019.

Bender, D. A. (2003). *Nutritional biochemistry of the vitamins*. Cambridge University Press.

Blazer, V. S. (1992). Nutrition and disease resistance in fish. *Annual Review of Fish Diseases, 2*, 309–323.

Boonyaratpalin, M., & Wanakowat, J. (1993). Effect of thiamin, riboflavin, pantothenic acid and inositol on growth, feed efficiency and mortality of juvenile seabass. *Colloques de l'INRA (France)*.

Brown, P. B. (1987). *Vitamin D requirement of juvenile channel catfish reared in calcium-free water*. Texas A&M University.

Burtle, G., & Lovell, R. (1989). Lack of response of channel catfish (*Ictalurus punctatus*) to dietary myo-inositol. *Canadian Journal of Fisheries and Aquatic Sciences, 46*(2), 218–222.

Carson, D. A., Carrera, C. J., Wasson, D. B., & Yamanaka, H. (1988). Programmed cell death and adenine deoxynucleotide metabolism in human lymphocytes. *Advances in Enzyme Regulation, 27*, 395–404.

Chavez de Martinez, M. (1990). Vitamin C requirement of the Mexsican native chiclid Chiclosoma uropthalmus (Gunter) fry for panthotenic acid and the pethological signs of deficiency. *Journal of Aquaculture and Fish Management, 21*, 145–146.

Chávez-Sánchez, M. C., Olvera-Novoa, M. A., Osuna-Durán, B., Abdo de la Parra, I., Abad-Rosales, S. M., & Martínez-Rodríguez, I. (2014). Ascorbic acid requirement and histopathological changes due to its deficiency in juvenile spotted rose snapper Lutjanus guttatus (Steindachner, 1869). *Aquaculture International, 22*(6), 1891–1909.

Chen, L., Feng, L., Jiang, W.-D., Jiang, J., Wu, P., Zhao, J. et al. (2015a). Intestinal immune function, antioxidant status and tight junction proteins mRNA expression in young grass carp (*Ctenopharyngodon idella*) fed riboflavin deficient diet. *Fish & Shellfish Immunology, 47*(1), 470–484.

Chen, Y.-J., Yuan, R.-M., Liu, Y.-J., Yang, H.-J., Liang, G.-Y., & Tian, L.-X. (2015b). Dietary vitamin C requirement and its effects on tissue antioxidant capacity of juvenile largemouth bass, Micropterus salmoides. *Aquaculture, 435*, 431–436.

Cho, C., & Woodward, B. (1990). Dietary pantothenic acid requirements of young rainbow trout (*Oncorhynchus mykiss*). *FASEB Journal, 4*, 3747.

Choopani, M., Mohammadiazarm, H., Rajabzadeh, E., & Salati, A. P. (2022). Growth and physiological response of juvenile common carp (*Cyprinus carpio*) to increased levels of dietary niacin. *Journal of Applied Aquaculture, 34*(1), 234–246.

Christodoulou, V., Bampidis, V., Lymberopoulos, A., Robinson, P., Ploumi, K., & Belibasaki, S. (2006). Effect of supplemental dietary biotin on performance of lactating ewes. *Animal Feed Science and Technology, 130*(3–4), 268–276.

Combs Jr, G. F., & McClung, J. P. (2016). *The vitamins: Fundamental aspects in nutrition and health*. Academic Press.

Corti, A., Casini, A. F., & Pompella, A. (2010). Cellular pathways for transport and efflux of ascorbate and dehydroascorbate. *Archives of Biochemistry and Biophysics, 500*(2), 107–115.

Council, N. R. (1993). *Nutrient requirements of fish*. National Academies Press.

Council, N. R. (2011). *Nutrient requirements of fish and shrimp*. National Academies Press.

Cowey, C., Adron, J., Knox, D., & Ball, G. (1975). Studies on the nutrition of marine flatfish. The thiamin requirement of turbot (Scophthalmus maximus). *British Journal of Nutrition, 34*(3), 383–390.

Cowey, C., Adron, J., & Youngson, A. (1983). The vitamin E requirement of rainbow trout (*Salmo gairdneri*) given diets containing polyunsaturated fatty acids derived from fish oil. *Aquaculture (Amsterdam, Netherlands), 30*(1–4), 85–93.

Cowey, C. B., & Woodward, B. (1993). The dietary requirement of young rainbow trout (*Oncorhynchus mykiss*) for folic acid. *The Journal of Nutrition, 123*(9), 1594–1600.

Craig, S. R., & Gatlin III, D. M. (1996). Dietary choline requirement of juvenile red drum (*Sciaenops ocellatus*). *The Journal of Nutrition, 126*(6), 1696–1700.

Dabrowski, K., & Guderley, H. (2003). Intermediary metabolism. In *Fish nutrition* (pp. 309–365). Elsevier.

Dakshinamurti, K. (2005). Biotin—A regulator of gene expression. *The Journal of Nutritional Biochemistry, 16*(7), 419–423.

Daniel, N., Muralidhar, A., Srivastava, P., Jain, K., Pani Prasad, K., Manish, J., & Sivaramakrishnan, T. (2018). Dietary ascorbic acid requirement for growth of striped catfish, *Pangasianodon hypophthalmus* (Sauvage, 1878) juveniles. *Aquaculture Nutrition, 24*(1), 616–624.

Darias, M., Mazurais, D., Koumoundouros, G., Cahu, C., & Zambonino-Infante, J. (2011). Overview of vitamin D and C requirements in fish and their influence on the skeletal system. *Aquaculture, 315*(1–2), 49–60.

Dawood, M. A., & Koshio, S. (2018). Vitamin C supplementation to optimize growth, health and stress resistance in aquatic animals. *Reviews in Aquaculture, 10*(2), 334–350.

de los Angeles Pereira, N., Sarasa, M. V., & Fernández-Gimenez, A. (2016). A comparative study of liposoluble vitamin effects in decapod crustacean with emphasis in argentine shrimp species. *Characteristics, Evolution and Conservation Strategies, 81*.

Deng, D.-F., Hemre, G.-I., & Wilson, R. P. (2002). Juvenile sunshine bass (Morone chrysops♀× Morone saxatilis♂) do not require dietary myo-inositol. *Aquaculture (Amsterdam, Netherlands), 213*(1–4), 387–393.

Deng, D.-F., & Wilson, R. P. (2003). Dietary riboflavin requirement of juvenile sunshine bass (Morone chrysops♀× Morone saxatilis♂). *Aquaculture, 218*(1–4), 695–701.

Duester, G. (2008). Retinoic acid synthesis and signaling during early organogenesis. *Cell, 134*(6), 921–931.

Duncan, P. L., & Lovell, R. (1994). Influence of vitamin C on the folate requirement of channel catfish, *Ictalurus punctatus*, for growth, hematopoiesis, and resistance to Edwardsiella ictaluri infection. *Aquaculture (Amsterdam, Netherlands), 127*(2–3), 233–244.

Dupree, H. K. (1966). *Vitamins essential for growth of channel catfish* (Vol. 7). US Department of the Interior, Fish and Wildlife Service, Bureau of Sport.

Ebi, I., Yong, A. S.-K., Lim, L.-S., & Shapawi, R. (2018). Dietary ascorbic acid requirement for the optimum growth performances and normal skeletal development in juvenile hybrid grouper, *Epinephelus fuscoguttatus× Epinephelus lanceolatus*. *Journal of King Saud University-Science, 30*(4), 493–499.

El Naggar, G. O., & Lovell, R. T. (1991). L-ascorbyl-2-monophosphate has equal antiscorbutic activity as L-ascorbic acid but L-ascorbyl-2-sulfate is inferior to L-ascorbic acid for channel catfish. *The Journal of Nutrition, 121*(10), 1622–1626.

El-Gawad, A., Eman, A., & Abdel Hamid, O. M. (2014). Effect of vitamin C dietary supplementation in reducing the alterations induced by fenitrothion in *Oreochromis niloticus*. *Fish Physiology and Biochemistry, 40*(3), 787–796.

Eo, J., & Lee, K.-J. (2008). Effect of dietary ascorbic acid on growth and non-specific immune responses of tiger puffer, Takifugu rubripes. *Fish & Shellfish Immunology, 25*(5), 611–616.

Eya, J. C. (1996). 'Broken-skull disease' in African catfish *Clarias gariepinus* is related to a dietary deficiency of ascorbic acid. *Journal of the World Aquaculture Society, 27*(4), 493–498.

Fatma Abidi, S., & A Khan, M. (2008). Dietary threonine requirement of fingerling Indian major carp, *Labeo rohita* (Hamilton). *Aquaculture Research, 39*(14), 1498–1505.

Feng, L., Li, S.-Q., Jiang, W.-D., Liu, Y., Jiang, J., Wu, P. et al. (2016). Deficiency of dietary niacin impaired intestinal mucosal immune function via regulating intestinal NF-κB, Nrf2 and MLCK signaling pathways in young grass carp (*Ctenopharyngodon idella*). *Fish & Shellfish Immunology, 49*, 177–193.

Gao, J., Koshio, S., Ishikawa, M., Yokoyama, S., Nguyen, B., & Mamauag, R. (2013). Effect of dietary oxidized fish oil and vitamin C supplementation on growth performance and reduction of oxidative stress in Red Sea Bream Pagrus major. *Aquaculture Nutrition, 19*(1), 35–44.

Gaylord, G. (1998). Re-evaluation of vitamin E supplementation of practical diets for channel catfish, *Ictalurus punctatus*, production. *Aquaculture Nutrition, 4*(2), 109–114.

Gerster, H. (1996). The importance of vitamin B 6 for development of the infant. Human medical and animal experiment studies. *Zeitschrift fur Ernahrungswissenschaft, 35*(4), 309–317.

Gong, W., Lei, W., Zhu, X., Yang, Y., Han, D., & Xie, S. (2014). Dietary myo-inositol requirement for juvenile gibel carp (*Carassius auratus gibelio*). *Aquaculture Nutrition, 20*(5), 514–519.

Gouillou-Coustans, M.-F., Bergot, P., & Kaushik, S. (1998). Dietary ascorbic acid needs of common carp (*Cyprinus carpio*) larvae. *Aquaculture (Amsterdam, Netherlands), 161*(1–4), 453–461.

Griffin, M. E., Wilson, K. A., White, M. R., & Brown, P. B. (1994). Dietary choline requirement of juvenile hybrid striped bass. *The Journal of Nutrition, 124*(9), 1685–1689.

Guimarães, I., Lim, C., Yildirim-Aksoy, M., Li, M., & Klesius, P. (2014). Effects of dietary levels of vitamin A on growth, hematology, immune response and resistance of Nile tilapia (*Oreochromis niloticus*) to *Streptococcus iniae*. *Animal Feed Science and Technology, 188*, 126–136.

Halver, J. E. (1957). Nutrition of salmonoid fishes: III. Water-soluble vitamin requirements of chinook salmon. *The Journal of Nutrition, 62*(2), 225–243.

Halver, J. E. (1982). The vitamins required for cultivated salmonids. *Comparative Biochemistry and Physiology Part B: Comparative Biochemistry, 73*(1), 43–50.

Halver, J. E. (2003). The vitamins. In *Fish nutrition* (pp. 61–141). Elsevier.

Halver, J. E., Ashley, L. M., & Smith, R. R. (1969). Ascorbic acid requirements of coho salmon and rainbow trout. *Transactions of the American Fisheries Society, 98*(4), 762–771.

Hamre, K. (2011). Metabolism, interactions, requirements and functions of vitamin E in fish. *Aquaculture Nutrition, 17*(1), 98–115.

Hang, Y., Hua, X., Li, X., Yi, W., & Cong, X. (2022). Effect of dietary levels of folic acid on growth performance, blood biochemistry, antioxidant capacity, and immunity of juvenile largemouth bass, Micropterus salmoides. *Journal of the World Aquaculture Society, 53*(4), 836–847.

Hansen, C. S., Jensen, J. S., Ridderstråle, M., Vistisen, D., Jørgensen, M. E., & Fleischer, J. (2017). Vitamin B12 deficiency is associated with cardiovascular autonomic neuropathy in patients with type 2 diabetes. *Journal of Diabetes and Its Complications, 31*(1), 202–208.

Hardy, R. W., & Kaushik, S. J. (2021). *Fish nutrition*. Academic Press.

Harrison, F. E., Green, R. J., Dawes, S. M., & May, J. M. (2010). Vitamin C distribution and retention in the mouse brain. *Brain Research, 1348*, 181–186.

He, P., Feng, L., Zhang, L., Mi, H.-F., Wu, P., Liu, Y. et al. (2022). The negative impacts of dietary biotin deficiency on antioxidant status, apoptosis and tight junction protein transcription of immune organs in grass carp (*Ctenopharyngodon idella*). *Aquaculture (Amsterdam, Netherlands), 547*, 737439.

He, W., Zhou, X. Q., Feng, L., Jiang, J., & Liu, Y. (2009). Dietary pyridoxine requirement of juvenile Jian carp (*Cyprinus carpio* var. Jian). *Aquaculture Nutrition, 15*(4), 402–408.

Hemre, G.-I., Deng, D.-F., Wilson, R. P., & Berntssen, M. H. (2004). Vitamin A metabolism and early biological responses in juvenile sunshine bass (*Morone chrysops*× *M. saxatilis*) fed graded levels of vitamin A. *Aquaculture (Amsterdam, Netherlands), 235*(1–4), 645–658.

Hernandez, L., Teshima, S. I., Ishikawa, M., Alam, S., Koshio, S., & Tanaka, Y. (2005). Dietary vitamin A requirements of juvenile Japanese flounder *Paralichthys olivaceus*. *Aquaculture Nutrition, 11*(1), 3–9.

Hien, D., & Doolgindachbaporn, S. (2011). Effect of niacin and folic acid in feed rations on growth and live weights of Green catfish (Mystus nemurus Valenciennes 1840). *Pakistan Journal of Biological Sciences, 14*(1), 64–68.

Hosokawa, H., Shimeno, S., & Takeda, M. (2001). Dietary choline requirement of fingerling yellowtail. *Aquaculture Science, 49*(2), 231–235.

Hu, C.-J., Chen, S.-M., Pan, C.-H., & Huang, C.-H. (2006). Effects of dietary vitamin A or β-carotene concentrations on growth of juvenile hybrid tilapia, *Oreochromis niloticus*× *O. aureus*. *Aquaculture (Amsterdam, Netherlands), 253*(1–4), 602–607.

Huang, C.-H., Lin, W.-Y., & Chu, J.-H. (2005). Dietary lipid level influences fatty acid profiles, tissue composition, and lipid peroxidation of soft-shelled turtle, *Pelodiscus sinensis*. *Comparative Biochemistry and Physiology Part A: Molecular & Integrative Physiology, 142*(3), 383–388.

Huang, F., Jiang, M., Wen, H., Wu, F., Liu, W., Tian, J., & Shao, H. (2016). Dietary vitamin C requirement of genetically improved farmed Tilapia, *Oreochromis Niloticus*. *Aquaculture Research, 47*(3), 689–697.

Huang, F., Wu, F., Zhang, S., Jiang, M., Liu, W., Tian, J. et al. (2017). Dietary vitamin C requirement of juvenile Chinese sucker (*Myxocyprinus asiaticus*). *Aquaculture Research, 48*(1), 37–46.

Huang, H. H., Feng, L., Liu, Y., Jiang, J., Jiang, W. D., Hu, K. et al. (2011). Effects of dietary thiamin supplement on growth, body composition and intestinal enzyme activities of juvenile Jian carp (*Cyprinus carpio* var. Jian). *Aquaculture Nutrition, 17*(2), e233–e240.

Huang, J., Tian, L., Du, Z., Yang, H., & Liu, Y. (2005). Pyridoxine deficiency of grouper, *Epinephelus coioides*: Physiological and biochemical alteration. *Fish Physiology and Biochemistry, 31*(4), 331–337.

Huang, J., Tian, L., Wu, X., Yang, H., & Liu, Y. (2010). Effects of dietary riboflavin levels on antioxidant defense of the juvenile grouper *Epinephelus coioides. Fish Physiology and Biochemistry, 36*(1), 55–62.

Hughes, S. G., Rumsey, G. L., & Nickum, J. G. (1981). Riboflavin requirement of fingerling rainbow trout. *The Progressive Fish-Culturist, 43*(4), 167–172.

Hung, S. S. (1989). Choline requirement of hatchery-produced juvenile white sturgeon (*Acipenser transmontanus*). *Aquaculture, 78*(2), 183–194.

Hung, S. S. (2017). Sturgeon, ACIPENSER spp. In *Handbook of nutrient requirements of finfish* (pp. 153–160). CRC Press.

Hung, S., Cho, C., & Slinger, S. (1980). Measurement of oxidation in fish oil and its effect on vitamin E nutrition of rainbow trout (*Salmo gairdneri*). *Canadian Journal of Fisheries and Aquatic Sciences, 37*(8), 1248–1253.

Ibiyo, L., Atteh, J., Omotosho, J., & Madu, C. (2007). Vitamin C (ascorbic acid) requirements of *Heterobranchus longifili*s fingerlings. *African Journal of Biotechnology, 6*(13).

Jauncey, K. (1982). The effects of varying dietary protein level on the growth, food conversion, protein utilization and body composition of juvenile tilapias (*Sarotherodon mossambicus*). *Aquaculture, 27*(1), 43–54.

Jhingran, V., & Pullin, R. (1988). A hatchery manual for common, Chinese and Indian major carps under field condition. *ICLARM Studies and Reviews, 11*.

Jia, X., Qian, P., Wu, C., Xie, Y., Yang, W., Song, R. et al. (2022). Effects of dietary pantothenic acid on growth, antioxidant ability and innate immune response in juvenile black carp. *Aquaculture Reports, 24*, 101131.

Jiang, M., Huang, F., Wen, H., Yang, C., Wu, F., Liu, W., & Tian, J. (2014a). Dietary niacin requirement of GIFT tilapia, *Oreochromis niloticus*, reared in freshwater. *Journal of the World Aquaculture Society, 45*(3), 333–341.

Jiang, M., Huang, F., Zhao, Z., Wen, H., Wu, F., Liu, W. et al. (2014b). Dietary thiamin requirement of juvenile grass carp, *Ctenopharyngodon idella. Journal of the World Aquaculture Society, 45*(4), 461–468.

Jiang, W. D., Feng, L., Liu, Y., Jiang, J., & Zhou, X. Q. (2009). Myo-inositol prevents oxidative damage, inhibits oxygen radical generation and increases antioxidant enzyme activities of juvenile Jian carp (*Cyprinus carpio* var. Jian). *Aquaculture Research, 40*(15), 1770–1776.

Jiang, W. D., Kuang, S. Y., Liu, Y., Jiang, J., Hu, K., Li, S. H. et al. (2013). Effects of myo-inositol on proliferation, differentiation, oxidative status and antioxidant capacity of carp enterocytes in primary culture. *Aquaculture Nutrition, 19*(1), 45–53.

Jiménez-Fernández, E., Ponce, M., Rodriguez-Rúa, A., Zuasti, E., Manchado, M., & Fernández-Díaz, C. (2015). Effect of dietary vitamin C level during early larval stages in Senegalese sole (Solea senegalensis). *Aquaculture, 443*, 65–76.

Jobling, M. (2012). National Research Council (NRC): Nutrient requirements of fish and shrimp. In: Springer.

Jurss, K., & Jonas, L. (1981). Electron microscopicand biochemical investigations on the pyridoxine deficiency of rainbow trout (*Salmo gairdneri* Richardson). *Zoologische Jahrbücher (Allg. Zool. Physiol. Terre), 85*, 185–196.

Karges, R., & Woodward, B. (1984). Development of lamellar epithelial hyperplasia in gills of pantothenic acid-deficient rainbow trout, *Salmo gairdneri* Richardson. *Journal of Fish Biology, 25*(1), 57–62.

Kashiwada, K.-I., Teshima, S.-I., & Kanazawa, A. (1970). Studies on the production of B vitamins by Intestinal. *Bulletin of the Japanese Society of Scientific Fisheries, 36*(4), 421.

Ketola, H. (1976). Choline metabolism and nutritional requirement of lake trout (Salvelinus namaycush). *Journal of Animal Science, 43*(2), 474–477.

Khan, Y. M., & Khan, M. A. (2019). Dietary biotin requirement of fingerling Catla catla (Hamilton) based on growth, feed conversion efficiency, and liver biotin concentration. *Journal of the World Aquaculture Society, 50*(3), 674–683.

Khan, Y. M., & Khan, M. A. (2020). Dietary folic acid requirement of fingerling Catla, Catla catla (Hamilton). *Aquaculture Nutrition, 26*(4), 1035–1045.

Khan, Y. M., & Khan, M. A. (2021a). Dietary niacin requirement of fingerling Indian major carp Catla catla Hamilton. *Aquaculture Nutrition, 27*(5), 1482–1493.

Khan, Y. M., & Khan, M. A. (2021b). Optimization of dietary pyridoxine improved growth performance, hematological indices, antioxidant capacity, intestinal enzyme activity, non-specific immune response, and liver pyridoxine concentration of fingerling major carp Catla catla (Hamilton). *Aquaculture (Amsterdam, Netherlands), 541*, 736815.

Khan, Y. M., & Khan, M. A. (2022). Effects of dietary pantothenic acid on growth performance, intestinal enzyme activity, non-specific immune response, antioxidant capacity, hematological parameters, carcass composition and liver pantothenic acid concentration of fingerling Catla, Catla catla (Hamilton). *Animal Feed Science and Technology, 285*, 115245.

Khosravi, S., Jang, J.-W., Rahimnejad, S., Song, J.-W., & Lee, K.-J. (2015a). Choline essentiality and its requirement in diets for juvenile parrot fish (*Oplegnathus fasciatus*). *Asian-Australasian Journal of Animal Sciences, 28*(5), 647–653.

Khosravi, S., Lim, S.-J., Rahimnejad, S., Kim, S.-S., Lee, B.-J., Kim, K.-W. et al. (2015b). Dietary myo-inositol requirement of parrot fish, *Oplegnathus fasciatus*. *Aquaculture (Amsterdam, Netherlands), 436*, 1–7.

Kissil, G. W., Cowey, C., Adron, J., & Richards, R. (1981). Pyridoxine requirements of the gilthead bream, Sparus aurata. *Aquaculture, 23*(1–4), 243–255.

Kocabas, A., & Gatlin Iii, D. (1999). Dietary vitamin E requirement of hybrid striped bass (Morone chrysops female × *M. saxatilis* male).

Krossøy, C., Waagbø, R., & Ørnsrud, R. (2011). Vitamin K in fish nutrition. *Aquaculture Nutrition, 17*(6), 585–594.

Lall, S., & Weerakoon, D. (1990). Vitamin B6 requirement of Atlantic salmon (Salmo salar). *FASEB Journal, 4*, 3749.

Landenberger, A., Kabil, H., Harshman, L. G., & Zempleni, J. (2004). Biotin deficiency decreases life span and fertility but increases stress resistance in *Drosophila melanogaster*. *The Journal of Nutritional Biochemistry, 15*(10), 591–600.

Lee, H. C., & Aarhus, R. (1991). ADP-ribosyl cyclase: An enzyme that cyclizes NAD+ into a calcium-mobilizing metabolite. *Cell Regulation, 2*(3), 203–209.

Li, J., Liang, X.-F., Tan, Q., Yuan, X., Liu, L., Zhou, Y., & Li, B. (2014). Effects of vitamin E on growth performance and antioxidant status in juvenile grass carp *Ctenopharyngodon idellus*. *Aquaculture (Amsterdam, Netherlands), 430*, 21–27.

Li, J., Zhang, L., Mai, K., Ai, Q., Wan, J., Zhang, C., & Zhang, J. (2010a). Estimation of dietary biotin requirement of Japanese seabass, *Lateolabrax japonicus* C. *Aquaculture Nutrition, 16*(3), 231–236.

Li, W., Zhou, X. Q., Feng, L., Liu, Y., & Jiang, J. (2010b). Effect of dietary riboflavin on growth, feed utilization, body composition and intestinal enzyme activities of juvenile Jian carp (*Cyprinus carpio* var. Jian). *Aquaculture Nutrition, 16*(2), 137–143.

Li, P., Hou, D., Zhao, H., Huang, W., Peng, K., & Cao, J. (2022). Dietary pyridoxine effect on growth performance, physiological metabolic parameters, intestinal enzymatic activities and antioxidant status of juvenile yellow catfish (*Pelteobagrus fulvidraco*). *Aquaculture Reports, 24*, 101153.

Li, S.-Q., Feng, L., Jiang, W.-D., Liu, Y., Jiang, J., Wu, P. et al. (2016a). Deficiency of dietary niacin impaired gill immunity and antioxidant capacity, and changes its tight junction proteins via regulating NF-κB, TOR, Nrf2 and MLCK signaling pathways in young grass carp (*Ctenopharyngodon idella*). *Fish & Shellfish Immunology, 55*, 212–222.

Li, X.-F., Wang, F., Qian, Y., Jiang, G.-Z., Zhang, D.-D., & Liu, W.-B. (2016b). Dietary vitamin B12 requirement of fingerling blunt snout bream Megalobrama amblycephala determined by growth performance, digestive and absorptive capability and status of the GH-IGF-I axis. *Aquaculture (Amsterdam, Netherlands), 464*, 647–653.

Li, X. F., Wang, T. J., Qian, Y., Jiang, G. Z., Zhang, D. D., & Liu, W. B. (2017). Dietary niacin requirement of juvenile blunt snout bream Megalobrama amblycephala based on a dose–response study. *Aquaculture Nutrition, 23*(6), 1410–1417.

Lim, C., & Lovell, R. T. (1978). Pathology of the vitamin C deficiency syndrome in channel catfish (*Ictalurus punctatus*). *The Journal of Nutrition, 108*(7), 1137–1146.

Lim, C., LeaMaster, B. R., & Brock, J. A. (1995). Pyridoxine requirement of fingerling red hybrid tilapia grown in seawater. *Journal of Applied Aquaculture, 5*(2), 49–60.

Lim, C., Yildirim-Aksoy, M., Barros, M. M., & Klesius, P. (2011). Thiamin requirement of Nile tilapia, *Oreochromis niloticus*. *Journal of the World Aquaculture Society, 42*(6), 824–833.

Limsuwan, T., & Lovell, R. T. (1981). Intestinal synthesis and absorption of vitamin B-12 in channel catfish. *The Journal of Nutrition, 111*(12), 2125–2132.

Lin, Y.-H., Lin, H.-Y., & Shiau, S.-Y. (2012). Estimation of dietary pantothenic acid requirement of grouper, *Epinephelus malabaricus* according to physiological and biochemical parameters. *Aquaculture, 324*, 92–96.

Lin, Y.-H., Wu, J.-Y., & Shiau, S.-Y. (2010). Dietary cobalt can promote gastrointestinal bacterial production of vitamin B12 in sufficient amounts to supply growth requirements of grouper, *Epinephelus malabaricus*. *Aquaculture, 302*(1–2), 89–93.

Liu, Y., Chi, L., Feng, L., Jiang, J., Jiang, W. D., Hu, K. et al. (2011). Effects of graded levels of dietary vitamin C on the growth, digestive capacity and intestinal microflora of juvenile Jian carp (*Cyprinus carpio* var. Jian). *Aquaculture Research, 42*(4), 534–548.

Lock, E. J., Waagbø, R., Wendelaar Bonga, S., & Flik, G. (2010). The significance of vitamin D for fish: A review. *Aquaculture Nutrition, 16*(1), 100–116.

Lovell, R., & Li, Y.-P. (1978). Essentiality of vitamin D in diets of channel catfish (*Ictalurus punctatus*). *Transactions of the American Fisheries Society, 107*(6), 809–811.

Lovell, R. T., & Limsuwan, T. (1982). Intestinal synthesis and dietary nonessentiality of vitamin B12 for Tilapia nilotica. *Transactions of the American Fisheries Society, 111*(4), 485–490.

Luo, C., Lin, Y., Miao, L., Liu, B., & Ge, X. (2022). The interactive effect of dietary inclusion levels and preparation methods of myo-inositol on growth performance, antioxidant function and lipid metabolism homeostasis of *Cyprinus carpio* var. jian. *Aquaculture Reports, 25*, 101249.

Mai, K., Xiao, L., Ai, Q., Wang, X., Xu, W., Zhang, W. et al. (2009). Dietary choline requirement for juvenile cobia, Rachycentron canadum. *Aquaculture, 289*(1–2), 124–128.

Mclaren, B. A., Keller, E., O'donnell, D. J., & Elvehjem, C. (1947). The nutrition of rainbow trout. 1. Studies of vitamin requirements. *Archives of Biochemistry, 15*, 169–178.

McMahon, R. J. (2002). Biotin in metabolism and molecular biology. *Annual Review of Nutrition, 22*, 221.

Merchie, G., Lavens, P., Storch, V., Übel, U., Nelis, H., De Leenheer, A., & Sorgeloos, P. (1996). Influence of dietary vitamin C dosage on turbot (Scophthalmus maximus) and European sea bass (Dicentrarchus labrax) nursery stages. *Comparative Biochemistry and Physiology Part A: Physiology, 114*(2), 123–133.

Miao, S., Zhang, W., Xu, W., & Mai, K. (2013). Dietary folic acid requirement of juvenile abalone Haliotis discus hannai Ino. *Aquaculture (Amsterdam, Netherlands), 400*, 73–76.

Misra, C., Das, B., Mukherjee, S., & Pradhan, J. (2007). Effects of dietary vitamin C on immunity, growth and survival of Indian major carp *Labeo rohita*, fingerlings. *Aquaculture Nutrition, 13*(1), 35–44.

Mohamed, J. S. (2001). Dietary pyridoxine requirement of the Indian catfish, *Heteropneustes fossilis*. *Aquaculture, 194*(3–4), 327–335.

Mohamed, J. S., & Ibrahim, A. (2001). Quantifying the dietary niacin requirement of the Indian catfish, *Heteropneustes fossilis* (Bloch), fingerlings. *Aquaculture Research, 32*(3), 157–162.

Mohamed, J. S., Ravisankar, B., & Ibrahim, A. (2000). Quantifying the dietary biotin requirement of the catfish, Clarias batrachus. *Aquaculture International, 8*(1), 9–18.

Mohamed, J. S., Sivaram, V., Roy, T. C., Marian, M. P., Murugadass, S., & Hussain, M. R. (2003). Dietary vitamin A requirement of juvenile greasy grouper (Epinephelus tauvina). *Aquaculture (Amsterdam, Netherlands), 219*(1–4), 693–701.

Mohd Khan, Y., & Khan, M. A. (2021). Effects of dietary cyanocobalamin on growth performance, non-specific immune response, antioxidant capacity, haematological parameters, body composition and liver cyanocobalamin concentration of fingerling major carp, Catla catla (Hamilton). *Aquaculture Nutrition, 27*(2), 604–614.

Mohd Khan, Y., & Khan, M. A. (2022). Dietary thiamin requirement of fingerling major carp Catla catla (Hamilton). *Journal of Animal Physiology and Animal Nutrition, 106*(4), 939–946.

Moren, M., Opstad, I., Berntssen, M., Infante, J.-L. Z., & Hamre, K. (2004). An optimum level of vitamin A supplements for Atlantic halibut (*Hippoglossus hippoglossus* L.) juveniles. *Aquaculture (Amsterdam, Netherlands), 235*(1–4), 587–599.

Morito, C. L., Conrad, D. H., & Hilton, J. W. (1986). The thiamin deficiency signs and requirement of rainbow trout (*Salmo gairdneri*, Richardson). *Fish Physiology and Biochemistry, 1*(2), 93–104.

Morris, P., Baker, R., & Davies, S. (1998). Nicotinic acid supplementation of diets for the African catfish, *Clarias gariepinus* (Burchell). *Aquaculture Research, 29*(11), 791–799.

Morris, P., Davies, S., & Lowe, D. (1995). Qualitative requirement for B vitamins in diets for the gilthead seabream (*Sparus aurata* L.). *Animal Science, 61*(2), 419–426.

Murai, T., & Andrews, J. W. (1974). Interactions of dietary α-tocopherol, oxidized menhaden oil and ethoxyquin on channel catfish (*Ictalurus punctatus*). *The Journal of Nutrition, 104*(11), 1416–1431.

Murai, T., & Andrews, J. W. (1979). Pantothenic acid requirements of channel catfish fingerlings. *The Journal of Nutrition, 109*(7), 1140–1142.

Natori, Y., & Oka, T. (1997). Vitamin B6 modulation of gene expression. *Nutrition Research, 17*(7), 1199–1207.

Ng, W.-K., Serrini, G., Zhang, Z., & Wilson, R. P. (1997). Niacin requirement and inability of tryptophan to act as a precursor of NAD+ in channel catfish, *Ictalurus punctatus*. *Aquaculture, 152*(1–4), 273–285.

Nsonga, A., Kang'Ombe, J., Mfitilodze, W., Soko, C., & Mtethiwa, A. (2009). Effect of varying levels of dietary vitamin C (ascorbic acid) on growth, survival and hematology of juvenile tilapia, Oreochromis karongae (Trewavas 1941) reared in aquaria. *Brazilian Journal of Aquatic Science and Technology, 13*(2), 17–23.

Ogino, C., Uki, N., Watanabe, T., Iida, Z., & Ando, K. (1970a). B vitamin requirements of carp. 4. Requirement for choline. *Bulletin of the Japanese Society of Scientific Fisheries, 36*(11), 1140–1146.

Ogino, C., Watanabe, T., Kakino, J., Iwanaga, N., & Mizuno, M. (1970b). B vitamin requirements of carp: III. Requirement of biotin. *Bulletin of the Japanese Society of Scientific Fisheries, 36*(7).

Oser, B. (1979). *Hawk's physiological chemistry* (pp. 1051–1053). Tata Mc Gray-Hall Pub. Co. Ltd.

Pan, S., Yan, X., Dong, X., Li, T., Suo, X., Liu, H. et al. (2022). Influence of Dietary Inositol Supplementation on Growth, Liver Histology, Lipid Metabolism, and Related Genes Expression on Juvenile Hybrid Grouper (♀ *Epinephelus fuscoguttatus*×♂ E. lanceolatu) Fed High-Lipid Diets. *Aquaculture Nutrition, 2022*.

Paul, B., Sarkar, S., & Mohanty, S. (2004). Dietary vitamin E requirement of mrigal, *Cirrhinus mrigala* fry. *Aquaculture (Amsterdam, Netherlands), 242*(1–4), 529–536.

Peng, L., & Gatlin Iii, D. (2009). Dietary vitamin E requirement of the red drum *Sciaenops ocellatus*. *Aquaculture Nutrition, 15*(3), 313–319.

Peres, H., Lim, C., & Klesius, P. H. (2004). Growth, chemical composition and resistance to Streptococcus iniae challenge of juvenile Nile tilapia (*Oreochromis niloticus*) fed graded levels of dietary inositol. *Aquaculture (Amsterdam, Netherlands), 235*(1–4), 423–432.

Phillips Jr, A. M., & Brockway, D. R. (1949). The niacin and biotin requirement of trout. *Transactions of the American Fisheries Society, 77*(1), 152–159.

Phillips Jr, A. M., Hammer, G. L., Edwards, J. P., & Hosking, H. F. (1964). Dry concentrates as complete trout foods for growth and egg production. *The Progressive Fish-Culturist, 26*(4), 155–159.

Poston, H. (1976). Optimum level of dietary biotin for growth, feed utilization, and swimming stamina of fingerling lake trout (Salvelinus namaycush). *Journal of the Fisheries Board of Canada, 33*(8), 1803–1806.

Poston, H. A., & Combs Jr, G. F. (1980). Nutritional implications of tryptophan catabolizing enzymes in several species of trout and salmon. *Proceedings of the Society for Experimental Biology and Medicine, 163*(4), 452–454.

Poston, H., & Wolfe, M. J. (1985). Niacin requirement for optimum growth, feed conversion and protection of rainbow trout, *Salmo gairdneri* Richardson, from ultraviolet-B irradiation. *Journal of Fish Diseases, 8*(5), 451–460.

Qian, Y., Li, X. F., Sun, C. X., Jiang, G. Z., Xu, W. N., Wang, Y., & Liu, W. B. (2014). Dietary biotin requirement of juvenile blunt snout bream, M egalobrama amblycephala. *Aquaculture Nutrition, 20*(6), 616–622.

Qian, Y., Li, X.-F., Zhang, D.-D., Cai, D.-S., Tian, H.-Y., & Liu, W.-B. (2015). Effects of dietary pantothenic acid on growth, intestinal function, anti-oxidative status and fatty acids synthesis of juvenile blunt snout bream Megalobrama amblycephala. *Plos One, 10*(3), e0119518.

Qian, Y., Tang, S., Zhang, M.-J., Liu, J., Xu, W.-N., & Zhang, J. (2022). Effects of pantothenic acid on growth performance, oxidative stress, and lipid content in high-fat-fed megalobrama amblycephala. *Aquaculture Nutrition, 2022.*

Qin, D., Dong, X., Tan, B., Yang, Q., Chi, S., Liu, H., & Zhang, S. (2017). Effects of dietary choline on growth performance, lipid deposition and hepatic lipid transport of grouper (E pinephelus coioides). *Aquaculture Nutrition, 23*(3), 453–459.

Raggi, T., Buentello, A., & Gatlin III, D. M. (2016). Characterization of pantothenic acid deficiency and the dietary requirement of juvenile hybrid striped bass, Morone chrysops× M. saxatilis. *Aquaculture (Amsterdam, Netherlands), 451*, 326–329.

Robinson, E., & Lovell, R. (1978). Essentiality of biotin for channel catfish (*Ictalurus punctatus*) fed lipid and lipid-free diets. *The Journal of Nutrition, 108*(10), 1600–1605.

Roem, A. J., Kohler, C. C., & Stickney, R. R. (1990). Vitamin E requirements of the blue tilapia, Oreochromis aureus (Steindachner), in relation to dietary lipid level. *Aquaculture (Amsterdam, Netherlands), 87*(2), 155–164.

Roosta, Z., Hajimoradloo, A., Ghorbani, R., & Hoseinifar, S. H. (2014). The effects of dietary vitamin C on mucosal immune responses and growth performance in Caspian roach (Rutilus rutilus caspicus) fry. *Fish physiology and biochemistry, 40*(5), 1601–1607.

Rumsey, G. L. (1991). Choline-betaine requirements of rainbow trout (*Oncorhynchus mykiss*). *Aquaculture (Amsterdam, Netherlands), 95*(1–2), 107–116.

Sandnes, K., Torrissen, O., & Waagbø, R. (1992). The minimum dietary requirement of vitamin C in Atlantic salmon (Salmo salar) fry using Ca ascorbate-2-monophosphate as dietary source. *Fish Physiology and Biochemistry, 10*(4), 315–319.

Sau, S., Paul, B., Mohanta, K., & Mohanty, S. (2004). Dietary vitamin E requirement, fish performance and carcass composition of rohu (*Labeo rohita*) fry. *Aquaculture (Amsterdam, Netherlands), 240*(1–4), 359–368.

Sealey, W. M., & Gatlin III, D. M. (1999). Dietary vitamin C requirement of hybrid striped bass Morone chrysops♀× M. saxatilis♂. *Journal of the World Aquaculture Society, 30*(3), 297–301.

Semba, R. D. (1998). The role of vitamin A and related retinoids in immune function. *Nutrition Reviews, 56*(1), S38-S48.

Semba, R. D., Lyles, C. M., Margolick, J. B., Caiaffa, W. T., Farzadegan, H., Cohn, S., & Vlahov, D. (1998). Vitamin A supplementation and human immunodeficiency virus load in injection drug users. *The Journal of Infectious Diseases, 177*(3), 611–616.

Sesay, D. F., Habte-Tsion, H.-M., Zhou, Q., Ren, M., Xie, J., Liu, B. et al. (2016). Effects of dietary folic acid on the growth, digestive enzyme activity, immune response and antioxidant enzyme activity of blunt snout bream (Megalobrama amblycephala) fingerling. *Aquaculture (Amsterdam, Netherlands), 452*, 142–150.

Sesay, D. F., Habte-Tsion, H.-M., Zhou, Q., Ren, M., Xie, J., Liu, B. et al. (2017). The effect of dietary folic acid on biochemical parameters and gene expression of three heat shock proteins (HSPs) of blunt snout bream (Megalobrama amblycephala) fingerling under acute high temperature stress. *Fish Physiology and Biochemistry, 43*(4), 923–940.

Shahkar, E., Yun, H., Kim, D.-J., Kim, S.-K., Lee, B. I., & Bai, S. C. (2015). Effects of dietary vitamin C levels on tissue ascorbic acid concentration, hematology, non-specific immune response and gonad histology in broodstock Japanese eel, Anguilla japonica. *Aquaculture, 438*, 115–121.

Shaik Mohamed, J. (2001). Dietary biotin requirement determined for Indian catfish, *Heteropneustes fossilis* (Bloch), fingerlings. *Aquaculture Research, 32*(9), 709–716.

Shao, L., Zhu, X., Yang, Y., Jin, J., Liu, H., Han, D., & Xie, S. (2016). Effects of dietary vitamin A on growth, hematology, digestion and lipometabolism of on-growing gibel carp (*Carassius auratus gibelio* var. CAS III). *Aquaculture (Amsterdam, Netherlands), 460*, 83–89.

Shiau, S., & Shiau, L. (2001). Re-evaluation of the vitamin E requirements of juvenile tilapia (*Oreochromis niloticus* × *O. aureus*). *Animal Science, 72*(3), 529–534.

Shiau, S.-Y., & Chin, Y.-H. (1999). Estimation of the dietary biotin requirement of juvenile hybrid tilapia, *Oreochromis niloticus× O. aureus*. *Aquaculture (Amsterdam, Netherlands), 170*(1), 71–78.

Shiau, S.-Y., & Hsieh, H.-L. (1997). Vitamin B6 requirements of tilapia *Oreochromis niloticus* × *O. aureus* fed two dietary protein concentrations. *Fisheries Science, 63*(6), 1002–1007.

Shiau, S.-Y., & Huang, S.-Y. (2001). Dietary folic acid requirement for maximum growth of juvenile tilapia *Oreochromis niloticus* × *O. aureus*. *Fisheries Science, 67*(4), 655–659.

Shiau, S.-Y., & Hwang, J.-Y. (1993). Vitamin D requirements of juvenile hybrid tilapia *Oreochromis niloticus* O. aureus*. *日本水産学会誌, 59*(3), 553–558.

Shiau, S.-Y., & Lo, P.-S. (2000). Dietary choline requirements of juvenile hybrid tilapia, *Oreochromis niloticus× O. aureus*. *The Journal of Nutrition, 130*(1), 100–103.

Shiau, S.-Y., & Lung, C.-Q. (1993). Estimation of the vitamin B12 requirement of the grass shrimp, *Penaeus monodon*. *Aquaculture (Amsterdam, Netherlands), 117*(1–2), 157–163.

Shiau, S.-Y., & Su, S.-L. (2004). Dietary inositol requirement for juvenile grass shrimp, *Penaeus monodon*. *Aquaculture, 241*(1–4), 1–8.

Shiau, S.-Y., & Su, S.-L. (2005). Juvenile tilapia (*Oreochromis niloticus× Oreochromis aureus*) requires dietary myo-inositol for maximal growth. *Aquaculture (Amsterdam, Netherlands), 243*(1–4), 273–277.

Shiau, S.-Y., & Suen, G.-S. (1992). Estimation of the niacin requirements for tilapia fed diets containing glucose or dextrin. *The Journal of Nutrition, 122*(10), 2030–2036.

Shirmohammad, F., Mehri, M., & Joezy-Shekalgorabi, S. (2016). A review on the role of inositol in aquaculture.

Soliman, A. K., & Wilson, R. P. (1992). Water-soluble vitamin requirements of tilapia. 1 pantothenic acid requirement of blue tilapia, Oreochromis aureus. *Aquaculture, 104*(1–2), 121–126.

Stephensen, C. B. (2001). Vitamin A, infection, and immune function. *Annual Review of Nutrition, 21*, 167.

Stickney, R. R., McGeachin, R. B., Lewis, D. H., Marks, J., Riggs, A., Sis, R. F. et al. (1984). Response of Tilapia aurea to dietary vitamin C. *Journal of the World Mariculture Society, 15*(1–4), 179–185.

Su, S.-L., & Shiau, S.-Y. (2004). Requirements of dietary myo-inositol of juvenile grouper, *Epinephelus malabaricus*. *臺灣水產學會刊, 31*(4), 311–317.

Sun, M., Li, X.-F., Ge, Y.-P., Zhang, L., Liu, B., & Liu, W.-B. (2022). Dietary thiamine requirement and its effects on glycolipid metabolism in oriental river prawn (*Macrobrachium nipponense*). *Aquaculture (Amsterdam, Netherlands), 550*, 737824.

Takeuchi, T., Watanabe, K., Satoh, S., & Watanabe, T. (1992). Requirement of grass carp fingerlings for *f¿*-Tocopherol. *Nippon Suisan Gakkaishi, 58*(9), 1743–1749.

Teixeira, C. P., Barros, M. M., Pezzato, L. E., Fernandes, J., Ademir Calvo, Koch, J. F. A., & Padovani, C. R. (2012). Growth performance of Nile tilapia, *Oreochromis niloticus*, fed diets containing levels of pyridoxine and haematological response under heat stress. *Aquaculture Research, 43*(8), 1081–1088.

Tomkins, A., & Hussey, G. (1989). Vitamin A, immunity and infection. *Nutrition Research Reviews, 2*(1), 17–28.

Tomlinson, D., Mülling, C., & Fakler, T. (2004). Invited review: Formation of keratins in the bovine claw: Roles of hormones, minerals, and vitamins in functional claw integrity. *Journal of Dairy Science, 87*(4), 797–809.

Trakatellis, A., Dimitriadou, A., & Trakatelli, M. (1997). Pyridoxine deficiency: New approaches in immunosuppression and chemotherapy. *Postgraduate Medical Journal, 73*(864), 617–622.

Twibell, R. G., & Brown, P. B. (2000). Dietary choline requirement of juvenile yellow perch (Perca flavescens). *The Journal of Nutrition, 130*(1), 95–99.

Villeneuve, L. A., Gisbert, E., Moriceau, J., Cahu, C. L., & Infante, J. L. Z. (2006). Intake of high levels of vitamin A and polyunsaturated fatty acids during different developmental periods modifies the expression of morphogenesis genes in European sea bass (*Dicentrarchus labrax*). *British Journal of Nutrition, 95*(4), 677–687.

Waagbø, R. (2010). Water-soluble vitamins in fish ontogeny. *Aquaculture Research, 41*(5), 733–744.

Wang, D., Zhao, L., & Tan, Y. (1995). Requirement of the fingerling grass carp (*Ctenopharyngodon idellus*) for choline. *Journal of Fisheries of China, 19*(2), 133–139.

Wang, X., Kim, K.-W., Bai, S. C., Huh, M.-D., & Cho, B.-Y. (2003). Effects of the different levels of dietary vitamin C on growth and tissue ascorbic acid changes in parrot fish (*Oplegnathus fasciatus*). *Aquaculture (Amsterdam, Netherlands), 215*(1–4), 203–211.

Webster, C. D., & Lim, C. (2002). Introduction to fish nutrition. In *Nutrient requirements and feeding of finfish for aquaculture* (pp. 1–27).

Wei, X., Hang, Y., Li, X., Hua, X., Cong, X., Yi, W., & Guo, X. (2023). Effects of dietary vitamin K3 levels on growth, coagulation, calcium content, and antioxidant capacity in largemouth bass, *Micropterus salmoides*. *Aquaculture and Fisheries, 8*(2), 159–165.

Wen, H., Yan, A., Gao, Q., Jiang, M., & Wei, Q. (2008). Dietary vitamin A requirement of juvenile Amur sturgeon (*Acipenser schrenckii*). *Journal of Applied Ichthyology, 24*(5), 534–538.

Wen, H., Zhao, Z., Jiang, M., Liu, A., Wu, F., & Liu, W. (2007). Dietary myo-inositol requirement for grass carp, *Ctrnopharyngodon idella* fingerling. *Journal of Fishery Sciences of China, 14*(5), 794–800.

Wen, Z. P., Zhou, X. Q., Feng, L., Jiang, J., & Liu, Y. (2009). Effect of dietary pantothenic acid supplement on growth, body composition and intestinal enzyme activities of juvenile Jian carp (*Cyprinus carpio* var. Jian). *Aquaculture Nutrition, 15*(5), 470–476.

Wilson, R. P., & Poe, W. E. (1973). Impaired collagen formation in the scorbutic channel catfish. *The Journal of Nutrition, 103*(9), 1359–1364.

Wilson, R. P., & Poe, W. E. (1988). Choline nutrition of fingerling channel catfish. *Aquaculture (Amsterdam, Netherlands), 68*(1), 65–71.

Wilson, R. P., Bowser, P. R., & Poe, W. E. (1983). Dietary pantothenic acid requirement of fingerling channel catfish. *The Journal of Nutrition, 113*(10), 2124–2128.

Wilson, R. P., Bowser, P. R., & Poe, W. E. (1984). Dietary vitamin E requirement of fingerling channel catfish. *The Journal of Nutrition, 114*(11), 2053–2058.

Woodward, B., & Frigg, M. (1989). Dietary biotin requirements of young rainbow trout (*Salmo gairdneri*) determined by weight gain, hepatic biotin concentration and maximal biotin-dependent enzyme activities in liver and white muscle. *The Journal of Nutrition, 119*(1), 54–60.

Wu, F., Huang, F., Wen, H., Jiang, M., Liu, W., Tian, J., & Yang, C. G. (2015). Vitamin C requirement of adult genetically improved farmed tilapia, *Oreochromis niloticus*. *Aquaculture International, 23*(5), 1203–1215.

Wu, F., Jiang, M., Wen, H., Liu, W., Tian, J., Yang, C.-g., & Huang, F. (2017). Dietary vitamin E effects on growth, fillet textural parameters, and antioxidant capacity of genetically improved farmed tilapia (GIFT), *Oreochromis niloticus*. *Aquaculture International, 25*(2), 991–1003.

Xia, M. H., Huang, X. L., Wang, H. L., Jin, M., Li, M., & Zhou, Q. C. (2015). Dietary niacin levels in practical diets for L itopenaeus vannamei to support maximum growth. *Aquaculture Nutrition, 21*(6), 853–860.

Xiang, X., Zhou, X. Q., Chen, G. F., Wu, P., & Zheng, Z. L. (2016). Effect of graded levels of dietary thiamine on the growth performance, body composition and haemato-biochemical parameters of juvenile S clizothorax prenanti. *Aquaculture Nutrition, 22*(3), 691–697.

Xiao, L., Mai, K., Ai, Q., Xu, W., Wang, X., Zhang, W., & Liufu, Z. (2010). Dietary ascorbic acid requirement of cobia, Rachycentron canadum Linneaus. *Aquaculture Nutrition, 16*(6), 582–589.

Yang, C., Jiang, M., Wen, H., Tian, J., Wu, F., Liu, W., & Yu, L. (2017a). Dietary vitamin B6 requirement of juvenile grass carp (*Ctenopharyngodon idella*).

Yang, Q., Ding, M., Tan, B., Dong, X., Chi, S., Zhang, S., & Liu, H. (2017b). Effects of dietary vitamin A on growth, feed utilization, lipid metabolism enzyme activities, and fatty acid synthase and hepatic lipase mRNA expression levels in the liver of juvenile orange spotted grouper, *Epinephelus coioides*. *Aquaculture (Amsterdam, Netherlands), 479*, 501–507.

Yazdani Sadati, M., Sayed Hassani, M., Pourkazemi, M., Shakourian, M., & Pourasadi, M. (2014). Influence of different levels of dietary choline on growth rate, body composition, Hematological indices and liver lipid of juvenile Siberian sturgeon *Acipenser baerii* Brandt, 1869. *Journal of Applied Ichthyology, 30*(6), 1632–1636.

Ying, Y., Bing, Y., Chen, D.-w., Gang, T., Mao, X.-b., Zheng, P., & Liu, J.-b. (2013). Effect of dietary folic acid supplementation on growth performance and hepatic protein metabolism in early-weaned intrauterine growth retardation piglets. *Journal of Integrative Agriculture, 12*(5), 862–868.

Yossa, R., Sarker, P. K., Mock, D. M., & Vandenberg, G. W. (2014). Dietary biotin requirement for growth of juvenile zebrafish D anio rerio (Hamilton-Buchanan). *Aquaculture Research, 45*(11), 1787–1797.

Yossa, R., Sarker, P. K., Mock, D. M., Lall, S. P., & Vandenberg, G. W. (2015). Current knowledge on biotin nutrition in fish and research perspectives. *Reviews in Aquaculture, 7*(1), 59–73.

Yuan, J., Feng, L., Jiang, W. D., Liu, Y., Jiang, J., Li, S. H. et al. (2016). Effects of dietary vitamin K levels on growth performance, enzyme activities and antioxidant status in the hepatopancreas and intestine of juvenile Jian carp (*Cyprinus carpio* var. Jian). *Aquaculture Nutrition, 22*(2), 352–366.

Zehra, S., & Khan, M. (2018). Dietary riboflavin requirement of fingerling *Channa punctatus* (Bloch) based on growth, conversion efficiencies, protein retention, liver riboflavin storage, RNA/DNA ratio and carcass composition. *Aquaculture Nutrition, 24*(1), 269–276.

Zehra, S., & Khan, M. A. (2012). Dietary vitamin C requirement of fingerling, *Cirrhinus mrigala* (Hamilton), based on growth, feed conversion, protein retention, hematological indices, and liver vitamin C concentration. *Journal of the World Aquaculture Society, 43*(5), 648–658.

Zehra, S., & Khan, M. A. (2017). Dietary thiamin and pyridoxine requirements of fingerling Indian major carp, *Cirrhinus mrigala* (Hamilton). *Aquaculture Research, 48*(9), 4945–4957.

Zehra, S., & Khan, M. A. (2018). Dietary pyridoxine requirement of fingerling *Channa punctatus* (Bloch) based on growth performance, liver pyridoxine concentration, and carcass composition. *Journal of Applied Aquaculture, 30*(3), 238–255.

Zehra, S., & Khan, M. A. (2019a). Effects of different levels of dietary cyanocobalamin on growth, liver cyanocobalamin concentration, antioxidant capacity, intestinal enzymes and non-specific immune response for optimum inclusion in the commercial feeds of fingerling *Channa punctatus* (Bloch). *Aquaculture (Amsterdam, Netherlands), 511*, 734272.

Zehra, S., & Khan, M. A. (2019b). Quantification of dietary inositol requirement for fingerling *Channa punctatus* (Bloch) based on growth, antioxidant status, hematological tools and liver inositol concentration. *Aquaculture (Amsterdam, Netherlands), 512*, 734280.

Zehra, S., & Khan, M. A. (2020). Dietary folic acid requirement of fingerling *Channa punctatus* (Bloch) based on growth, protein productive value and liver folic acid concentrations. *Animal Feed Science and Technology, 262*, 114397.

Zehra, S., & Khan, M. A. (2021). Dietary vitamin C requirement based on growth performance, non-specific immune response, antioxidant capacity, and liver vitamin C concentration of fingerling *Channa punctatus* (Bloch). *Animal Feed Science and Technology, 280*, 115058.

Zhao, S., Feng, L., Liu, Y., Kuang, S. Y., Tang, L., Jiang, J. et al. (2012). Effects of dietary biotin supplement on growth, body composition, intestinal enzyme activities and microbiota of juvenile Jian carp (*Cyprinus carpio* var. Jian). *Aquaculture Nutrition, 18*(4), 400–410.

Zhao, Y., Zhao, J., Zhang, Y., & Gao, J. (2017). Effects of different dietary vitamin C supplementations on growth performance, mucus immune responses and antioxidant status of loach (*Misgurnus anguillicaudatus* Cantor) juveniles. *Aquaculture Research, 48*(8), 4112–4123.

Zhao, Z., Wen, H., Wu, F., Liu, A., Jiang, M., & Liu, W. (2008). Dietary folic acid requirement for grass carp fingerling, *Ctenopharyngodon idella*. *Journal of Shanghai Fisheries University, 17*, 187–192.

Zheng, X., Jiang, W. D., Feng, L., Wu, P., Liu, Y., Jiang, J. et al. (2022). Effects of dietary pyridoxine on the skin immunity, tight junctions, antioxidants and apoptosis of grass carp (*Ctenopharyngodon idella*) infected with Aeromonas hydrophila. *Aquaculture Research, 53*(4), 1582–1596.

Zhou, Q. C., Wang, L. G., Wang, H. L., Wang, T., Elmada, C. Z., & Xie, F. J. (2013). Dietary vitamin E could improve growth performance, lipid peroxidation and non-specific immune responses for juvenile cobia (*Rachycentron canadum*). *Aquaculture Nutrition, 19*(3), 421–429.

Zhu, Z., Li, R., Yang, Q., Tan, B., Dong, X., Chi, S. et al. (2022). The potential advantage of myo-inositol as a dietary supplement for juvenile hybrid grouper (Brown-Marbled grouper *epinephelus fuscoguttatus*♀× Giant Grouper *Epinephelus lanceolatus*♂). *North American Journal of Aquaculture, 84*(1), 74–82.

CHAPTER 19

Hatchery System and Production of African Catfish in Aquaculture

YASSINE MUBARK ALI GAMEREDINN

Agricultural Engineer, Aquaculture Expert and Specialist, Secretary General of the Sudanese Fisheries Sciences Society, Sudan, Africa

ABSTRACT

African catfish is considered one of the important commercial breeding fish in Africa and in many countries of the world, due to its distinguished characteristics and many positive characteristics that helped in the success of its breeding. It is one of the best species for artificial breeding fish in that it has the ability to withstand difficult environmental conditions such as the low level of dissolved oxygen. In water and constant changes in water temperatures, in addition to the ease of acclimatization to different breeding conditions; the possibility of breeding them naturally or artificially, which provides the quantities required for breeding, with the ease of distinguishing between males and females through the apparent shape of the genitals, which facilitates spawning operations; it has the ability to grow in high densities. Under the intensive breeding system, the possibility of feeding them on cheap food, which reduces production costs, and their growth and feed conversion rates are considered high. They have high resistance to diseases, especially fungal and bacterial, due to the abundance of mucus secreted on the skin. It allows the manufacturing process, and besides all that, being a predatory fish, its use in polyculture with tilapia fish helps in controlling unwanted frequent reproduction of tilapia fry.

Recent Advances in Sustainable Aquaculture and Fisheries. Shabir Ahmad Dar, Imtiyaz Qayoom, Gohar Bilal Wani, Adnan Abubakr, and Seemab Zehra (Eds.)

DOI: 10.1201/9781998511372-19

19.1 INTRODUCTION

Several hundred catfish species (order: *Siluriformes*) belonging to 11 families inhabit Africa, but there is a comprehensive study conducted by Teugels and identified them with 32 species. The best species according to Teugels (1982 a, b, 1984), Boulenger (1911), and David (1935) is *Clarias gariepinus.*

The African catfish is one of the important fisheries for commercial farming in warm regions and is distributed in the Nile River regions, West Africa, and Algeria to South Africa, as well as in the continent of Asia. The fish is one of the commercially promising, due to its distinguished characteristics and numerous positive traits, such as it can withstand low levels of dissolved oxygen in the water and fluctuations in water temperatures. Moreover, it readily adapts to different breeding conditions, allowing breeding in both natural and artificial environments, which enables the production of desired quantities. African catfish are widely farmed in sub-Saharan Africa still accounts for nearly 91% of its global production and Nigeria is the largest producer in the region. African catfish is resilient and feeds on a variety of foods in the aquatic environment. It grows very fast, making it an excellent species in fish farming. In Africa, aquaculture is particularly significant given the continent's rapidly growing population, widespread food insecurity, and the economic reliance of millions of people on fisheries and aquatic resources. Despite this reliance, Africa's contribution to global aquaculture remains modest, accounting for only about 2.7% of total production. This low output is largely due to limited investment, resulting in an underdeveloped infrastructure and insufficient technological advancement across the continent . However, some African countries like Egypt, Nigeria, Ghana, and Uganda have made notable strides in expanding their aquaculture sub-sector industries. (Gerald Kwikiriza et al. 2025). They can grow in high densities under the intensive breeding system; the possibility of feeding them on cheap food reduces production costs. They have high resistance to diseases, especially fungal and bacterial, due to the abundance of mucus secreted on the skin.

The African catfish is particularly adaptable and, in addition to gill breathing, has also developed the ability to utilize oxygen from the air. The 50 genes that control this mechanism have now been identified, which can further advance evolutionary research and enable exciting research approaches to this physiological peculiarity. Following the reference genome for the pike perch, the FBN has now created the basic prerequisite for modern breeding approaches for this species with the African catfish, another economically relevant aquaculture fish species. (Nguinkal et al. 2024)

19.2 CHARACTERISTICS OF AFRICAN CATFISH

African catfish have an elongated body, four barbels, large dorsal fin with (61–80) soft fin rays, and the anal fin is also long (45–65) soft fin rays and four pairs of oral tentacles (whiskers). The eyes are small; the pelvic and pectoral fins are also small. The pectoral fin spine has a serrated outer edge (serrated). The body color is dark gray or black on the dorsal side, while the ventral side is creamy white. The body is smooth (free from scales). African catfish farmed in freshwater, with a temperature range of 28–30°C, which is the optimum for growth, while it resists thermal changes between 8 and 35°C. These fish have good resistance to environmental changes, especially in muddy water and low oxygen (muddy water). These fish can crawl in shallow waters, especially at night, using their powerful pectoral fins to navigate to another environment.

FIGURE 19.1 African catfish.

Source: Public domain

19.2.1 FOOD AND FEEDING HABITS

It is a bottom feeder; it is a carnivorous fish and feeds on a wide variety of food items including detritus, fish, zooplankton, macrophytes, insects, phytoplankton, and gastropods.

19.2.2 REPRODUCTION

Fertilization in these fish is external, which is usually during summer in shallow waters where there are weeds and aquatic plants between which eggs are laid.

19.3 IMPORTANT FEATURES OF AFRICAN CATFISH

1. High growth rate and feed conversion factor.
2. Tolerance to difficult environmental conditions, especially low oxygen and heat.
3. The mucous substance on the skin increases its resistance to diseases.
4. Ease of hatching under captivity conditions, naturally and artificially.
5. Their relative fecundity is high (70 eggs/g body weight).
6. They can be reared at higher densities than other fish.
7. They can be fed at large ages on low-cost foods.
8. They are used as a biological control to check the indiscriminate hatching of tilapia.

19.4 ENVIRONMENTAL NEEDS

19.4.1 TEMPERATURE

From 24 to 32°C, the optimum is 29°C, it tolerates up to 5°C, and the optimum for hatching and embryo development is from 25 to 27°C.

Factors	Temperature °C	DO mg/L	Salinity g/L	pH	Hardness mg/L
Optimal range	24–32	0.2–6	8–10	6.5–9	50–100

19.4.2 DISSOLVED OXYGEN

The optimum dissolved oxygen required is in the range of 4–5 mg/L at a depth of 6 inches from the water's surface, however, the fry needs about 6 mg/L due to the incompleteness of the accessory respiration organ and dies quickly at 0.1 mg/L. Large fish can withstand severe depression up to 40 h out of water, but they need a good aeration source in intensive culture, and the lethal oxygen limit is 0.2–0.8 mg/L.

19.4.3 SALINITY

Catfish is one of the most sensitive freshwater fish to salinity. It grows well in freshwater and can tolerate brackish water up to 8–10 g/L. It may live in salinity of up to 29 g/L without growth or reproduction, and the ideal salinity should not exceed 5 g/L.

19.4.4 PH

The optimal pH required is from 7 to 8.5 or 6.5 to 9.

19.5 NATURAL REPRODUCTION OF AFRICAN CATFISH

African catfish (male and female) attain sexual maturity in nature at the age of 2–3 years and a length greater than 40 cm. Under captive conditions (ponds), they attain sexual maturity at the age of 1 year and a weight of 200–300 g. The GSI reaches its maximum value in May and is 1.9% for males and 15.4% for females. Catfish lay eggs once a year during the rainy season, floods, and the increase in the water column in rivers, as these conditions stimulate mothers to ovulate and fertilize. The eggs are laid in batches (15–50 batches), which stick to aquatic plants and grass located at a distance of 25–30 cm from the surface of the water, or in holes in embankments at the same depth, then the male fertilizes the eggs. The survival rate is very low, reaching a rate of 1–2% due to the characteristic of predation in mothers and super-growth fry.

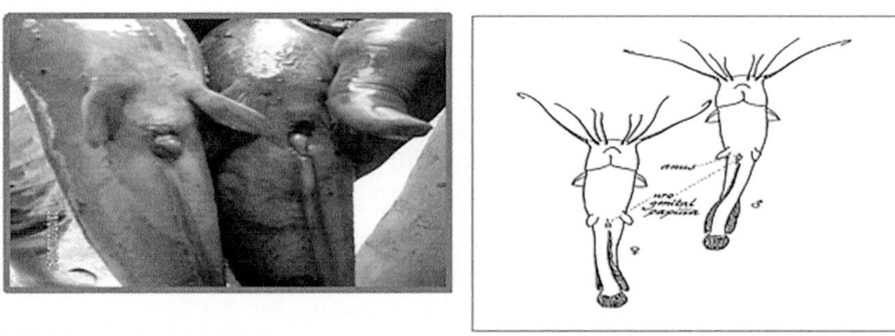

FIGURE 19.2 Male and female.

Source: https://www.fao.org/4/w3595e/w3595e04.htm

19.6 INDUCED BREEDING OF CATFISH

Catfish are characterized by high growth rates, tolerance to difficult environmental conditions, and their ability to survive for a long time after being caught, which is an important factor in their marketing. Therefore, it was necessary to pay attention to the hatching of these fish due to the unavailability of their fry from natural sources, and this requires shedding light on how to artificially hatch catfish and care for the fry to obtain an intensive production of fry and fingerlings that cover the needs of fish farms and lead to higher productivity. Since catfish mothers do not lay their eggs naturally in fish ponds as happens in nature, it is required to induce them to ovulate artificially by hormonal treatment by injecting pituitary extract.

1. The steps for hatching are summarized as:Selection of brooders
2. Collecting and extracting the pituitary gland
3. Injection of females
4. Preparing the semen
5. Extracting eggs from injected mothers
6. Fertilization of eggs

19.6.1 NATURAL ARTIFICIAL SPWAING AND BROODER SELECTION CRITERIA

In induced natural spawning, hatcheries must prepare spawning tanks and proceed by inducing female broodstock with a synthetic or natural Depending on the age and condition of the males, induce and then place them in the same tank as the females before spawning. To avoid contaminating the tanks, disinfect the broodstock using a 0.5% iodine solution before induction and follow immediately after they are acclimatized. Once vitellogenesis is complete, the females will shedd their eggs on the kakaban (a nest made from sacks) and the males will simultaneously fertilize them. Carefully remove both broodstock once spawning is complete, usually within 2 hours after it starts, i.e. when eggs are visible on the kakaban or the floor of the tank. Place at least two kakabans in the tank, and avoid using females that are too small or too big to prevent predation instead of courtship. Ideally, use males of similar age and size as the females, at least 12 months old and weighing 1–2 kg. For induced natural spawning, it is not necessary to kill or dissect the males Induced natural spawning is not as popular as induced artificial spawning because it is not possible to determine the quality of the sperm beforehand, so there is no room for replacement. Increasing the ratio of males to females in the tank can be detrimental because the males could fight for dominance while the females are laying their unfertilized eggs. Aggressive males can injure the females and vice versa. The main disadvantage of this method of breeding catfish is that it is not possible to determine the weight and, by extension, the number of eggs from the females and ascertain if an additional male is required. Weighing eggs can only be done after removing the females and weighing them to determine the difference from their initial weight before spawning. However, this is only for recordkeeping purposes, as there is no room for intervention.

In hatcheries, African catfish do not reproduce naturally in captivity, so human intervention is required to aid the process. Female African catfish require stimulation of the pituitary gland with a hormone, either natural or synthetic, so that their eggs mature completely. However, it is not possible to strip the sperm of the males because they possess micro fibrils at the end of the lobes of the testicles, which prevents the free flow of milt. As a result, they have to be dissected to induce artificial spawning.

To start you must collection of mature females with smooth, distended bellies, age 7–10 months, weight 200–500 g, and disinfected with formalin concentration of 50–150 ppm for 3 h.

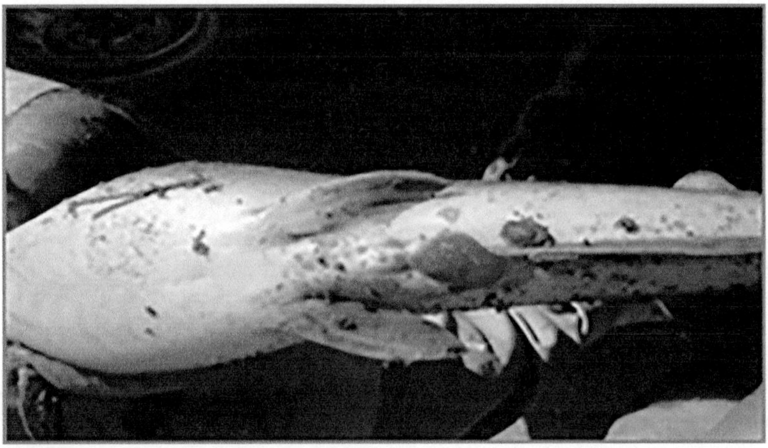

FIGURE 19.3 Mature female.

19.6.2 COLLECTING AND EXTRACTING THE PITUITARY GLAND

The pituitary gland is collected by sacrificing the brooders, the head is cut off, the lower jaw is separated with a knife, and the skull cap is spread from the top with a fine, sharp saw. The pituitary gland appears on the abdominal part of the brain, like a grain of white rice stained with a light violet color. It can be used fresh or dry after removing the water and fat from it using acetone. One per gland, changing it several times after 10 min, then after 8 h, 24 h, dried on filter paper and kept in a desiccator.

FIGURE 19.4 The use of the extracted hormone.

The dose varies according to the type of hormone, but the optimal rate is 4 mg per 1 kg of female fish, and the dose is reduced by adding saline solution by 5 times the dose of the hormone, as in the case of the pituitary gland it must be crushed and dipped in a saline solution of 10 mL for each gland of a fish weighing 1 kg to be injected into a fish a female weighing 1 kg.

19.6.3 FEMALE INJECTION

Females are injected at a rate of 4 mg per kg of fish. The needle is placed at a depth of 2–2.5 mm and at an angle of 45 in the dorsal muscle towards the tail. Then female fish are placed in the basin for ovulation, which takes 7–21 h, depending on the temperature.

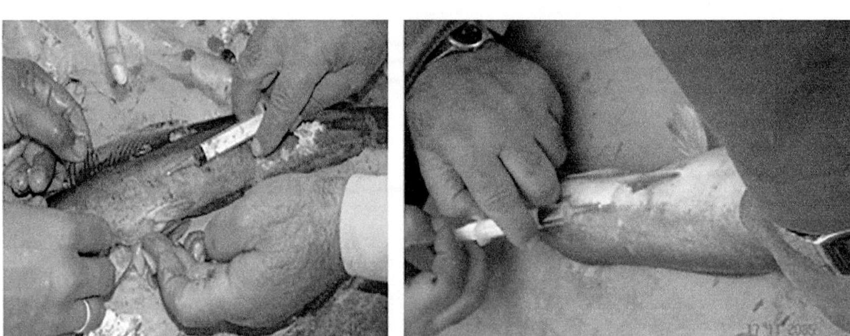

FIGURE 19.5 Female injection.

The dose varies according to the type of hormone, but the optimal rate is 4 mg per 1 kg of female fish, and the dose is reduced by adding saline solution by 5 times the dose of the hormone, as in the case of the pituitary gland it must be crushed and dipped in a saline solution of 10 mL for each gland of a fish weighing 1 kg to be injected into a fish a female weighing 1 kg, then the mothers are placed in the basin for ovulation, which takes 7–21 h, depending on the temperature.

19.6.4 PREPARATION OF SPERM SUSPENSION

Mature males with pink genitals are selected. The abdomen is cut open and the testicles are removed gently, dried well, cut with scissors on a dry nylon cloth, squeezed with fingers, and washed with 50 mL of saline in a cup for every 5 males. It can be used directly for fertilization or placed in the refrigerator at 4°C for 48 h. After an hour, to ensure the viability of the sperm, a drop of semen is placed under the microscope, and a drop of water is placed on it. If it moves for 30 s, it is active.

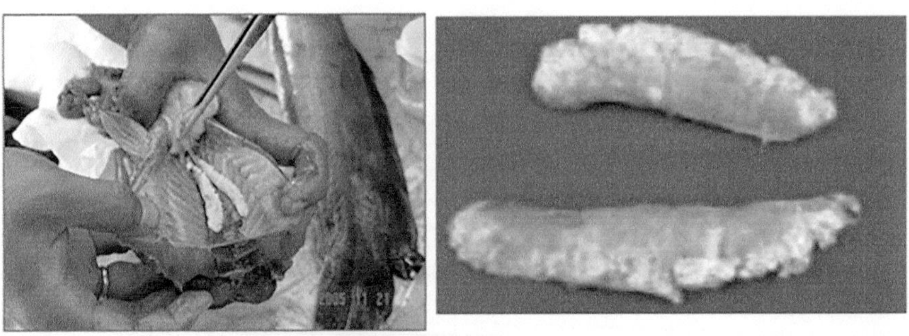

FIGURE 19.6 Preparation of sperm suspension.

19.7 OVULATION AND FERTILIZATION

After 8–10 h, the female is pressed to release eggs (1000–2000 eggs depending on the size and age of the mother) are fertilized with the sperm suspension. Eggs are mixed thoroughly with the help of a bird feather and a small quantity of water is added to the egg mass for activation of the sperms and successful fertilization. After 2–3 min of mixing, the eggs are washed thoroughly in running water and shifted to the flow-through incubation tray. The incubation unit is well disinfected with a formalin bath with a concentration of 50–150 ppm for 3 h and 200 g of eggs are scattered, with about 140,000 eggs produced by 4 females, with an average of 500 g for the female gently in the form of a thin layer on the floor of the incubation tray after disinfecting it with 150 L of water and the color of the viable eggs are yellow to brown, and the unfit eggs are white in color. The water stream is regulated at a rate of 1–3 L per min. The incubation period takes 20–57 h, depending on the temperature. The hatching rate is 50–80%. The eggs are still broken by siphoning, and the larvae feed (0.03 mg) on the yolk sac. After 3 days, the percentage of survival after the yolk sac is 90%.

FIGURE 19.7 Striping of the injected female.

FIGURE 19.8 Mixing sperm with egg.

19.8 INCUBATION

A total of 4000–8000 catfish eggs are incubated, as they depend in their early stages on natural food, especially zooplankton, up to a size of 7–8 g. After this size, it begins the process of preying on the smaller fry, so it is transferred to the ponds. It is reported that the highest growth rate of catfish is during the first three months of its life until the age of one year, after which the growth rate gradually decreases.

FIGURE 19.9 Incubation and rearing eggs.

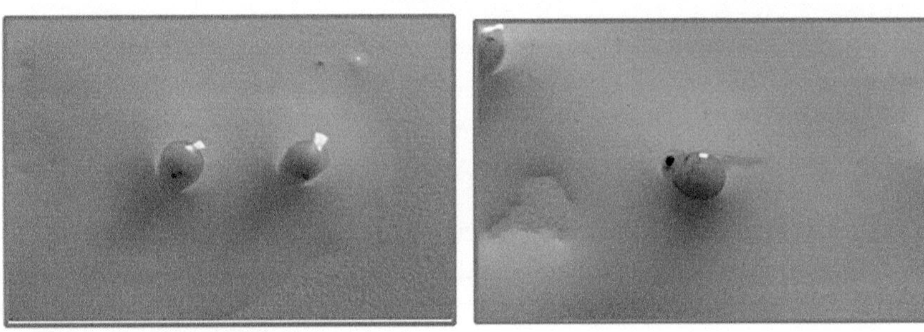

Figure 19.10 Fertilized eggs.

19.9 LARVAL FEEDING

The larvae should be reared in the indoor tanks for about 15–18 days at the stocking density of 1000–1500/m^2. Care should be taken to minimize disturbance during cleaning or refilling the tank. Zooplankton such as Copepod, Daphnia, Moina, Artemia nauplii, molluscan meat, tubifex, and egg custard are often provided as larval feed after yolk sac absorption (from the fourth day onwards). There may be a phenomenon of predation among the larvae, which leads to a difference in sizes, and, therefore, feeding must be intensified.

FIGURE 19.11 Fry of African catfish.

19.10 POLYCULTURE OF AFRICAN CATFISH

Catfish production is considered a secondary product, as it is included in the crop composition in mixed breeding, at a rate of 8–10% of the total number in the pond, and in large sizes up to 500 g/fish, and to get rid of tilapia fry resulting from random spawning, where catfish are stored in breeding ponds. With a period of not less than three months of storing tilapia fish, we find that the best housing ratio between Nile tilapia and catfish is 1:4, respectively.

19.11 MONOCULTURE OF AFRICAN CATFISH

In this system, catfish fingerlings are stocked at high rates in the earthen pond, and the basis relies on balanced artificial feeding (25–40%). Therefore, this type of rearing requires the provision of such food, as well as specialized management. The rearing period in this system extends from six months to a year, depending on the target marketing weight as well as the type of product to be marketed (whole fish or pure meat). Therefore, stocking rates vary according to the marketing size to be reached. For example, if the desired marketing size of catfish is 200 g, then the stocking rate for fingerlings with an average body weight of 1–5 g is 10 fingerlings/m^2. In the case of the desire to breed catfish for sizes greater than 500 g, then in this case the stocking rates of 2 fingerlings/m^2 are reduced. Catfish feeding begins at high rates that may reach 10% of the weight of fingerlings, and decreases with the growth of the fish until it reaches 3% for large fish. In each period, consideration must be given to sorting the sizes of the fish because their variation leads to the phenomenon of predation and weight differences in sizes.

FIGURE 19.12 Monoculture of African catfish.

19.12 CONCLUSION

Because the African catfish is an excellent fish for fish farming and have the advantage of resisting in difficult conditions and have the ability to grow fast and give excellent sizes, there was a need for its artificial hatching because it does not spawn naturally in ponds, so the hormone is used to stimulate it and fertilize the eggs and incubate them until the larvae hatch and are rearing. The production of African catfish depends on the /high production of hatcheries to increase the production of these to contribute to food security in Africa and Asia.

KEYWORDS

- **breeding**
- **mucus**
- **polyculture**
- **reproduction**
- **salinity**
- **spawning**
- **species**
- **temperature**

REFERENCES

Boulenger, G. A. (1911). *Catalogue of the freshwater fishes of Africa* (Vol. 2, pp. 529). London: British Museum (Natural History).

David, L. (1935). Die entwicklung der Clariiden und ihre verbreitung. *Revue de Zoologie et de Botanique Africaines*, *28*(1), 77–147.

Gerald Kwikiriza et al. (2025). Enhancing African Catfish (*Clarias gariepinus*) Aquaculture in Uganda: Insights into Hatchery Propagation, Population Suitability, and Broodstock Management. Fishes, journal published by MDPI, 10(6), 290.

Isa .SI, (2023). Best management practices for hatcheries culturing African catfish. Penang, Malaysia: World Fish. Manual: 2023–37.

Nguinkal et al., (2024), new study at the FBN decodes the genome of the African catfish. IDW magazine.

Teugels, G. G. (1982a). A systematic outline of the African species of the genus Clarias (Pisces, Clariidae), with an annotated bibliography. *Annales du Musée royal de l'Afrique centrale, Sciences Zoologiques*, *236*, 1–249.

Teugels, G. G. (1982b). Preliminary data of a systematic outline of the African species of the genus Clarias (Pisces, Clariidae). *Revue de Zoologie Africaine, 96*(4), 731–748.

Teugels, G. G. (1984). The nomenclature of African Clarias species used in aquaculture. *Aquaculture*, *38*(4), 373–374.

CHAPTER 20

Applications of Nanotechnology in Improving Aquaculture

QUSEEN MUSHTAQ RESHI[1,*], ISHTIYAQ AHMAD[2], IMTIAZ AHMED[1], and NAVEED NABI[1]

[1]*Department of Zoology, Fish Nutrition Research Laboratory, University of Kashmir, Srinagar, Jammu and Kashmir, India*

[2]*Division of Fish Genetics and Biotechnology, Rangil Sher-e-Kashmir University of Agricultural Sciences and Technology, Jammu and Kashmir, India*

ABSTRACT

Aquaculture is critical to global food production. The aquaculture sector is currently undergoing new scientific and technological breakthroughs in order to deliver high-quality end products through advancements in nanoscience. The application of nutritional supplements in shellfish and fish for better health management, adding value, and stress reduction is a big frontier of aquaculture studies. Fish, mollusc, and crustacean farming is the world's leading animal production sector, and the highly integrated fish farming industry may be among the first to incorporate and commercialize nanotechnology products. However, the possible advantages of nanotechnology for aquaculture as well as agriculture must be adjusted against concerns regarding water, soil, and the environment. It is preferable to increase productivity through nanotechnology-driven precision farming and maximize output while minimizing input efforts via better oversight and target-oriented action. Holistic strategies have the ability

Recent Advances in Sustainable Aquaculture and Fisheries. Shabir Ahmad Dar, Imtiyaz Qayoom, Gohar Bilal Wani, Adnan Abubakr, and Seemab Zehra (Eds.)

DOI: 10.1201/9781998511372-20

to enhance food production by incorporating breakthrough technologies and disciplines such as biopharmaceuticals integrated with nanoscience to address existing biological impediments that are presently restricting additional development. This chapter details breakthroughs in nanotechnology in general, as well as these novel possible future avenues that are envisaged to have a significant impact on fisheries and aquaculture.

20.1 INTRODUCTION

Aquaculture refers to the fish and seafood farming, and it is one of the most rapidly expanding segments of the animal food manufacturing industry (Can et al., 2011; Sekhon et al., 2014). Aquaculture, one of the fastest-growing food-producing industries, is seen to have a good potential to complement capture fisheries in order to meet the world's need for fish as food (Gabriel et al., 2022). The demand for food fish has been increasing over the past seven decades (annual consumption rate at 3.1%), at a rate that is almost twice as fast as the yearly growth in the global population (1.6%) (FAO, 2020). The advantages of aquaculture, such as the availability of easily accessible resources, food security, economic development, and community empowerment, may have contributed to the radical look for practical ways to lessen harmful effects rather than discourage the exercise (Gabriel et al., 2022). Nanotechnology, which is among the new technologies, is dealing with the design, synthesis, characterization, and as well as the use of materials with a practical organization in the nanometric scale (Silva, 2010; Gabriel et al., 2022). The impact of nanoparticles (NPs) on aquatic organisms is not fully understood yet. Nanoscience allows scientists to evaluate, modify, and make products at the nanometer range, and that is one billionth of a meter and confers unique chemical and physical characteristics to the compounds (Singh and Dutta, 2020). NPs are used in a variety of forms and shapes, including nanospheres (Donbrow, 1991), carbon nanotubes (Reilly, 2007), nanocapsules (Torchilin, 2006), and dendrimers (Aulenta et al., 2003; Gillies and Frechet, 2005; Wu et al., 2015), and others have been observed with numerous benefits, including dose reduction, tissue-specific aiming, minimization in harmful or subsequent negative impacts, and high drug bioactivity and effectiveness (Toyokawa et al., 2008; Xu et al., 2018; Shah and Mraz, 2020). As of now, a diverse range of materials have already been efficaciously utilized to fabricate these NPs; however, polymeric NPs have been extensively investigated as drug delivery carriers due to their numerous advantages, such as the ability to protect drugs from degradation, improve drug utilization efficiency, and control rate of drug release (Fan et al., 2012; Shah and Mraz, 2020). Some of the aspects include transfer of genes (Murata et al., 1998), water filtration (Khosravi-Katuli et al., 2017), nutrient delivery (Ashouri et al., 2015), nutraceuticals (Aklakur et al., 2016), delivery of vaccines (Rajeshkumar et al., 2009), and nutrient delivery (Lavertu et al., 2006). Industrial aquaculture practices have emerged as a fast-budding industry in both developing as well as developed countries. The world in the last 20 years has witnessed nearly fourfold escalation in industrial aquaculture (Naylor and Burke, 2005; Banerjee et al., 2014). There is a significant gap to be filled in modernization for disease treatment, drug usage, water quality management, and manufacturing of suited fish for improved outcomes, profitability driven by both nutrigenomic and epigenetic interactions, improved breeding excellence through effective implementation of spawning and maturation inducing agents, nutraceutical delivery for significant growth promotion and reduction in culture time, utilization of impactful auto-transgenics and vaccine in this area (Aklakur et al., 2016; Shah and Mraz, 2020).

Fish farmers may gain from the use of nanoscience in terms of production, development of innovative products in food preservation, processing, and packaging (Sekhon et al., 2014). Re-circulatory aquaculture systems (RAS) are one example of the intensive technologies that have replaced widespread aquaculture as a means of bridging the supply and demand gap for food fish. High fish stocking densities in intensive systems have been demonstrated to stress farmed fish, which negatively impacts fish performance and wellbeing (Gabriel et al., 2022). Disturbing arrival of chemical compounds into the aquaculture systems remains a major issue for fish production as well as public health. To address these limitations, a comprehensive strategy of understanding, deploying and integrating new science and technology approaches in sustaining a desirable aquaculture is considered necessary (Shah and Mraz, 2020). Of numerous recent scientific developments, nanotechnology is gaining prominence as the unique technology and research platform for the upcoming generation of agri-food system in aspects of transformation and development (Kumari et al., 2014; Rodrigues et al., 2017; Shah and Mraz, 2020). Nanotechnology has a wide range of applications in the aquaculture and seafood industries

(Can et al., 2011; Sekhon et al., 2014). Effect of different NPs on various species of experimental fish is given in Table 20.1. Involvement of nanotechnology can also help to reduce the concerns associated with different hydrobiological challenges of the aquaculture system. The increasing use of nanomaterials in aquaculture with favorable findings has been widely reported in research platforms, although its long-term impact on life forms and the atmosphere must be investigated further in field applications (Singh and Dutta, 2020).

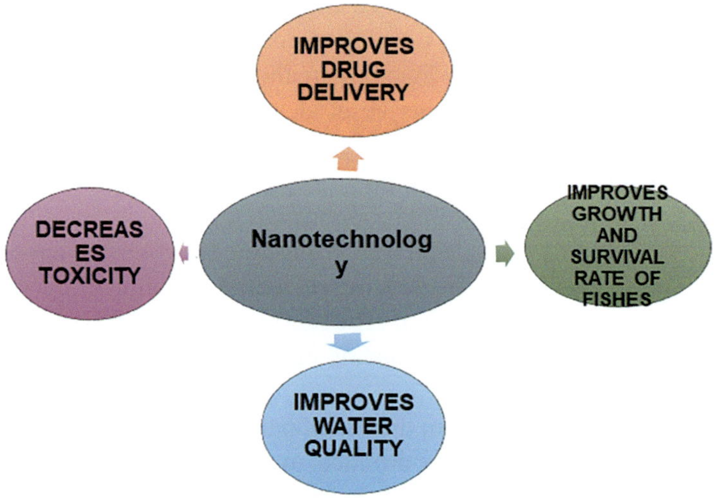

FIGURE 20.1 Schematic representation showing influence of nanoscience on aquaculture.

20.2 ROLE OF NANOTECHNOLOGY IN IMPROVING WATER QUALITY

Water quality is important for animal health. Fish culture system is invariably influenced by water quality issues. Water quality degradation causes poor outcomes and harm to fish and other aquatic organisms. In aquaculture, particularly in semi-intensive practices, the diet fed to animals is usually rich in proteins and fertilizers, to enrich natural feeds to achieve optimum growth of cultured species and augment production. Nevertheless, management of fecal matter and uneaten food particles remains an impending issue which repeatedly affects the water quality in culture systems (Ninh et al., 2016; Gabriel et al., 2022). Amongst numerous water-related issues confronting intensive aquaculture, nitrogenous waste

TABLE 20.1 Effect of Different Nanoparticles on Various Fish Species.

Source of Nanoparticles	Experimental Fish	Observations	References
Azadirachta indica	*Cirrhinus mrigala*	Immunological measures' functional activity was considerably increased in fish treated with these nanoparticles. They may have immunomodulatory and antimicrobial effects, it was determined.	Rather et al. (2017)
Ginger	*Cyprinus carpio*	Using ginger nanoparticles as an effective formulation to protect common carp fingerlings against motile *Aeromonas septicaemia.*	Korni and Khalil (2017)
Curcuma longa	*Labeo rohita*	Substantially increased serum bactericidal activity, superoxide anion generation, lysozyme activity, and protection against *Aeromonas hydrophila.*	Sahu et al. (2008)
Curcumin	*Oreochromis mossambicus*	Improvement of the digestive enzymes' functions. Influenced how GH in the brain and factors like IGF-1 and IGF-2 in the muscle were expressed.	Midhun et al. (2016)
Iron	*Oncorhynchus mykiss*	An iron (Fe) nanoparticle- and probiotic-supplemented meal dramatically improved the growth performance of rainbow trout.	Mohammadi and Tukmechi (2015)
Selenium	*Cyprinus carpio*	In comparison to a control group, common carp's growth and antioxidant defense system significantly improved when exposed to nano-selenium.	Ashouri et al. (2015)
Copper	*Pagrus major*	The addition of dietary copper nanoparticles enhanced greatly the aquatic animals' biochemical status, growth, metabolic and digestive enzyme activities as well as nonspecific immune response.	El Basuini et al. (2017)

accumulation is said to be the most pernicious and harmful (Singh and Dutta, 2020). The separation of nitrogen waste from the aquatic environment is a difficult task (Handy and Poxton, 1993). Moreover, waste buildup entices pathogenic organisms such as algae, bacteria, and viruses. To act, a novel approach is being used, one of the most noticeable being the application of nanotechnology (Singh and Dutta, 2020). Very often water quality in aquaculture system deteriorates because of fecal matter and enrichment due to organic matter and nutrients (Sikder et al., 2016), by which it has a negative impact on fish health, growth making the fish prone to stress and various infectious diseases (Boyd and Tucker, 1998; Boyd, 2001; Gabriel et al., 2022). Nanotechnology has a wide range of applications in aquaculture, including water treatment, pond sterilization, detection and elimination of aquatic diseases, prompt delivery of nutrient levels and medicines (including hormones and vaccines), and augmentation of fish potential in assimilating such drugs (Bhattacharyya et al., 2015; Huang et al., 2015; Shah and Mraz, 2020).

Hence, proper monitoring of water quality is fundamental for aquaculture operations (Gabriel et al., 2022). These NPs can alter the pharmacological characteristics of active ingredients and have increased reactivity (Jiang et al., 2019; Gabriel et al., 2022). Research on this technique for numerous aquaculture applications is extensive. A nanotechnology-based treatment could indeed help to reduce the challenge of nitrogen-containing waste accumulation of an aquaculture system (Singh and Dutta, 2020). Nanotechnology is used to purify water by utilizing nanoscopic materials such as alumina fibres and carbon nanotubes for nanofiltration. Nanofiltration is a comparatively recent membrane filtration methodology that is primarily used to remove solids and pathogenic organisms from surface as well as fresh groundwater (Sekhon, 2014). Although the direct use of silver-based NPs in water to control a fungal disease was found to be toxic to young trout, however, a water filter covered with silver-based NPs managed to prevent fungal infections in rainbow trout reared in indoor setups (Sci dev. net, 2022).

20.3 INFLUENCE OF NANOTECHNOLOGY IN FISH NUTRITION

Usually, fish have been fed by being given food in the form of pellets, by catch, or fish offal. The main considerations in the formulation of the pellet are the daily nutritional requirements of fish for key elements such as proteins, carbohydrates, lipids, minerals, and vitamins (Gabriel et al., 2022; Ahmed and Ahmad, 2020; Ahmad et al., 2021a, b, 2022). Nutritionists have recently used nanotechnology to develop a number of delivery strategies for micronutrients, including encapsulation, protection, and controlled release. Therefore, nanotechnology has a significant potential to improve nutritional assessment and bioavailability measurements (Gabriel et al., 2022; Ahmad et al., 2021). For instance, the understanding of nutritional and biomolecular interactions in particular tissues is improved by the ultra-sensitive detection of nutrients and metabolites. The gastrointestinal tract has consistently been the primary and most significant channel for feed/food delivery principles, including for NPs, in nutritional studies (Gabriel et al., 2022). There are several different ways that NPs can enter the gastrointestinal tract, including the following: (1) ingestion or swallow pathway: ingestion directly from food and water as well as from the administration of therapeutic nano-drugs; (2) inhalation pathway: after clearing the respiratory tract, inhaled NPs can be swallowed and enter the gastrointestinal tract; and (3) oral pathway: oral or smart delivery into gastrointestinal tract, in which particle uptake in the gastrointestinal tract (Gabriel et al., 2022).

Nanotechnology has a plethora of possibilities in the feed industry, such as: (i) minor modifications of natural ingredients to improve taste, palatability, and sensory improvement like flavor, color, and contour; (ii) boosting nutrient quality of food products by balancing active components like nutritional supplements in feed formulations, packaging, and new product development to prolong storage; and (iii) intensifying systemic absorption of vital nutrients (Food Safety Authority of Ireland, 2008). The Russian Academy of Sciences observed that when sturgeon and young carp were fed iron NPs, their growth rates increased by 24 and 30%, respectively (ETC, 2004). Various selenium sources (nano-Se and selenomethionine) augmented in the normal diet enhanced the total weight, relative gain rate, antioxidant level capacity, muscle Se concentration and glutathione peroxidase (GSH-Px) activities *of Carassius auratus gibelio*. Furthermore, nano-Se showed up to be significantly efficient at elevating muscle selenium content as compared to organic selenomethionine (Zhou et al., 2009). Likewise, the nano level scale delivery of these nutritional supplements improved the performance and growth of the tested fish. The faster the absorption via the gastrointestinal tract barrier to reach the blood, followed by diffusion through gastrointestinal tract mucus to reach the cells of the intestinal lining, is achieved by smaller particle size (Hoet et al., 2004; Gabriel et al., 2022). Pollutants and food waste in fisheries are impending issues caused by decreased food stabilization, inadequate buoyancy or texture of the pellet (Handy and Poxton, 1993; Handy, 2012). The physical characteristics of food pellets can be significantly altered by low concentrations of nanomaterials (Handy, 2012). One significant concept is that NPs will improve aquafeeds by boosting the fraction of fish feed ingredients that traverse the gut tissue and enter the fish, as opposed to passing unused via the fish digestive system (Handy, 2012; Gabriel et al., 2022). By enhancing the nutrients' solubility and protecting the fish gut, the delivery systems of NPs specifically aim to improve the bioavailability, bioaccessibility, and consequently effectiveness of the nutrients. The specific components of the delivery systems include micronutrients that are contained within NPs made of surfactants, lipids, proteins, and/ or carbohydrates (Joye et al., 2014; Gabriel et al., 2022). These systems' small particle size is superior to traditional delivery methods in a number of ways, including greater bioavailability, increased stability against aggregation and gravitational separation, and increased optical clarity (Joye et al., 2014; Gabriel et al., 2022). Young carp and sturgeon grew at a faster rate after being fed iron NPs (Sekhon et al., 2014). A diet supplemented with nanoselenium improved antioxidants, final weight, relative gain rate, muscle selenium concentrations, and glutathione peroxidase activities in *Carassius auratus*. Nanoselenium was discovered to be more beneficial at elevating selenium content of muscle in comparison to selenomethionine. The nano level transfer of these nutraceuticals was found to improve both growth and performance of the fish tested (Rather et al., 2011). Furthermore, different feed nanoformulations aid in maintaining optimum feed consistency and taste as well (FOE, 2008).

To enhance fish health and development performance, immuno-modulatory substances such as phenolic compounds, vitamins, and minerals are being added to aquafeeds more frequently (Gabriel et al., 2022). However, because of their low bioavailability, these nutraceuticals are frequently difficult to incorporate into feeds (Gabriel et al., 2022). This problem might be resolved by encapsulating the bioactive components. The bioavailability of encapsulated bioactive compounds rises with decreasing particle size, allowing for quicker digestion and absorption. Additionally, NPs can be designed to travel through particular parts of the digestive tract before releasing their payload at a specific location, potentially increasing their immune-nutritional advantages (Jafari and McClements, 2017; Gabriel et al., 2022).

Since fat-soluble vitamins are less soluble or to and have exhibit decreased assimilation efficiency across the gut of fishes, so such nutrients can be encapsulated in NPs (Handy, 2012). NPs have the potential to change the physical characteristics of fish food in alongside increasing the stability and bioactivity of food additives. Nanoencapsulation technology has indeed been proposed for vitamins, carotenoids, fatty acids and trace minerals, with the primary objective of increasing bioavailability (Acosta, 2009; Bouwmeester et al., 2009). Casein-derived nano-micelles have been recommended as a vehicle for water-insoluble additives such as vitamin D2

(Semo et al., 2007; Handy, 2012). Nanoscale mineral supplements could act as a supply of trace metals without causing extensive fecal losses usually accompanied with mineral salts (Carriquiriborde et al., 2004; Handy, 2012).

Better understanding of feed nanoscale materials enables nano-delivery of bioactive/nutrient in feedstuffs or *in vivo* in fish. NPs with the potency to be utilized for this objective include nanofibers, nanoclays, and nanotubes. Nanochips, nanoimaging, nanofilters, and nanosensors are examples of nanoproducts that could find use. Similarly, the prospective nano-delivery systems are nanomachines, nanocapsules, nanoballs, nanocochleates, nanodevices, and nanorobots (Thulasi et al., 2013). Nanotechnology has already been utilized in the food manufacturing (Chaudhry et al., 2008; Tiede et al., 2008), and research interests have to do with how NPs affect the texture, structure, as well as food's quality, in addition to potential technological aspects in producing food, handling, traceability, transportation and traceability and storage (Handy, 2012). Nanoparticle mineral additives could be a source of trace metals without the substantial fecal matter negative effects associated with inorganic salts (e.g., Fe salts; Carriquiriborde et al., 2004). Nanomaterials could also serve as an alternate solution to biological kinds of food additives, where anti-nutritional factors components (inadvertent agrochemicals, toxic materials, etc.) in the active component can or sometimes be a concern (Berntssen et al., 2010). Romero-Perez et al. (2010) reported improvement in selenium absorption in ruminants by giving sodium selenite in nanoform. Nanomaterials may also provide a substitute to natural types of food nutrients, in which anti-nutritional factors such as toxicants, inadvertent agrochemicals, etc.) with the active component could pose a challenge (Berntssen et al., 2010).

20.4 AQUACULTURE DISEASE CONTROL USING NANOTECHNOLOGY

Disease outbreaks are a massive impediment to the sustainability and development of aquaculture. The aquaculture sector (particularly intensive and super-intensive commercial farms) has suffered significant economic losses as a result of epidemics caused by a wide range of infectious organisms (including parasites, viruses, and bacteria) (Huang et al., 2015; Shinn et al., 2015; Tandel et al., 2017; Ahmad and Kaur, 2017, 2018; Ahmad et al., 2019, 2021). Numerous antibiotics used in aquaculture have been reported to be ineffective against most common pathogenic bacteria such as *Yersinia ruckeri*, *Aeromonas hydrophila*, *A. salmonicida*, *Pseudomonas*, *Vibrio*, *Edwardsiella*, and *Listeria* species (Sørum, 2008; Swain et al., 2014). Overuse of these chemicals in aquaculture may be toxic to other organisms, including humans and the environment (Malheiros et al., 2020). Antimicrobial agents, probiotics, and biopharmaceutical/nutritional supplements are now administered via injection system or feed as a safety precaution and whenever symptoms appear. Nanosized tools could be able to identify and treat health issues and infections. Monitoring the effect of vaccines, probiotics, chemicals, and hormones is possible with the help of a smart delivery system that has multipurpose properties such as time controlled and preprogrammed delivery (Murata et al., 1998). Nowadays, nanoscience is the new approach with the ability to be used as antimicrobial agents, inoculations, and disease-causing agent identification tools in fish farming (Shaalan et al., 2016). Various metal NPs (both chemically synthesized and biological) have been proposed as alternative antibacterial agents, with the potential to eliminate or reduce the use of traditional antibiotics in aquaculture.

Vaccination is one technique which has been used in efforts to resolve the disease issues associated with aquaculture (Rather et al., 2011). Utilization of oil emulsions could have serious consequences because some shellfish and fish exhibit abhorrent side effects. In this regard, the utilization of nano-particle carriers of vaccine antigens such as poly-lactide-co-glycolide acid (PLGA) and chitosan (Eldridge et al., 1990; Rather et al., 2011) in conjunction with mild inflammatory elicitors may provide both shellfish and fish with a high level of defense against bacterial diseases and certain viral diseases with live attenuated side effects (Rather et al., 2011). Likewise, oral treatment of these vaccines and site-specific discharge of the active substance for vaccination would then decrease the price and exertion of disease prevention, implementation of vaccine and drug delivery, resulting in sustainable food production at the very same expense of feed intake (Rather et al., 2011). The use of nanocapsules containing NPs allows for mass vaccination in fish. These would be difficult to digest and degrade. These nanocapsules contain simple strand DNA, which is assimilated into the fish cells when applied to water. The ultrasonographic technique is used to tear the capsules, releasing the DNA and inducing an immune response in the fish as a result of the vaccination (Rather et al., 2011).

Vaccines have been used extensively in aquaculture as a protective mechanism against pathogenic organisms so as to safeguard the host from infection caused by these pathogenic organisms. In the fishing industry, the most trustworthy and efficient methods of vaccination are either oral administration or via injection (Shah and Mraz, 2020). The latter, a conventional combination therapy process, necessitates the preparation of vaccines with oil/water dosage forms, which has a number of negative consequences. In general, these formulations, along with the practice guidelines, may every once in a while result in fish mortality (Ji et al., 2015). To address these issues, the scientific world has recently suggested a nano-delivery system as an alternate plan for vaccine delivery in fish that is not only perfectly safe but also more effective (Shah and Mraz, 2020). To date, various encapsulation methods have been created and tested in this context. Alginate particles were identified as exploratory candidates for oral-based delivery of vaccines to aquatic animals (Joosten et al., 1997). For use in fish, alginate particles are usually created by process of emulsification (Leal et al., 2010; Ana et al., 2010; Shah and Mraz, 2020) that has proved as one of the top methods for NP preparation and is easily scalable as well (Reis et al., 2017), and to a minor extent by supplementary techniques such as the orifice-ionic gelation and the spray method (BI, 2010). Information from diverse researchers offers alginate as weight and survival supporter of fish (Fujiki et al., 1994; Chiu et al., 2008), an antigen adjuvant (Tafaghodi et al., 2007; Borges et al., 2008). Also, administration of alginate has also depicted elevated immune-stimulant response of *Cyprinus carpio* and *Epinephelus fuscoguttatus* (Cheng et al., 2008; Huttenhuis et al., 2006; Yeh et al., 2008), also enhanced defense of *Scophthalmus maximus* in opposition to *V. anguillarum* (Skjermo and Bergh, 2004), and brown-marbled grouper and *Epinephelus coioides* against Streptococcus and iridovirus (Cheng et al., 2008; Yeh et al., 2008). Chitosan is usually found in the exoskeleton of insects and crustaceans and is a naturally occurring abundant biopolymer. Chitosan-based formulations are of great importance as bio-nanosensor, vehicles of carrying drugs, coatings and in various medical disciplines such as surgical operations and dentistry (Dutta et al., 2004; Sogias et al., 2008).

20.5 NANOTECHNOLOGY FOR IMPROVED PACKAGING AND SHELF LIFE

Presently, there are several practical benefits in both shellfish and fish production. However, the spoilage tendency of fresh fish has long been a source of important concern and, therefore, any such packaging that can prolong the shelf-life of fresh fish products and by-products would serve to be beneficial (Handy, 2012). This can be accomplished in a variety of ways. First of all, numerous nanopolymers as well as surface coatings are available that offer effective packaging (De Azeredo, 2009), which may reduce the likelihood of soft tissue damage or structural loss to packed fish (Handy, 2012). Unlike some conventional plastics which are non-biodegradable, nanopackaging can be made from natural nanoscopic polymeric materials like chitosan, starch, and cellulose particles (De Azeredo, 2009; Thompson et al., 2004). The meat industry has also been recommended to use both strong and light nanopackaging (Lee, 2010).

Microbial activity also causes fresh fish products to spoil. Antimicrobial and antifungal surfaces can be used to create efficient packaging (De Azeredo, 2009; Moraru et al., 2003; Handy, 2012). NPs could be of great use in order to trace gases such as carbon dioxide that are known to be released by deteriorating fish fillet (Von Bultzingsl owen et al., 2002; Handy, 2012). Fish oils are indeed commonly used in the food sector, and nanoencapsulation using maltodextrin in conjunction with a surface-active biopolymer of whey protein or starch concentrate has been proposed as an approach to maintain the quality, flavor, and aroma of fish oil used in the food processing industry (Jafari et al., 2008).

20.6 GENE TRANSFER

The creation of novel delivery systems for gene transfer symbolizes a pivotal moment in the therapies of numerous genetic disorders. Moreover, developing a reliable and secure delivery mechanism remains a crucial obstacle to achieving successful gene therapy (Rather et al., 2011). Because of their security, consistency, and capacity to be manufactured in massive numbers, non-viral delivery mechanisms have indeed been propagated as choices to viral vectors (Tomlinson and Rolland, 1996). Chitosan (CS), a biocompatible, natural biodegradable, and non-toxic natural polymer extracted from shells of crustaceans, has been studied extensively for pharmaceutical and medical applications, particularly in artificial organs, targeted therapy, drug transfer, gene transfer protein delivery, and so on (Sakai et al., 2002; Lavertu et al., 2006; Wei et al., 2007; Shah and Mraz, 2020). Some approaches use DNA complexes with peptide, protein, lipid or polymeric carriers, as well as ligands competent of aiming DNA structures to receptor molecules on target tissue and binding sites proficient of channeling subcellular DNA traffic to the nucleus. The composition of compounds between DNA and chitosan has produced promising outcomes (Rather et al., 2011). Even though chitosan improves therapeutic efficacy, adding proper ligands to the DNA-chitosan complex appears to improve gene delivery through use of receptor-mediated endocytosis. Such research suggests that chitosan has similar effect to certain other chemically synthesized vectors without the corresponding toxic effects, indicating that it may become a successful gene vehicle for *in vivo* assays (Rather et al., 2011). Thus, there are endless possibilities for nanoscience to offer a significant role in agriculture, fish production, and other aquaculture practices.

There is a wealth of scientific literature expressing a broad analysis of nanotechnological applications in fisheries and aquaculture (Ji et al., 2015; Luis et al., 2017; Masoomi Dezfooli et al., 2018; Shah and Mraz, 2020). Nevertheless, despite its effectiveness, there is a risk that it will make a contribution to aquaculture pollution, which is currently either unspecified or neglected (Mahboob et al., 2014; Huang et al., 2015). Besides, the overuse of antibiotics for the treatment of many diseases, as well as the use of other synthetic substances as growth promoters, does have a negative impact on the aquatic ecosystem (Awad and Awaad, 2017; Baldissera et al., 2018). Such procedures, when combined, cause alarming situations such as impairment in growth and reproduction, death rates, and chemical pathways in both adult and embryo stages of fish, that can eventually lead to massive financial losses of the fishery industry (Khan et al., 2015; Shah and Mraz, 2020). In addition, they elevate possible problems about human health and environmental safety (Seaton et al., 2009; Purohit et al., 2017). As a result, this discussion shows that there are still discrepancies in terms of novel techniques of enumeration in nanoscience that are safe and secure and more environmentally friendly, particularly for lipophilic bioactive compounds that can be employed as natural cures rather than synthetic materials.

20.7 CONCLUSION

Researchers suggest that in the coming years, there will be innumerable nanoscale agroformulations with elevated bioactivity, effectiveness, and specificity. Despite their minimal necessity, the integration of nutraceuticals is often more expensive. As a result, it must be utilized so that waste is minimized in order to maximize efficiency and ensure that the final item is commercially feasible. The progression of nano-delivery systems for such compounds may resolve the commercial issues associated with their use in aquaculture. The potential for using nanomaterials to deliver such supplements in fish meal as well as neutrogenomics studies is enormous. Nanosensors/nanobiosensors for pathogen detection, health monitoring, nanoporous zeolites for slow-release and efficient dosage of nutrients and drugs for fish, nanocapsules for chemical delivery, nanocomposites for plastic coating materials used in packaging, antimicrobial nanoemulsions have revolutionized aquaculture. Nanotechnology has enormous potential applications and benefits in enhancing production and protection against diseases as well. While much more technology diffusion is required to improve the prospective use of nanostructured materials in fisheries, there are currently numerous evidences of hope of this technology's future implementation in fish healthcare, aquaculture water management, live-stock breeding, harvest, and pre- and post-harvest innovations.

KEYWORDS

- **aquaculture**
- **fish**
- **fish health**
- **nanoscience**
- **nitrogen**
- **nutrition**
- **organism**
- **population**
- **stocking**
- **toxicity**

REFERENCES

Acosta, E. (2009). Bioavailability of nanoparticles in nutrient and nutraceutical delivery. *Current Opinion in Colloid and Interface Science, 14*, 3–15.

Ahmad, I., Ahmed, I., Reshi, Q. M., Jan, K., Gupta, A., Dar, S. A., Molnar, K., Kaur, H., Malla, B. A., & Andrabi, S. M. (2021). Morphological, histopathological and molecular characterization of *Myxobolus szekelyianus* n. sp. (Cnidaria: Myxosporea: Myxobolidae) causing acute gill disease in *Schizothorax esocinus* (Heckel, 1838) from River Jhelum of Kashmir Himalayan region, India. *Aquaculture Research, 52*, 6537–6549.

Ahmad, I., & Kaur, H. (2017). Redescription and histopathology of two species of myxozoans infecting gills of fingerlings of Indian major carps. *Journal of FisheriesSciences.com, 11*(1), 1–10.

Ahmad, I., & Kaur, H. (2018). Prevalence, site and tissue preference of myxozoan parasites infecting gills of cultured fingerlings of Indian major carps in District Fatehgarh Sahib, Punjab (India). *Journal of Parasitic Diseases, 42*, 559–569.

Ahmed, I., & Ahmad, I. (2020). Effect of dietary protein levels on growth performance, hematological profile and biochemical composition of fingerlings rainbow trout, *Oncorhynchus mykiss* reared in Indian Himalayan region. *Aquaculture Reports, 16*, 100268.

Ahmed, I., Ahmad, I., Dar, S. A., Awas, M., Kaur, H., Ganie, B. A., & Shah, B. A. (2019). *Myxobolus himalayaensis* sp. nov. (Cnidaria: Myxozoa) parasiting *Schizothorax richardsonii* (Cyprinidae: Schizothoracinae) from River Poonch in North West Himalaya, India. *Aquaculture Reports, 14*, 100192.

Ahmad, I., Ahmed, I., Dar, N. A. (2021a). Effects of dietary leucine levels on growth performance, hemato-biochemical parameters, liver profile, intestinal enzyme activities and target of rapamycin signaling pathway related gene expression in rainbow trout, *Oncorhynchus mykiss* fingerlings. *Aquaculture Nutrition, 27,* 1837–1852.

Ahmad, I., Ahmed, I., & Dar, N. A. (2021b). Dietary valine improved growth, immunity, enzymatic activities and expression of TOR signaling cascade genes in rainbow trout, *Oncorhynchus mykiss* fingerlings. *Scientific Reports, 11*, 22089.

Ahmad, I., Ahmed, I., & Dar, N. A. (2022). Effects of dietary isoleucine on growth performance, enzymatic activities, antioxidant properties and expression of TOR related genes in rainbow trout, *Oncorhynchus mykiss* fingerlings. *Aquaculture Research,* 1–17.

Aklakur, M., Rather, M. A., & Kumar, N. (2016). Nanodelivery: An emerging avenue for nutraceuticals and drug delivery. *Critical Reviews in Food Science and Nutrition, 56*, 2352–2361.

Ana, I., Saint-Jean, S. R., Perez-Prieto, S. I. (2010). Immunogenic and protective effects of an oral DNA vaccine against infectious pancreatic necrosis virus in fish. *Fish and Shellfish Immunology, 28*, 562–570.

Ashouri, S., Keyvanshokooh, S., Salati, A. P. et al. (2015). Effects of different levels of dietary selenium nanoparticles on growth performance, muscle composition, blood biochemical profiles and antioxidant status of common carp (*Cyprinus carpio*). *Aquaculture, 446*, 25–29.

Aulenta, F., Hayes, W., & Rannard, S. (2003). Dendrimers: A new class of nanoscopic containers and delivery devices. *European Polymer Journal, 39*, 1741–1771.

Awad, E., & Awaad, A. (2017). Role of medicinal plants on growth performance and immune status in fish. *Fish and Shellfish Immunology, 67*, 40–54.

Baldissera, M. D., Souza, C. F., Zeppenfeld, C. C., Descovi, S., Machado, V. S., Santos, R. C. V. et al. (2018). Efficacy of dietary curcumin supplementation as bactericidal for silver catfish against *Streptococcus agalactiae. Microbial Pathogenesis, 116*, 237–240.

Banerjee, S., Kumar, V., & Pathak, A. N. (2014). Drug effects on aquaculture and it's remediation. *Journal of Drug Delivery and Therapeutics, 4*(1), 117–122.

Berntssen, M. H. G., Julshamn, K., & Lundebye, A. K. (2010). Chemical contaminants in aquafeeds and Atlantic salmon (*Salmo salar*) following the use of traditional-versus alternative feed ingredients. *Chemosphere, 78*, 637–646.

Bhattacharyya, A., Reddy, J. S., Hasan M. M., Adeyemi, M., Mary, R. R., Naika, H. R. (2015). Nanotechnology: A unique future technology in aquaculture for food security. *International Journal of Bioassay, 4*, 4115–4416.

Bi, D. (2010). Oral vaccination against lactococcosis in rainbow trout (Oncorhynchus mykiss) using sodium alginate and poly(lactide-co-glycolide) carrier. *Kafkas Universitesi Veteriner Fakultesi Dergisi, 16*, S211–S217.

Borges, O., Silva, M., de Sousa, A., Borchard, G., Junginger, H. E., & Cordeiro-da-Silva, A. (2008). Alginate coated chitosan nanoparticles are an effective subcutaneous adjuvant for hepatitis B surface antigen. *International Immunopharmacology, 8*, 1773–1780.

Bouwmeester, H., Dekkers, S., Noordam, M. Y., Hagens, W. I., Bulder, A. S., de Heer, C., ten Voorde, S. E. C. G., Wijnhoven, S. W. P., Marvin, H. J. P., & Sips, A. J. A. M. (2009). Review of health safety aspects of nanotechnologies in food production. *Regulatory Toxicology and Pharmacology, 53*, 52–62.

Boyd, C. E. (2001, June 27–30). Decision support systems for water resources management. In *AWRA/UCOWR Summer Specialty Conference, Snowbird, Utah* (p. 153).

Boyd, C. E., & Tucker, C. S. (1998). Ecology of aquaculture ponds. In *Pond aquaculture water quality management*. Boston, MA: Springer. https://doi.org/10.1007/978-1-4615-5407-3_2.

Can, E., Kizak, V., Kayim, M., Can, S.S., Kutlu, B., Ates, M., ... & Demirtas, N. (2011). Nanotechnological applications in aquaculture-seafood industries and adverse effects of nanoparticles on environment. *Journal of Materials Science and Engineering, 5*(5), 605–609.

Carriquiriborde, P., Handy, R. D., & Davies, S. J. (2004). Physiological modulation of iron metabolism in rainbow trout (*Oncorhynchus mykiss*) fed low and high iron diets. *Journal of Experimental Biology, 207*, 75–86.

Chaudhry, Q., Scotter, M., Blackburn, J., Ross, B., Boxall, A., Castle, L., Aitken, R., & Watkins, R. (2008). Applications and implications of nanotechnologies for the food sector. *Food Additives and Contaminants, 25*(3), 241–258.

Cheng, A. C., Chen, Y. Y., & Chen, J. C. (2008). Dietary administration of sodium alginate and j-carrageenan enhances the innate immune response of brown-marbled grouper *Epinephelus fuscoguttatus* and its resistance against *Vibrio alginolyticus*. *Veterinary Immunology and Immunopathology, 121*, 206–215.

Chiu, S. T., Tsai, R. T., Hsu, J. P., Liu, C. H., & Cheng, W. (2008). Dietary sodium alginate administration to enhance the non-specific immune responses, and disease resistance of the juvenile grouper *Epinephelus fuscoguttatus*. *Aquaculture, 277*, 66–72.

De Azeredo, H. M. C. (2009). Nanocomposites for food packaging applications. *Food Research International, 42*, 1240–1253.

Donbrow, M. (1991). *Microcapsules and nanoparticles in medicine and pharmacy*. Boca Raton, FL: CRC Press.

Dutta, P. K., Dutta, J., & Tripathi, V. (2004). Chitin and chitosan: Chemistry, properties and applications. *Journal of Scientific and Industrial Research, 63*, 20–31.

ElBasuini, M. F., El-Hais, A. M., Dawood, M. A. O., et al. (2017). Effects of dietary copper nanoparticles and vitamin C supplementations on growth performance, immune response and stress resistance of red sea bream, *Pagrus major*. *Aquaculture Nutrition, 23*, 1329–1340.

Eldridge, J. H., Hammond, C. J., Meulbroek, J. A., Staas, J. K., Gilley, R. M., & Tice, T. R. (1990). Controlled vaccine release in gut-associated lymphoid tissues. I. Orally administered biodegradable microspheres target the Peyer's patches. *Journal of Controlled Release, 11*, 205–214.

ETC (Action Group on Erosion, Technology and Concentration). (2003). Down on the farm: The impact of nanoscale technologies on food and agriculture. http://www.etcgroup.org/en/materials/publications.html?pub_id=80).

Fan, W., Yan, W., Xu, Z., & Ni, H. (2012). Erythrocytes load of low molecular weight chitosan nanoparticles as a potential vascular drug delivery system. *Colloids and Surfaces B: Biointerfaces, 95*, 258–265.

FAO. (2020). *The state of the world fisheries and aquaculture 2020*. Rome: FAO.

FOE. (2008). Out of the laboratory and onto our plates: Nanotechnology in food and agriculture. [http://www.foe.org/].

Food Safety Authority of Ireland. (2008). The relevance of food safety of applications of nanotechnology in the food and feed industries. Court lower abbey street, Abbey, Dublin 1. www.fsai.ie.

Fujiki, K., Matsuyama, H., & Yano, T. (1994). Protective effect of sodium alginates against bacterial infection in common carp, *Cyprinus carpio* L. *Journal of Fish Diseases, 17*, 349–355.

Gabriel, N. N., Habte-Tsion, H. M., & Haulofu, M. (2022). Perspectives of nanotechnology in aquaculture: Fish nutrition, disease, and water treatment. In *Emerging nanomaterials for advanced technologies* (pp. 463–485). Cham: Springer.

Gillies, E., & Frechet, J. (2005). Dendrimers and dendritic polymers in drug delivery. *Drug Discovery Today, 10*, 35–43.

Handy, R. D. (2012). FSBI briefing paper: Nanotechnology in fisheries and aquaculture. *Fisheries Society of the British Isles*, 1–29.

Handy, R. D., & Poxton, M. G. (1993). Nitrogen pollution in mariculture – toxicity and excretion of nitrogenous compounds by marine fish. *Reviews in Fish Biology and Fisheries, 3*, 205–241.

Hoet, P., Bruske-Hohlfeld, I., & Salata, O. (2004). Nanoparticles-known and unknown health risks. *Journal of Nanobiotechnology, 2*, 12–27.

Huang, S., Wang, L., Liu, L., et al. (2015). Nanotechnology in agriculture, livestock, and aquaculture in China. A review. *Agronomy for Sustainable Development, 35*, 369–400.

Huttenhuis, H. B. T., Ribeiro, A. S. P., Bowden, T. J., Bavel, V. C., Taverne-Thiele, A. J., Rombout, J. H. W. M. (2006). The effect of oral immuno-stimulation in juvenile carp (*Cyprinus carpio* L.). *Fish and Shellfish Immunology, 21*, 261–271.

Jafari, S. M., & McClements, D. J. (2017). Nanotechnology approaches for increasing nutrient. Bioavailability. *Advances in Food and Nutrition Research, 81*, 1–30.

Jafari, S. M., Assaidpoor, E., Bhandari, B., & He, Y. H. (2008). Nano-particle encapsulation of fish oil by spray drying. *Food Research International, 41*, 172–183.

Ji, J., Torrealba, D., Ruyra, A., & Roher, N. (2015). Nanodelivery systems as new tools for immunostimulant or vaccine administration: Targeting the fish immune system. *Biology, 4*, 664–696.

Jiang, Y., Chekuri, S., Fang, R. H., et al. (2019). Engineering biological interactions on the nanoscale. *Current Opinion in Biotechnology, 55*, 1–8.

Joosten, P., Tiemersma, E., Threels, A., Dhieux-Caumartin, C., & Rombout, J. (1997). Oral vaccination of fish against *Vibrio anguillarum* using alginate microparticles. *Immunological Aspects of Oral Vaccination in Fish, 56*, 471–485.

Joye, I. J., Davidov-Pardo, G., & McClements, D. J. (2014). Nanotechnology for increased micronutrient bioavailability. *Trends in Food Science and Technology, 40*, 168–182.

Khan, M. S., Jabeen, F., Qureshi, N. A., Asghar, M. S., Shakeel, M., & Noureen, A. (2015). Toxicity of silver nanoparticles in fish: A critical review. *Journal of Biological Environment Science, 6*, 211–227.

Khosravi-Katuli, K., Prato, E., Lofrano, G., et al. (2017). Effects of nanoparticles in species of aquaculture interest. *Environmental Science and Pollution Research, 24*, 17326–17346.

Korni, F. M. M., & Khalil, F. (2017). Effect of ginger and its nanoparticles on growth performance, cognition capability, immunity and prevention of motile *Aeromonas septicaemia* in *Cyprinus carpio* fingerlings. *Aquaculture Nutrition, 23*, 1492–1499.

Kumari, A., & Yadav, S. K. (2014). Nanotechnology in agri-food sector. *Critical Reviews in Food Science and Nutrition, 54*, 975–984.

Lavertu, M., Methot, S., Tran-Khanh, N., & Buschmann, M. D. (2006). High efficiency gene transfer using chitosan/DNA nanoparticles with specific combinations of molecular weight and degree of deacetylation. *Biomaterials, 27*, 4815–4824.

Leal, C. A. G., Carvalho-Castro, G. A., Sacchetin, P. S. C., Lopes, C. O., Moraes, A. M., & Figueiredo, H. C. P. (2010). Oral and parenteral vaccines against *Flavobacterium columnare*: Evaluation of humoral immune response by ELISA and *in vivo* efficiency in Nile tilapia (*Oreochromis niloticus*). *Aquaculture International, 18*, 657–666.

Lee, K. T. (2010). Quality and safety aspects of meat products as affected by various physical manipulations of packaging materials. *Meat Science, 86*, 138–150.

Luis, A. I. S., Campos, E. V. R., de Oliveira, J. L., & Fraceto, L. F. (2017). Trends in aquaculture sciences: From now to use of nanotechnology for disease control. *Reviews in Aquaculture, 11*, 119–132.

Mahboob, S., Al-Balawi, H. F., Al-Misned, F., Al-Quraishy, S., & Ahmad, Z. (2014). Tissue metal distribution and risk assessment for important fish species from Saudi Arabia. *Bulletin of Environmental Contamination and Toxicology, 92*(1), 61–66.

Malheiros, D. F., Sarquis, I. R., Ferreira, I. M., et al. (2020). Nanoemulsions with oleoresin of *Copaifera reticulata* (Leguminosae) improve anthelmintic efficacy in the control of monogenean parasites when compared to oleoresin without nanoformulation. *Journal of Fish Diseases, 43*, 687–695.

Dezfooli, M. S., Gutierrez-Maddox, N., Alfaro, A., & Seyfoddin, A. (2018). Encapsulation for delivering bioactives in aquaculture. *Reviews in Aquaculture, 10*, 1–30.

Midhun, S. J., Arun, D., Edatt, L., et al. (2016). Modulation of digestive enzymes, GH, IGF-1 and IGF-2 genes in the teleost, Tilapia (*Oreochromis mossambicus*) by dietary curcumin. *Aquaculture International, 24*, 1277–1286.

Mohammadi, N., & Tukmechi, A. (2015). The effects of iron nanoparticles in combination with *Lactobacillus casei* on growth parameters and probiotic counts in rainbow trout (*Oncorhynchus mykiss*) intestine. *Journal of Veterinary Research, 70*, 47–53.

Moraru, C., Panchapakesan, C., Huang, Q., Takhistov, P., Liu, S., & Kokini, J. (2003). Nanotechnology: A new frontier in food science. *Food Technology, 57*, 24–29.

Murata, J., Ohya, Y., & Ouchi, T. (1998). Design of quaternary chitosan conjugate having antennary galactose residues as a gene delivery tool. *Carbohydrate Polymers, 32*(2), 105–109.

Naylor, R., & Burke, M. (2005). Aquaculture and ocean resources: Raising tigers of the sea. *Annual Review of Environment and Resources, 30*, 185–218.

Ninh, N. T. H., Dung, N. M., & Cuong, H. N. (2016, November 12). Water quality management for sustainable aquaculture production in the Mekong Delta. In *International Conference of the Mekong, Salween and Red River: Sharing Knowledge and Perspectives Across Borders, Faculty of Political Science*. Bangkok: Chulalongkorn University.

Purohit, R., Mittal, A., Dalela, S., Warudkar, V., Purohit, K., & Purohit, S. (2017). Social, environmental and ethical impacts of nanotechnology. *Materials Today: Proceedings, 4*, 5461–5467.

Rajeshkumar, S., Venkatesan, C., Sarathi, M., et al. (2009). Oral delivery of DNA construct using chitosan nanoparticles to protect the shrimp from white spot syndrome virus (WSSV). *Fish Shellfish and Immunology, 26*, 429–437.

Rather, M. A., Bhat, I. A., Sharma, N., et al. (2017). Synthesis and characterization of *Azadirachta indica* constructed silver nanoparticles and their immunomodulatory activity in fish. *Aquaculture Research, 48*, 3742–3754.

Rather, M. A., Sharma, R., Aklakur, M., Ahmad, S., Kumar, N., Khan, M., & Ramya, V. L. (2011). Nanotechnology: A novel tool for aquaculture and fisheries development. A prospective mini-review. *Fisheries and Aquaculture Journal, 16*(1–5), 3.

Reilly, R. M. (2007) Carbon nanotubes: Potential benefits and risks of nanotechnology in nuclear medicine. *The Journal of Nuclear Medicine, 48*, 1039.

Reis, C. P., Neufeld, R. J., & Veiga, F. (2017). Preparation of drug-loaded polymeric nanoparticles. In L. P. Balogh (Ed.), *Nanomedicine in cancer* (pp. 197–240). Singapore: Pan Stanford Publishing Pte Ltd.

Rodrigues, S. M., Demokritou, P., Dokoozlian, N., Hendren, C. O., Karn, B., Mauter, M. S., et al. (2017). Nanotechnology for sustainable food production: Promising opportunities and scientific challenges. *Environmental Science: Nano, 4*, 767–781.

Romero-Perez, A., Garcia-Garcia, E., Zavaleta-Mancera, A., Ramirez-Bribiesca, J. E., Revilla-Vazquez, A., Hernandez-Calva, L. M., Lopez-Arellano, R., & Cruz-Monterrosa, R. G. (2010). Designing and evaluation of sodium selenite nanoparticles *in vitro* to improve selenium absorption in ruminants. *Veterinary Research Communications, 34*, 71–79.

Sahu, S., Das, B. K., Mishra, B. K., Pradhan, J., Samal, S. K., & Sarangi, N. (2008). Effect of dietary *Curcuma longaon* enzymatic and immunological profiles of rohu, *Labeo rohita* (Ham.), infected with *Aeromonas hydrophila*. *Aquaculture Research, 39*(16), 1720–1730.

Sakai, S, Ono, T, Ijima, H, & Kawakami, K. (2002). *In vitro* and *in vivo* evaluation of alginate/sol-gel synthesized aminopropyl- silicate/alginate membrane for bioartificial pancreas. *Biomaterials, 23*, 4177–4183.

SciDev.Net. (2013). Nanotech filter may protect farmed fish from fungus. London, UK: SciDev.Net. Available from: http://www.scidev.net/global/news/

nanotech-filter-may-protect-farmed-fish-from-fungus.html (accessed December 4, 2022)

Seaton, A., Tran, L., Aitken, R., Donaldson, K. (2009). Nanoparticles, human health hazard and regulation. *Journal of the Royal Society Interface, 7*, S119–S129.

Sekhon, B. S. (2014). Nanotechnology in agri-food production: An overview. *Nanotechnology, Science and Applications*, *7*, 31.

Semo, E., Kesselman, E., Danino, D., & Livney, Y. D. (2007). Casein micelle as a natural nano-capsular vehicle for nutraceuticals. *Food Hydrocolloids*, *21*, 936–942.

Shaalan, M., Saleh, M., El-Mahdy, M., et al. (2016) Recent progress in applications of nanoparticles in fish medicine: A review. *Nanomedicine Nanotechnology*, *12*, 701–710.

Shah, B. R., & Mraz, J. (2020). Advances in nanotechnology for sustainable aquaculture and fisheries. *Reviews in Aquaculture*, *12*(2), 925–942.

Shinn, A. P., Pratoomyot, J., Bron, J. E., et al. (2015). Economic costs of protistan and metazoan parasites to global mariculture. *Parasitology, 142*, 196.

Sikder, M. N. A., Min, W. W., Ziyad, A. O., et al. (2016). Sustainable treatment of aquaculture effluents in future-A review. *International Research Journal of Advanced Engineering and Science, 1*, 190–193.

Silva, A. A. (2010). Nanotechnology applications and approaches for neuroregeneration and drug delivery to the central nervous system. *Annals of the New York Academy of Sciences, 1199*, 221–230.

Singh, M. K., & Dutta, M. P. (2020). Scope of nanotechnology in aquaculture: An overview. *Research Journal of Agricultural Sciences, 11*(5), 1203–1208.

Skjermo, J., & Bergh, Ø. (2004). High-M alginate immunostimulation of Atlantic halibut (*Hippoglossus hippoglossus* L.) larvae using Artemia for delivery, increases resistance against vibriosis. *Aquaculture, 238*, 107–113.

Sogias, I. A., Williams, A. C., & Khutoryanskiy, V. V. (2008). Why is chitosan mucoadhesive? *Biomacromolecules, 9*, 1837–1842.

Sørum, H. (2008). Antibiotic resistance associated with veterinary drug use in fish farms. In Ø. Lie (Ed.), *Improving farmed fish quality and safety. Food science, technology, and nutrition* (pp. 157–182). Woodhead Publishing.

Swain, P., Nayak, S. K., Sasmal, A., et al. (2014). Antimicrobial activity of metal-based nanoparticles against microbes associated with diseases in aquaculture. *World Journal of Microbiology and Biotechnology*, *30*, 2491–2502.

Tafaghodi, M., Sajadi Tabasi, S. A., & Payan, M. (2007). Alginate microsphere as a delivery system and adjuvant for autoclaved *Leishmania major* and *Quillaja saponin*: Preparation and characterization. *Iranian Journal of Pharmaceutical Sciences, 3*, 61–68.

Tandel, G. M., John, K. R., George, M. R., et al. (2017). Current status of viral diseases in Indian shrimp aquaculture. *Acta Virologica, 61*, 131–137.

Thompson, R. C., Olsen, Y., Mitchell, R. P., Davis, A., Rowland, S. J., John, A. W. G., McGonigle, D., & Russell, A. E. (2004). Lost at sea: Where is all the plastic? *Science, 204*, 838.

Thulasi, A., Rajendran, D., Jash, S., et al. (2013). Nanobiotechnology in animal nutrition. In K. T. Sampath, J. Ghosh, & R. Bhatta (Eds.), *Animal nutrition and reproductive physiology*. New Delhi: Satish Serial Publishing House.

Tiede, K., Boxall, A. B. A., Tear, S. P., Lewis, J., David, H. & Hassellov, M. (2008). Detection and characterization of engineered nanoparticles in food and the environment. *Food Additives and Contaminants: Part A*, *25*, 7795–7821.

Tomlinson, E., & Rolland, A. P. (1996). Controllable gene therapy pharmaceutics of non-viral gene delivery systems. *Journal of Controlled Release, 1*, 357–372.

Torchilin, V. P. (2006). *Nanoparticulates as drug carriers*. London, UK: Imperial College Press.

Toyokawa, H., Nakao, A., Bailey, R. J., Nalesnik, M. A., Kaizu, T., Lemoine, J. L., et al. (2008). Relative contribution of direct and indirect allorecognition in developing tolerance after liver transplantation. *Liver Transplantation, 14*, 346–357.

Von Bültzingslöwen, C., Mcevoy, A. K., McDonagh, C., MacCraith, B. D., Klimant, I., Krause, C., & Wolfbeis, O. S. (2002). Sol-gel based optical carbon dioxide sensor employing dual luminophore referencing for application in food packaging technology. *Analyst, 127*, 1478–1483.

Wei, H., Zhang, X. Z., Chen, W. Q., Cheng, S. X., Zhuo, R. X. (2007). Self-assembled thermosensitive micelles based on poly (L-lactide-star block-N-isopropylacrylamide) for drug delivery. *Journal of Biomedical Materials Research Part A, 83A*, 980–989.

Wu, L. P., Ficker, M., Christensen, J. B., Trohopoulos, P. N., & Moghimi, S. M. (2015). Dendrimers in medicine: Therapeutic concepts and pharmaceutical challenges. *Bioconjugate Chemistry, 26*, 1198–1211.

Xu, W., Huang, K., Jin, W., Luo, D., Liu, H., Li, Y., et al. (2018). Catalytic and anti-bacterial properties of biosynthesized silver nanoparticles using native inulin. *RSC Advances, 8*, 28746–28752.

Yeh, S. P., Chang, C. A., Chang, C. Y., Liu, C. H., & Cheng, W. (2008). Dietary sodium alginate administration affects fingerling growth and resistance to Streptococcus sp., and iridovirus, and juvenile non-specific immune responses of the orange-spotted grouper *Epinephelus coioides*. *Fish and Shellfish Immunology, 25*, 19–27.

Zhou, X., Wang, Y., Gu, Q., & Li, W. (2009). Effects of different dietary selenium sources (selenium nanoparticle and selenomethionine) on growth performance, muscle composition and glutathione peroxidase enzyme activity of crucian carp (*Carassius auratus gibelio*). *Aquaculture, 291*, 78–81.

CHAPTER 21

Nanotechnology as an Emerging Diagnostic and Therapeutic Tool in Fish Health Management

MEHAK HAFEEZ[1,*], IRFAN AHMAD[1], KEEZIA KHURSHID[2], and SHAHZIYA SHAH[3]

[1]*Division of Fish Genetics and Biotechnology, Faculty of Fisheries, Sher-e-Kashmir University of Agricultural Sciences and Technology of Kashmir, Rangil, Ganderbal, Jammu and Kashmir, India*

[2]*Division of Aquatic Animal Health Management, Faculty of Fisheries, Sher-e-Kashmir University of Agricultural Sciences and Technology of Kashmir, (SKUAST-K), Rangil, Ganderbal, Jammu and Kashmir, India*

[3]*Division of Fisheries Resource Management, Faculty of Fisheries, Sher-e-Kashmir University of Agricultural Sciences and Technology of Kashmir, Rangil, Ganderbal, Jammu and Kashmir, India*

*Corresponding author

ABSTRACT

The science of nanotechnology has colossal prospects of innovative advancements for the sustainable development of aquaculture in terms of economic stability and global food security. At present, several kinds of nanotechnology-based applications are being used in aquaculture systems to improve efficiency, increase productivity, reduce environmental impacts, and lower costs and monetary losses. Contemporary attempts have been made in the field of finfish and shellfish health management by employing nanoparticles (NP)-based drugs, vaccines, medicines,

Recent Advances in Sustainable Aquaculture and Fisheries. Shabir Ahmad Dar, Imtiyaz Qayoom, Gohar Bilal Wani, Adnan Abubakr, and Seemab Zehra (Eds.)

DOI: 10.1201/9781998511372-21

and biosensors. These NP-based technologies have made it possible for aquaculturists to have easy and timely detection, prevention, control, and subsequent monitoring of various fish diseases. These nanotechnologies have proved to act as a significant tool to improve fish and shellfish health, thereby enhancing aquaculture production. However, there are certain risks associated with the application of NPs on fish and human health and natural ecosystems. The related concerns of NPs are balanced with their potential benefits and these ought not preclude us from applying nanotechnology-based approaches in the aquaculture sector. Toxicological studies so far suggest that NPs are less harmful than other commonly employed chemicals in veterinary and fisheries. But their use should be based on regulatory guidelines, linked with careful monitoring.

21.1 INTRODUCTION

The term "nanotechnology" was coined by Professor Norio Taniguchi of Tokyo Science University and by definition is the applied branch of science that deals with the development and utilization of the so-called nanoparticles (NPs). NPs are a special category of particles having a size in the range of 1–100 nm. The word "nano" is Greek in origin and means "dwarf." NPs are normally found in nature as ultrafine nano-sized particles but they can also be assembled and specifically engineered for an explicit intent (Oberdörster et al., 2005). On the nano-scale dimension, all the attributes (thermal, magnetic, catalytic, and electric) of a substance alter which countenances these NPs to entail remarkable scientific utilities. Due to these unique characteristics of NPs, they can be used in many new ways that otherwise are not conceivable for substances having large-sized particles in nature. Over the last couple of decades, the field of nanotechnology has grown significantly and has revolutionized nearly every branch of science and technology.

In biomedical science, NPs are employed for the diagnosis and treatment of several life-threatening diseases, the development of vaccines especially against viral pathogens, the controlled delivery of life-saving drugs, and the delivery of target genes during gene therapy (Cavalieri et al., 2014). In the field of aquaculture, NPs are utilized in both direct (fish health and fish nutrition) as well as indirect (pond-water treatment, fish packaging, fish tagging, and barcoding) ways. This chapter essentially deals with the direct application of nanotechnology in fish health management, with an emphasis on disease diagnosis and therapeutics.

21.2 NANOTECHNOLOGY IN DISEASE CONTROL AND DIAGNOSIS

At present, the global fish production is approximately 177.8 million tons, which is expected to increase to 181 million tons, of which over 60% will be contributed by aquaculture alone by the year 2030 (World Bank, 2013; Park et al., 2018; FAO, 2020). The recent intensification in aquaculture activities to cater to the ever-growing global food (fish protein) demand has resulted in an exponential increase in fish production. However, simultaneously it has led to an increased incidence of many deadly diseases and the entry of novel pathogens into the aquaculture systems present worldwide, thus challenging the stability and the sustainability of the sector (Pulkkinen et al., 2010). Conventionally, the majority of fish farmers apply hazardous chemicals or regular broad-spectrum antibiotics for the on-farm treatment of fish diseases. But the overuse of antibiotics for the prevention and control of diseases, at present, is an emerging global concern. Several studies have proven the existence of "superbugs," microbes having very high antibiotic resistance and mortality rate, which ultimately end in the death of the infected organism (Hafeez et al., 2021). However, nanotechnology can be a critical scientific tool to overcome the shortcomings of antibiotic abuse in aquaculture. NPs can help in minimizing the dosage of antibiotics while maintaining their therapeutic efficacy for longer periods and, in turn, subsiding their side effects. NPs when conjugated with antibiotic-resistant bacteria can be administered as vaccines for disease prevention, thus operating as potential antibacterial agents (Shaalan et al., 2016).

There is another key hindrance to the use of traditional drugs/vaccines via feed and injections to keep the fish healthy and that is their nonbioavailability. In this regard, NP-conjugated drugs have a remarkable potential to safeguard fish species from diseases because of their antiviral, antiparasitic, antifungal, and antibacterial characteristics. These encapsulated drugs/vaccines are highly target-specific which subsequently increases their chances of absorption by cells. Furthermore, NP-conjugated drugs have additional advantages such as less leaching of active drug ingredients into the water, prevention of drug oxidation, drug preservation for longer storage, high thermal stability, and easy and quick production process (Aklakur et al., 2015). To maintain bioavailability, nanocapsules resistant to digestion and enzymatic degradation have been successfully used for mass vaccination of fish. These nanocapsules contain short DNA strands that are absorbed by the target fish species directly from the surrounding waters. It is reported that NP-coatings such as poly-lactide-co-glycolide acid (PLGA) or chitosan in vaccines confer an amplified immunity to finfish and shellfish species against bacterial and viral diseases (Dubey et al., 2016). NP-encapsulated DNA vaccines against bacterial infection in Asian carp and white spot syndrome virus (WSSV) in shrimp have been developed (Rajeshkumar et al., 2009). For shrimp immunization by immersion and feeding techniques, polyanhydride NPs have been used for vaccine encapsulation and controlled release of antigens (Ross et al., 2014). Silica-based NPs are suitable for the delivery of drugs due to their porous nature and high adsorption capacity (Stromme et al., 2009). Lately, NPs have been utilized for the delivery of probiotics in fish to improve their health status. A novel NP-tagged probiotic, G-Clear Plus, was recently developed to prevent white gut disease caused by *Vibrio* spp. in fish (Moges et al., 2020).

It is reported that NPs containing silver metal (Ag) possess antimicrobial properties. The Ag^+ ions released from these NPs interact electrostatically with the cell membrane and cell wall components of the bacteria which results in the formation of pores and ultimately the death of bacterial cells (Lara et al., 2010; Huang et al., 2011; Mathur et al., 2014). These cations also link to the membrane ion channels to disrupt ion transport, interfere with the protein assembly of ribosomes, alter the enzymatic activity, and cause DNA breakage in microbial cells. AgNPs are effective against methicillin-resistant *Staphylococcus aureus* (Jeong et al., 2005). Aspera-constructed AgNPs were observed to have antibacterial potential against *Aeromonas hydrophila* infections in catla (*Catla catla*) (Antony et al., 2013). Additionally, Rather et al. (2016) insinuated the antibacterial and immunomodulatory role of *Azadirachta indica*-derived AgNPs against *A. hydrophila* infection in Mrigal (*Cirrhinus mrigala*), where higher resistance and survival rate were identified in the fish treated with these NPs compared to the control group when infected with the bacteria. Nanocomposites of chitosan with silver were reported to have antimicrobial action against the fish pathogen, *Aliivibrio salmonicida* (Dananjaya et al., 2016).

AgNPs derived from the bacterium *Aeromonas sobria* act as an antibacterial agent against a wide range of animal (including fish and human) pathogens, namely, *Providencia rettgeri*, *Hafnia alvei*, *Citrobacter braakii*, *A. hydrophila*, *Escherichia hermannii*, *Escherichia coli*, *Escherichia cloacae*, and *Morganella morganii* (Erdem et al., 2018). Several studies have shown that the AgNP nutraceuticals improved resistance in fish against pathogens such as *Lactococcus garvieae*, *Vibrio harveyi*, *Streptococcus iniae*, and *Vibrio parahaemolyticus* (Vaseeharan et al., 2009; Swain et al., 2014). These nutraceuticals also have antiparasitic and antifungal activity against *Ichthyophthirius multifiliis*, *Aphanomyces invadans*, and *Fusarium oxysporum* (Daniel et al., 2016). Johari et al. (2015) used AgNPs to treat the rainbow trout eggs affected with the fungal infection caused by *Saprolegnia*. Shaalan et al. (2017) found that AgNPs and zinc oxide NPs inhibited the growth and survival of *Aeromonas salmonicida*, *A. invadans*, and *Yersinia ruckeri* in the infected fish.

As studied and tested in numerous fish species, NPs containing copper metal (Cu) have also been found to act as antimicrobial agents against many fish pathogens. Titanium dioxide NPs display robust *in vitro* antibacterial activity and this property has been proved by their strong immunomodulatory action on fish neutrophils (Cheng et al., 2009; Jovanovic et al., 2011). When calcium phosphate NPs are conjugated with the S-layer proteins of *A. hydrophila*, they stimulate primary as well as secondary immune responses in infection and, thus, provide a broad immunity against bacterial disease (Behera and Swain, 2011). NPs derived from ginger are capable of inhibiting the bacterial infection known as Motile Aeromonas Septicemia (MAS) in the Asian carp fingerlings (Korni and Khalil, 2017).

Over the past 10 years, there has been an increasing trend in the aquaculture industry toward the use of nano-constructed diagnostic kits for disease detection (pathogen identification). These NP-based kits can detect pathogens in both aquatic ecosystems as well as inhabiting fish fauna which leads to subsequent disease monitoring and control. At present, nanobiosensors are being developed which can detect, even trace, quantities of bacteria, parasites, viruses, and pollutants such as heavy metals in water and fish (Chen et al., 2016). Electrical nanosensors are more likely to detect single virus particles (Patolsky et al., 2004). These sensors become exceptionally beneficial to prevent disease outbreaks and consequent economic losses in large commercial aquaculture systems. Using these sensors, the etiological agent is identified before it spreads frenziedly and corresponding treatment is applied to control the same. In this way, nanotechnology acts as a potent tool for early disease detection and pathogen eradication.

The most commonly used NPs for disease diagnosis are gold (Au) NPs, which have been helpful for the detection of *A. salmonicida* and *Aeromonas invadans* in many fish species to date (Saleh et al., 2011; Kuan et al., 2013). IgG-capped AuNPs are reported to detect *Staphylococcus pyrogenes* and *S. aureus* in fish (Roy et al., 2012). AuNPs conjugated with *A. salmonicida* antibodies have been successfully employed to detect bacterial furunculosis infection in fish (Saleh et al., 2012). Guo et al. (2016) created an immunomagnetic microfluidic chip containing indium-tin-oxide NPs for the detection of *S. aureus*. Compared to the universal colony counting method, this system was superior in terms of less detection time and high sensitivity and specificity, without actually cultivating the colonies (Guo et al., 2016). For detecting yellow head virus (YHV) and WSSV in shrimps, a colorimetric method employing AuNPs and loop-mediated isothermal amplification was developed (Jaroenram et al., 2012; Seetang-Nun et al., 2013). A similar approach was used to develop a sensor containing AuNPs to detect spring viremia of carp virus (SVCV) and DNA herpesvirus 3 in fish (Saleh et al., 2012; Saleh and El-Matbouli, 2015). For the detection of nervous necrosis virus (NNV), two methods were reported, viz., using magnetic NPs coated with rabbit anti-NNV antibodies and developing a biosensor with AuNPs (Yang et al., 2012; Toubanaki et al., 2015).

21.3 NANOTECHNOLOGY IN DRUG/VACCINE DELIVERY

In aquacultural systems, generally, the drugs are administered in three ways of immersion, oral intake, and injection. Compared to the other two, the oral drug/vaccine through the feed is especially suitable for targeted delivery using two key approaches either known as nano-encapsulated or nano-coated drug delivery exploiting nanotechnological advances or receptor/ligand-specific delivery (Aklakur et al., 2015). In organisms, including fish, the targeting of the vast majority of medications remains ineffective owing to their rapid distribution due to which only a fraction of them reaches the target site in the body. Yet, by conjugating nanoshells with high molecular-weight polymers, it is possible to get over this constraint and augment the bioavailability of drugs by slowing down their renal elimination and stretching their circulation period, thereby improving the responsiveness (Dragojevic et al., 2015). In India and other countries with significant aquacultural production, the use of nanotechnology in developing drugs/vaccines is currently on the rise. For drug/vaccine delivery, the employed NPs should be safe, biocompatible, and biodegradable and should protect the conjugated molecules for a prolonged time (De Jong and Borm, 2008). Nowadays, various types of NPs are used for the same purpose, of which polymeric NPs such as chitosan and PLGA are most safe and least toxic to living organisms. In novel fish and shellfish vaccines, NPs have been successfully used as either adjuvants or vectors for their targeted delivery and action. NPs help in shielding drugs or vaccine molecules from enzymatic destruction, directing them to their target sites, and releasing them gradually so that the effects last longer and can pass through any internal living barrier inside the host organism owing to their smaller size. However, the nano-drug/vaccine delivery has some critical concerns, including the cost of production, feed palatability and acceptance by target fish species, the toxic response by fish species, and environmental impacts (Aklakur et al., 2015).

21.3.1 POLYMER-BASED NANOCARRIERS

Polymeric NPs are widely used drug/vaccine delivery systems because of their fewer side effects and biodegradable property. These NPs have great potential to adsorb, entrap, disperse, or attach molecules of size <1 μm in their lattice (Gonçalves et al., 2018). Antigen-conjugated NPs shield and lead the antigen to reach the target immune cells, where they trigger the host immune system to generate a specific reaction chain, thus working as an immuno-stimulant (Zhao et al., 2014). Polymeric NPs are grouped into two categories, namely, nanospheres (matrix system) and nanocapsules (reservoir-type vesicular system). Based on the means of production, these NPs can be naturally formed such as chitosan, hyaluronic acid, and alginates or synthetically derived such as PLGA. Polymeric NPs have higher loading efficiency, higher biocompatibility and biodegradability, fairly good stability, and higher drug/vaccine protection ability and show controlled release of active ingredients when compared with liposomes (Katata-Seru et al., 2019).

Chitosan is a naturally occurring polysaccharide of β-linked D-glucosamine and N-acetyl-D-glucosamine, found in the exoskeleton of crustaceans and insects (Elieh-Ali-Komi and Hamblin, 2016). Owing to its distinctive properties such as nontoxicity, biodegradability, bio-adhesiveness, and biocompatibility, its formulated emulsions are used for drug delivery in aquaculture species (Shah and Mraz, 2020). Chitosan NPs can gradually release the conjugated drug and maintain it for longer durations inside the host body. Chitosan NPs conjugated with vitamin C when given to rainbow trout demonstrated the controlled release of the vitamin up to 48 h after administration. It was also found that the innate immunity and nonspecific defense factors were also elevated in the treated fish which is primarily due to the synergistic effects of vitamin C and the chitosan polymer (Alishahi et al., 2011). It has also been reported that chitosan NPs boost mucosal immunity in fish when taken orally through the feed. Vaccines based on chitosan NP-carrier and drugs with chitosan NPs coating were found to improve immunity and resistance in rainbow trout against environmental stress and bacterial and viral infections (Muruganandam et al., 2019).

Tian et al. (2008) developed an oral DNA vaccine having chitosan as the carrier, which was successful in fighting against lymphocystis disease virus (LCDV) in Japanese halibut (*Paralichthys olivaceus*). Another oral DNA vaccine was successfully developed by Kumar et al. (2008) using chitosan and chitosan/tripolyphosphate NPs as adjuvant as well as a delivery agent against the bacterial infection caused by *Vibrio anguillarum* in Asian sea bass (*Lates calcarifer*). Shrimps, when vaccinated with chitosan-encapsulated DNA vaccine having VP28 gene of WSSV, showed a significant rise of 85% in the survival rate during challenge tests while there was 100% mortality in the control. But when the challenge tests were repeated after 1–4 weeks of vaccination, the survival rate reduced to 50%, the reason being the lack of immune memory function in the crustaceans (Rajeshkumar et al., 2009). A recombinant nanovaccine based on chitosan was found to be effective in generating a protective immune response in black sea bream (*Acanthopagrus schlegelii*), against *V. parahaemolyticus* (Li et al., 2013). The administration of nodavirus DNA vaccine encapsulated within chitosan NPs was reported to improve immunity and survivability in freshwater prawns (Ramya et al., 2014). Chitosan-based NP vaccine conjugated with the inactivated virus was efficient in developing resistance against infectious salmon anemia virus (ISAV), where after vaccination, the survival rate was witnessed to increase up to 77% in the affected fish species (Rivas-Aravena et al., 2015). For vaccine delivery in zebra fish (*Danio rerio*) against viral hemorrhagic septicemia virus (VHSV), chitosan NPs were found to be highly successful (Kavaliauskis et al., 2016). Zheng et al. (2016) were efficacious in preparing an oral vaccine against reddish-body iridovirus infection in turbot (*Scophthalmus maximus*). When chitosan-coated membrane vesicles (cMVs) from the fish pathogen, *Piscirickettsia salmonis* were injected into adult zebrafish (*D. rerio*), it was found to induce a better immune response with improved survival rate in the species (Tandberg et al., 2018). It is reported that the inactivated VHSV antigen encapsulated within the chitosan NPs can generate antiviral immunity against the same virus in olive flounder (*P. olivaceus*) when administered via the mucosal route (Kole et al., 2019). Chitosan-coated mucoadhesive nanovaccine, given by immersion technique to tilapia (*Oreochromis* spp.) was proven to be efficient against columnaris disease (Kitiyodom et al., 2019). Leya et al. (2020) synthesized chitosan NPs-conjugated bicistronic DNA vaccine encoding *Edwardsiella tarda* antigen glyceraldehyde-3-phosphate dehydrogenase and immune adjuvant IFN-γ, which led to increased immunity against Edwardsiellosis infection in rohu (*Labeo rohita*) when given through feed and bath treatment.

Alginate is a natural polymer of β-D-mannuronic acid and α-L-guluronic acid present in certain species of brown algae and bacteria (Shah and Mraz, 2020). Alginate polymer is reported to act as an antigen adjuvant and growth promoter in fish (Fujiki et al., 1994; Tafaghodi et al., 2007; Borges et al., 2008; Chiu et al., 2008). The administration of alginate microparticles has proven to be an effective immunostimulant for several fish species such as brown-marbled grouper (*Epinephelus fuscoguttatus*) and orange-spotted grouper (*Epinephelus coioides*) against iridovirus and *Streptococcus* spp. and turbot (*S. maximus*) against *V. anguillarum* (Skjermo and Bergh, 2004; Huttenhuis et al., 2006; Cheng et al., 2008; Yeh et al., 2008; Fundueanu et al., 2010). Alginate microparticles were employed to deliver the antigens against *Ichthyophthirius multifilis* in rainbow trout and *Flavobacterium columnare* in Nile tilapia (Leal et al., 2010; Heidarieh et al., 2015). Similarly, for the oral administration of plasmid DNA against infectious pancreatic necrosis virus (IPNV) in brown and rainbow trouts, alginate microparticles were utilized as the delivery vehicle (Ana et al., 2010). These particles when conjugated to *L. garvieae* antigens were reported to elicit an efficient defensive response against the infection in rainbow trout (Ana et al., 2010). When alginate-chitosan micro/NPs were combined with *S. iniae* and *Y. ruckeri* lipopolysaccharides, these oral vaccines improved the immune response in rainbow trout against the diseases and increased the survival rate after challenge tests (Halimi et al., 2019; Dezfuly et al., 2020).

Several studies have reported the usage of the synthetic polymer, PLGA as a fish vaccine carrier against infectious hematopoietic necrosis virus (IHNV), IPNV, and *Streptococcus agalactiae* in rainbow trout, Atlantic salmon, and tilapia, respectively (Adomako et al., 2012; Munangandu et al., 2012; Zhang et al., 2015; Dubey et al., 2016). PLGA-encapsulated antigens from *A. hydrophila* acted as a vaccine in rohu, which was effective in generating potent immunostimulatory and antibody formation responses against the bacteria even after 42 days post-treatment (Behera et al., 2010). A highly protective immune reaction against the opportunistic protozoic pathogen, *Uronema marinum*, which infects fish such as flounder and grouper, was successfully evoked by its antigens encapsulated by PLGA and the resulting response persisted for a month in the vaccinated fish (Harikrishnan et al., 2012a, b). Another vaccine was developed for flounder in which PLGA NPs were conjugated to a plasmid that codes for the major capsid protein (MCP) of the LCDV. The findings indicated that after 4 weeks of vaccination, antibodies against the virus reached peak levels and that the PLGA-encapsulated vaccine was more successful in controlling the infection than the nonencapsulated treatment (Tian and Yu, 2011). PLGA microparticles have been used for the delivery of formalin-killed *A. hydrophila* antigens to pond loach (*Misgurnus anguillicaudatus*) and common carp (*Cyprinus carpio*). During the challenge test, the treated fish revealed greater survival and better innate and adaptive immune responses against the infection (Yun et al., 2017).

21.3.2 LIPOSOME NANOCARRIERS

Liposomes are another widely used drug delivery vehicle nowadays for drugs that are hydrophilic as well as lipophilic in nature (Ji et al., 2015). Liposomes are sphere-like vesicles, composed of an aqueous compartment surrounded by either bilayer or multilayer phospholipids in which all the layers are auto-sealing. These are usually derived from soy, milk, eggs, and nonionic surfactants using different methods such as membrane extrusion, detergent depletion, high-pressure homogenization, lipid layer hydration, sonication, ethanol injection, micro-fluidization, and reverse phase evaporation (Dima et al., 2015). The method employed for producing the liposomes depends on certain key factors such as the type of liposomes to be synthesized, the type of phospholipids used, and the interaction of the dispersion medium with the used phospholipids. Among all the methods of liposome production, the ethanol method has been upheld to be a superior encapsulating method for the production of soy phospholipid-based liposomes. Drug encapsulation in liposomes is done mainly by two methods, first is passive loading where the drug is encapsulated during the process of vesicle formation and second is active loading where the drug is entrapped after vesicle formation. Hydrophilic drugs are loaded inside the core of liposomes by mixing them with a buffer that hydrates the thin lipid film during liposome formation, while lipophilic drugs are loaded into the lipid bilayers by mixing them with other liposome components during the preparation of thin, dry lipid films (Pandey et al., 2016). Liposome nanocarriers are biocompatible and biodegradable; display higher bioavailability, targeted delivery, and sustained release of active ingredients in drugs; and can encapsulate more than one compound (Singh et al., 2012). Liposomes also safeguard the conjugated drugs from degradation by external factors such as light, moisture, oxygen, and temperature (Surve et al., 2018). Nonetheless, liposomes have certain disadvantages in terms of low storage stability and poor encapsulation efficiency (Dima et al., 2015; Shin et al., 2015). However, these concerns regarding the usage of liposomes can be bypassed by suitable coating.

Studies show that the orally administered liposome-encapsulated *A. salmonicida* antigens significantly boost the immune response in common carp (*C. carpio*) against bacterial septicemia. In the liposome-treated carp group, the fish showed fewer skin ulcers and the survival rate improved by up to 83% (Irie et al., 2005). A similarly enhanced immune response and resistance to *A. salmonicida* was also seen in rainbow trout when vaccinated with the bacterial lipopolysaccharide antigens coated with liposomes. An oral vaccine was developed by Yasumoto et al. (2006) for common carp using the formalin-treated inactivated koi herpesvirus coated with liposomes. When formalin-killed *V. harveyi* antigens encapsulated within liposomes were used to treat the bacterial infection in grouper, it was discovered that the morality rate was lowest in the fish group treated with encapsulated antigens rather than the noncoated antigens (Harikrishnan et al., 2012b). To control various fish pathogens, especially viruses, liposomes were utilized to deliver cinnamaldehyde (cinnamon extract) and melittin (an antimicrobial peptide). It was revealed that the liposome-coated drugs were more effective in inducing a swift and elevated immune response against diseases (Falco et al., 2013; Faikoh et al., 2014). In recent times, Malheiros et al. (2020) reported that the nanoemulsions of oleoresin from *Copaifera reticulata* (Leguminosae) improved its anthelmintic efficacy in the control of monogenean parasites on the gills of Tambaqui (*Colossoma macropomum*), compared to oleoresin without nanoformulation.

21.3.3 CARBON NANOTUBE CARRIERS

Carbon nanotubes are made of rolled-up graphene sheets, cylinder-shaped and exist in either single-walled form or multiwalled form (concentric sheets wrapped one above the other), both with different properties. The multiwall carbon nanotubes were first discovered in the year 1991 while the single-wall carbon nanotubes were reported in 1993 (Iijima, 1991; Iijima and Ichihashi, 1993). The applications of carbon nanotubes encompass many scientific areas such as biomedicine, nanotechnology, engineering, and biotechnology. Over the past few years, these nanoparticles have grabbed the significant interest of the scientific community due to their unique properties to be exploited for therapeutics, disease diagnosis, and drug/vaccine delivery (Sanginario et al., 2017). One of the principal barriers to the usage of carbon nanotubes in biological systems is their lack of solubility in aqueous media, nonbiocompatibility, and toxicity. If the particle size of the nanotubes is <100 nm, they are potentially harmful to organisms. Inside the host body, these nanoparticles activate the inflammatory and immunological responses, alter the protein structure, cause toxicity in the lungs, relocate from their site of deposition, and can successfully evade the immune phagocytic cells. However, researchers are trying to modify the external surface of carbon nanotubes by functionalization with diverse hydrophilic molecules to enhance their solubility and biocompatibility (Wu et al., 2004). There are many reports on the synthesis of modified carbon nanotubes that are water soluble, biocompatible, and nontoxic to animals (Eatemadi et al., 2014).

In the aquaculture sector, both single-wall and multiwall carbon nanotubes, have been successfully used for drug/vaccine delivery in safe dosages. For the intramuscular transfer of plasmid encoding the VP7 protein of grass carp reovirus (GCRV) in grass carp (*Ctenopharyngodon idella*), a single-wall carbon nanotube was used as a carrier. The findings revealed a higher immune response and resistance against the virus in the nanotube-conjugated treatment group than in the nonconjugated one (Zhu et al., 2015). Similar results were obtained in recent studies on grass carp using these nanotubes for vaccine delivery (Qiu et al., 2020; Zhu et al., 2020). Single-wall carbon nanotubes have also been successfully used for vaccine delivery against infectious spleen and kidney necrosis virus (ISKNV) in mandarin fish (Zhao et al., 2020). To provide passive immunity in fish species against rhabdovirus infection, carbon nanotubes were employed for viral antigen delivery (Zhang et al., 2020).

21.4 NANOTECHNOLOGY-ASSOCIATED CONCERNS

At present, the possible toxicity of NPs to biological systems is a budding global public health hazard. NP pollutants are now a new form of environmental contamination and studies are being conducted to evaluate their impending harmful effects (Moore, 2006; Shah and Mraz, 2020). Because of their extremely small size, NPs can easily penetrate the cells via the cell membranes and cause genotoxicity. Being highly reactive, the intrinsic NPs form reactive oxygen species and free radicals which are the principal mechanisms of NP biotoxicity. NPs cause inflammation, oxidative stress, modifications in protein/enzyme structures, and DNA damage. It has been validated that NPs produce DNA mutation and major structural damage to mitochondria, which could even result in cell death (Majumder and Dash, 2017; Vicari et al., 2018; Meghani et al., 2020).

There are numerous reports on the toxicity of NPs in aquatic fauna (Asharani et al., 2008; Griffitt et al., 2008; Matranga and Corsi, 2012; Garcia-Negrete et al., 2013; Canesi et al., 2015; Vignardi et al., 2015; Swain et al., 2016; Canesi et al., 2017; Sendra et al., 2017a, b, c; 2018a, b; Aouini et al., 2018; Volland et al., 2018; Naguib et al., 2020; Sayed et al., 2020). Nanotoxicology studies elucidate the regulatory aspects and biodistribution of NPs and also assess their threats on molecular as well as environmental levels (Bello and Leong, 2017). More studies are required regarding the sustainable use of nanotechnology in aquaculture, especially on how the engineered NPs can bioaccumulate and whether they could become noxious to fish in the long run (Skjolding et al., 2016; Sibaja-Luis et al., 2019). Risk analyses show that possible human health risks are likely to be associated with the long-term consumption of fish having 1 ng or lower quantities of NPs. NPs pose a relatively low risk for the majority of environmental niches but the inhabiting organisms could be at greater risk (Giese et al., 2018). This is only a mere prediction since it is still unknown how NPs behave, disperse, and dispose of in key ecosystems such as oceans.

21.5 CONCLUSION

Nanotechnology undoubtedly plays a pivotal role in the sustainable development of the global aquaculture sector. One of the promising applications of this field is related to fish health management for disease diagnosis and therapeutics. Conventional medications such as antibiotics have many disadvantages, especially the emergence of antibiotic-resistant strains. This problem is subjugated by the use of metallic NPs such as Ag, Au, Zn, Ti, and Cu, which exhibit significant antimicrobial activity. AgNPs have proved to be the most effective and efficient against bacteria, viruses, fungi, and other fish pathogens. Nanomedicine is an evolving part of nanotechnology and new technological advances in this field help in improving fish health. Generally, the pharmaceuticals used in veterinary and fisheries sectors have poor water stability, with a major portion of the active ingredients getting leached out into the water and washed away, which leads to ineffective results. Now, nanomedicines containing NPs such as chitosan, PLGA, liposomes, CNTs, and so forth, are being exploited to act as new drug/vaccine delivery agents in aquaculture. At present, nano-vaccination is a new attempt to enhance the immunogenicity of vaccines using NPs as carriers or adjuvants. Since NPs and fish pathogens share a similar size scale, nano-vaccines have the ability to trigger cellular and humoral immune responses, thereby activating the fish immune system. The novel nanobiosensors are being engineered, which can detect even minute concentrations of infectious agents and pollutants in aquatic ecosystems. In the immediate future, these sensors will play a crucial role in fish disease diagnosis. Thus, nanotechnology in aquaculture has become a comprehensive tool for the early detection, prevention, control, and subsequent monitoring of fish diseases. Currently, the chief limiting factor in the wider implementation of NPs in fisheries is their high cost of production.

On the other hand, the usage of nanotechnology in aquaculture also calls for a broad evaluation of the related potential negative impacts; therefore, a careful analysis of the life cycle and shelf life of NPs, combined with an assessment of their potential health and environmental risks, including exposure, release and deposition, must be performed. Nanotoxicologists have raised some major concerns over the *in vivo* application of NPs, especially those having a size <40 nm as they can easily penetrate the cells and their fate in the host body remains a mystery. A few studies have been done to understand the potential toxicity associated with the use of NPs on biological systems, including plants, humans, fish, and other invertebrates so as to determine their safe dosages for disease treatment. However, further research needs to be carried out in this regard. Nonetheless, such trepidations must not stop us from employing nanotechnological tools in aquaculture but in a controlled manner with vigilant scrutiny in order to avail the utmost benefits with the least possible perils.

KEYWORDS

- **aquatic health**
- **fish**
- **nanobiosensors**
- **nanoparticles**
- **nano-vaccines**
- **toxicity**

REFERENCES

Adomako, M. O., St-Hilaire, S., Zheng, Y. X., Eley, J., Marcum, R. D., Sealey, W. M., Donahower, B. C., Lapatra, S. E., & Sheridan, P. P. (2012). Oral DNA vaccination of rainbow trout, *Oncorhynchus mykiss* (Walbaum), against infectious haematopoietic necrosis virus using PLGA [Poly(D,L-Lactic-Co-Glycolic Acid)] nanoparticles. *Journal of Fish Diseases, 35*(3), 203–214.

Aklakur, M., Rather, M. A., & Kumar, N. (2015). Nanodelivery: An emerging avenue for nutraceuticals and drug delivery. *Critical Reviews in Food Science and Nutrition, 56*(14), 2352–2361.

Alishahi, A., Mirvaghefi, A., Tehrani, M., Farahmand, H., Koshio, S., Dorkoosh, F., & Elsabee, M. Z. (2011). Chitosan nanoparticle to carry vitamin C through the gastrointestinal tract and induce the non-specific immunity system of rainbow trout (*Oncorhynchus mykiss*). *Carbohydrate Polymers, 86*(1), 142–146.

Ana, I., Saint-Jean, S. R., & Pérez-Prieto, S. I. (2010). Immunogenic and protective effects of an oral DNA vaccine against infectious pancreatic necrosis virus in fish. *Fish & Shellfish Immunology, 28*, 562–570.

Antony, J. J., Ni, M., Durairaj, S., Ganesapandy, P., Palanivel, K., Seenivasan, K., Arunachalam, S., & Shanmugam, A. (2013). Antimicrobial activity of Leucas aspera engineered silver NPs against *Aeromonashydrophila* in infected Catla. *Colloids and Surfaces B: Biointerfaces, 109*, 20–24.

Aouini, F., Trombini, C., Volland, V., Elcafsi, M., & Blasco, J. (2018). Assessing lead toxicity in the clam *Ruditapes philippinarum*: Bioaccumulation and biochemical responses. *Ecotoxicology and Environmental Safety, 158*, 193–203.

Asharani, P. V., Wu, Y. L., Gong, Z., & Valiyaveettil, S. (2008). Toxicity of silver nanoparticles in zebrafish models. *Nanotechnology, 19*(25), 255102.

Behera, T., Nanda, P. K., Mohanty, C., Mohapatra, D., Swain, P., Das, B. K., Routray, P., Mishra, B. K., & Sahoo, S. K. (2010). Parenteral immunization of fish, *Labeo rohita* with Poly D, L-lactide-co-glycolic acid (PLGA) encapsulated antigen microparticles promotes innate and adaptive immune responses. *Fish & Shellfish Immunology, 28*(2), 320–325.

Behera, T., & Swain, P. (2011). Antigen adsorbed calcium phosphate nanoparticles stimulate both innate and adaptive immune response in fish, *Labeo rohita* H. *Cellular Immunology, 271*(2), 350–359.

Bello, D., & Leong, D. T. (2017). A decade of nanotoxicology: Assessing the impact on human health and the environment. *NanoImpact, 7*, 15–16.

Borges, O., Silva, M., de Sousa, A., Borchard, G., Junginger, H. E., & Cordeiro da Silva, A. (2008). Alginate coated chitosan nanoparticles are an effective subcutaneous adjuvant for hepatitis B surface antigen. *International Immunopharmacology, 8*(13-14), 1773–1780.

Canesi, L., Balbi, T., Fabbri, R., Damonte, G., Salis, A., Volland, M., & Blasco, J. (2017). Biomolecular coronas in invertebrate species: Implications in the environmental impact of nanoparticles. *Nanoimpact, 8*, 89–98.

Canesi, L., Ciacci, C., & Balbi, T. (2015). Interactive effects of nanoparticles with other contaminants in aquatic organisms: Friend or foe? *Marine Environmental Research, 111*, 128–134.

Cavalieri, F., Tortora, M., Stringaro, A., Colone, M., & Baldassarri, L. (2014). Nanomedicines for antimicrobial interventions. *Journal of Hospital Infection, 88*(4), 183–190.

Chen, B., Zou, L., Wu, Z., & Sun, M. (2016). The application of quantum dots in aquaculture pollution detection. *Toxicological and Environmental Chemistry, 98*(3-4), 385–394.

Cheng, A. C., Chen, Y. Y., & Chen, J. C. (2008). Dietary administration of sodium alginate and κ-carrageenan enhances the innate immune response of Brown-marbled grouper *Epinephelus fuscoguttatus* and its resistance against *Vibrio alginolyticus*. *Veterinary Immunology and Immunopathology, 121*, 206–215.

Cheng, T. C., Yao, K. S., Yeh, N., Chang, C. I., Hsu, H. C., Chien, Y. T., & Chang, C. Y. (2009). Visible light activated bactericidal effect of TiO2/Fe3O4 magnetic particles on fish pathogens. *Surface and Coatings Technology, 204*(6-7), 1141–1144.

Chiu, S. T., Tsai, R. T., Hsu, J. P., Liu, C. H., & Cheng, W. (2008). Dietary sodium alginate administration to enhance the non-specific immune responses, and disease resistance of the juvenile grouper *Epinephelus fuscoguttatus*. *Aquaculture, 277*(1-2), 66–72.

Dananjaya, S., Godahewa, G., Jayasooriya, R., Lee, J., & De Zoysa, M. (2016). Antimicrobial effects of chitosan silver nano composites (CAgNCs) on fish pathogenic *Aliivibrio (Vibrio) salmonicida*. *Aquaculture, 450*, 422–430.

Daniel, S. C. G. K., Sironmani, T. A., & Dinakaran, S. (2016). Nano formulations as curative and protective agent for fish diseases: Studies on red spot and white spot diseases of ornamental gold fish *Carassius auratus*. *International Journal of Fisheries and Aquatic Studies, 4*, 255–261.

De Jong, W. H., & Borm, P. J. (2008). Drug delivery and NPs: Applications and hazards. *International Journal of Nanomedicine, 3*(2), 133.

Dezfuly, Z. T., Alishahi, M., Ghorbanpoor, M., Tabandeh, M. R., & Mesbah, M. (2020). Immunogenicity and protective efficacy of *Yersinia ruckeri* lipopolysaccharide (LPS), encapsulated by alginate-chitosan micro/nanoparticles in rainbow trout (*Oncorhyncus mykiss*). *Fish & Shellfish Immunology, 104*, 25–35.

Dima, S., Dima, C., & Iordachescu, G. (2015). Encapsulation of functional lipophilic food and drug biocomponents. *Food Engineering Reviews, 7*, 417–438.

Dragojevic, S., Ryu, J. S., & Raucher, D. (2015). Polymer-based prodrugs: improving tumor targeting and the solubility of small molecule drugs in cancer therapy. *Molecules, 20*, 21750–21769.

Dubey, S., Avadhani, K., Mutalik, S., Sivadasan, S. M., Maiti, B., Paul, J., Girisha, S. K., Venugopal, M. N., Mutoloki, S., Evensen, O., Karunasagar, I., & Munangandu, H. M. (2016). *Aeromonashydrophila* OmpW PLGA nanoparticle oral vaccine shows a dose-dependent protective immunity in rohu (*Labeo rohita*). *Vaccine, 4*, 21.

Eatemadi, A., Daraee, H., Karimkhanloo, H., Kouhi, M., Zarghami, N., Akbarzadeh, A., Abasi, M., Hanifehpour, Y., & Joo, S. W. (2014). Carbon nanotubes: Properties, synthesis, purification, and medical applications. *Nanoscale Research Letters, 9*(1), 393.

Elieh-Ali-Komi, D., & Hamblin, M. R. (2016). Chitin and chitosan: Production and application of versatile biomedical nanomaterials. *International Journal of Advanced Research, 4*(3), 411–427.

Erdem, B., Dayangaç, A., Kıray, E., & Duygu, D. (2018). Biosynthesis of silver nanoparticles from *Aeromonas sobria* and antibacterial activity against fish pathogens. *International Journal of Environmental Science and Technology, 16*, 5125–5130.

Faikoh, E. N., Hong, Y. H., & Hu, S. Y. (2014). Liposome-encapsulated cinnamaldehyde enhances zebrafish (*Danio rerio*) immunity and survival when challenged with *Vibrio vulnificus* and *Streptococcus agalactiae*. *Fish & Shellfish Immunology, 38*, 15–24.

Falco, A., Barrajón-Catalán, E., Menéndez-Gutiérrez, M. P., Coll, J., Micol, V., & Estepa, A. (2013). Melittinloaded immunoliposomes against viral surface proteins, a new approach to antiviral therapy. *Antiviral Research, 97*, 218–221.

FAO. (2020). The State of World Fisheries and Aquaculture. Sustainability in action. Rome.

Fujiki, K., Matsuyama, H., & Yano, T. (1994). Protective effect of sodium alginates against bacterial infection in common carp, *Cyprinus carpio* L. *Journal of Fish Diseases, 17*(4), 349–355.

Fundueanu, G., Constantin, M., Ascenzi, P., & Simionescu, B. C. (2010). An intelligent multicompartmental system based on thermo-sensitive starch microspheres for temperature-controlled release of drugs. *Biomedical Microdevices, 12*, 693–704.

Garcia-Negrete, C. A., Blasco, J., Volland, M., Rojas, T. C., Hampel, M., Lapresta Fernandez, A., Jimenez-de Haro, M. C., Soto, M., & Fernandez, A. (2013). Behaviour of Au-citrate nanoparticles in seawater and accumulation in bivalves at environmentally relevant concentrations. *Environmental Pollution, 174*, 134–141.

Giese, B., Klaessig, F., Park, B., Kaegi, R., Steinfeldt, M., Wigger, H., von Gleich, A., & Gottschalk, F. (2018). Risks, release and concentrations of engineered nanomaterial in the environment. *Scientific Reports, 8*, 1565.

Gonçalves, R. F., Martins, J. T., Duarte, C. M., Vicente, A. A., & Pinheiro, A. C. (2018). Advances in nutraceutical delivery systems: from formulation design for bioavailability enhancement to efficacy and safety evaluation. *Trends in Food Science Technology, 78*, 270–291.

Griffitt, R. J., Luo, J., Gao, J., Bonzongo, J. C., & Barber, D. S. (2008). Effects of particle composition and species on toxicity of metallic nanomaterials in aquatic organisms. *Environmental Toxicology & Chemistry, 27*(9), 1972–1978.

Guo, J., Zhang, R., & Yang, N. (2016). An Immunomagnetic-nanoparticle-based microfluidic system for rapid separation and detection of aquaculture pathogens. *Journal of Computational and Theoretical Nanoscience, 13*, 2226–2231.

Hafeez, M., Ahmad, I., Qureshi, S., Kashoo, Z., Farooq, S., Wani, S., Asmi, O., Shah, F., Shah, S., & Razak, N. (2021). Resistant *Clostridium perfringens* isolates from freshwater fishes of Dal Lake, Kashmir: A public health concern. *National Academy Science Letters, 44*, 397–400.

Halimi, M., Alishahi, M., Abbaspour, M. R., Ghorbanpoor, M., & Tabandeh, M. R. (2019). Valuable method for production of oral vaccine by using alginate and chitosan against *Lactococcus garvieae/Streptococcus iniae* in rainbow trout (*Oncorhynchus mykiss*). *Fish & Shellfish Immunology, 90*, 431–439.

Harikrishnan, R., Balasundaram, C., & Heo, M. S. (2012a). Poly D, L-lactide-co-glycolic acid (PLGA)-encapsulated vaccine on immune system in *Epinephelus bruneus* against *Uronema marinum*. *Experimental Parasitology, 131*, 325–332.

Harikrishnan, R., Kim, J. S., Balasundaram, C., & Heo, M. S. (2012b). Vaccination effect of liposomes entrapped whole cell bacterial vaccine on immune response and disease protection in *Epinephelusbruneus* against *Vibrio harveyi*. *Aquaculture, 342*, 69–74.

Heidarieh, M., Moodi, S., Katuli, K. K., & Unger, H. (2015). Biochemical effects of encapsulated Radiovaccine via alginate NPs as strategy for booster in immunized rainbow trout against *Ichthyophthirius multifiliis*. *Acta Scientiae Veterinari, 43*, 1330.

Huang, C. M., Chen, C. H., Pornpsattananangkul, D., Zhang, L., Chan, M., Hsieh, M. F., & Zhang, L. (2011). Eradication of drug resistant *Staphylococcus aureus* by liposomal oleic acids. *Biomaterials, 32*(1), 214–221.

Huttenhuis, H. B. T., Ribeiro, A. S. P., Bowden, T. J., Van Bavel, C., Taverne-Thiele, A. J., & Rombout, J. H. W. M. (2006). The effect of oral immuno-stimulation in juvenile carp (*Cyprinus carpio* L.). *Fish & Shellfish Immunology, 21*(3), 261–271.

Iijima, S. (1991). Helical microtubules of graphitic carbon. *Nature, 9*(6348), 56–58.

Iijima, S., & Ichihashi, T. (1993). Single-shell carbon nanotubes of 1-nm diameter. *Nature, 363*, 603–605.

Irie, T., Watarai, S., Iwasaki, T., & Kodama, H. (2005). Protection against experimental *Aeromonas salmonicida* infection in carp by oral immunisation with bacterial antigen entrapped liposomes. *Fish & Shellfish Immunology, 18*, 235–242.

Jaroenram, W., Arunrut, N., & Kiatpathomchai, W. (2012). Rapid and sensitive detection of shrimp yellow head virus using loop-mediated isothermal amplification and a colorogenic nanogold hybridization probe. *Journal of Virological Methods, 186*(1), 36–42.

Jeong, S. H., Hwang, Y. H., & Yi, S. C. (2005). Antibacterial properties of padded PP/PE nonwovens incorporating nano-sized silver colloids. *Journal of Materials Science, 40*(20), 5413–5418.

Ji, J., Torrealba, D., Ruyra, À., & Roher, N. (2015). Nanodelivery systems as new tools for immunostimulant or vaccine administration targeting the fish immune system. *Biology, 4*(4), 664–696.

Johari, S. A., Kalbassi, M. R., Soltani, M., & Yu, I. J. (2015). Study of fungicidal properties of colloidal silver NPs (AgNPs) on trout egg pathogen, *Saprolegnia* sp. *International Journal of Aquatic Biology, 3*, 191–119.

Jovanovic, B., Anastasova, L., Rowe, E. W., Zhang, Y., Clapp, A. R., & Palic, D. (2011). Effects of nanosized titanium dioxide on innate immune system of fathead minnow (*Pimephales promelas* Rafinesque, 1820). *Ecotoxicology and Environmental Safety, 74*(4), 675–683.

Katata-Seru, L., Ramalapa, B., & Tshweu, L. (2019). Nanoformulated delivery systems of essential nutraceuticals and their applications. In *Nanoemulsions - Properties, fabrications and applications*. IntechOpen.

Kavaliauskis, A., Arnemo, M., Speth, M., Lagos, L., Rishovd, A. L., Estepa, A., Griffiths, G., & Gjøen, T. (2016). Protective effect of a recombinant VHSV-G vaccine using poly(I:C) loaded nanoparticles as an adjuvant in zebrafish (*Danio rerio*) infection model. *Developmental and Comparative Immunology, 61*, 248–257.

Kitiyodom, S., Yata, T., Yostawornkul, J., Kaewmalun, S., Nittayasut, N., Suktham, K., Surassmo, S., Namdee, K., Rodkhum, C., & Pirarat, N. (2019). Enhanced efficacy of immersion vaccination in tilapia against columnaris disease by chitosan-coated "pathogen-like" mucoadhesive nanovaccines. *Fish & Shellfish Immunology, 95*, 213–219.

Kole, S., Qadiri, S. S. N., Shin, S. M., Kim, W. S., Lee, J., & Jung, S. J. (2019). Nanoencapsulation of inactivated-viral vaccine using chitosan nanoparticles: Evaluation of its protective efficacy and immune modulatory effects in olive flounder (*Paralichthys olivaceus*) against viral haemorrhagic septicaemia virus (VHSV) infection. *Fish & Shellfish Immunology, 91*, 136147.

Korni, F. M. M., & Khalil, F. (2017). Effect of ginger and its nanoparticles on growth performance, cognition capability, immunity and prevention of motile Aeromonas septicaemia in *Cyprinus carpio* fingerlings. *Aquaculture Nutrition, 23*(6), 1492–1499.

Kuan, G. C., Sheng, L. P., Rijiravanich, P., Marimuthu, K., Ravichandran, M., Yin, L. S., Lertanantawong, B., & Surareungchai, W. (2013). Goldnanoparticle based electrochemical DNA sensor for the detection of fish pathogen *Aphanomyces invadans*. *Talanta, 117*, 312–317.

Kumar, S. R., Ahmed, V. I., Parameswaran, V., Sudhakaran, R., Babu, V. S., & Hameed, A. S. (2008). Potential use of chitosan nanoparticles for oral delivery of DNA vaccine in Asian sea bass (*Lates calcarifer*) to protect from *Vibrio* (*Listonella*) *anguillarum*. *Fish & Shellfish Immunology, 25*(1-2), 47–56.

Lara, H. H., Ayala-Núnez, N. V., Turrent, L. C. I., & Padilla, C. R. (2010). Bactericidal effect of silver NPs against multidrug-resistant bacteria. *World Journal of Microbiology and Biotechnology, 26*, 615–621.

Leal, C. A. G., Carvalho-Castro, G. A., Sacchetin, P. S. C., Lopes, C. O., Moraes, Â. M., & Figueiredo, H. C. P. (2010). Oral and parenteral vaccines against *Flavobacterium columnare*: evaluation of humoral immune response by ELISA and *in vivo* efficiency in Nile tilapia (*Oreochromis niloticus*). *Aquaculture International, 18*, 657–666.

Leya, T., Ahmad, I., Sharma, R., Tripathi, G., Kurcheti, P. P., Rajendran, K. V., & Bedekar, M. K. (2020). Bicistronic DNA vaccine macromolecule complexed with poly lactic-co-glycolic acid-chitosan NPs enhanced the mucosal immunity of *Labeo rohita* against *Edwardsiella tarda* infection. *International Journal of Biological Macromolecules, 156*, 928–937.

Li, L., Lin, S. L., Deng, L., & Liu, Z. G. (2013). Potential use of chitosan NPs for oral delivery of DNA vaccine in black seabream *Acanthopagrus schlegelii* Bleeker to protect from *Vibrio parahaemolyticus*. *Journal of Fish Diseases, 36*(12), 987995.

Majumder, D., & Dash, G. (2017). Application of nanotechnology in fisheries and aquaculture. *Everyman Science, 51*(6), 358–364.

Malheiros, D. F., Sarquis, I. R., Ferreira, I. M., Mathews, P. D., Mertins, O., & Tavares-Dias, M. (2020). Nanoemulsions with oleoresin of *Copaifera reticulata* (Leguminosae) improve antihelmintic efficacy in the control of monogenean parasites when compared to oleoresin without nanoformulation. *Journal of Fish Diseases, 43*, 687–695.

Mathur, A., Kushwaha, A., Dalakoti, V., Dalakoti, G., & Singh, D. S. (2014). Green synthesis of silver NPs using medicinal plant and its characterization. *Pharmaceutical Letters, 5*, 118–122.

Matranga, V., & Corsi, I. (2012). Toxic effects of engineered nanoparticles in the marine environment: Model organisms and molecular approaches. *Marine Environmental Research, 76*, 32–40.

Meghani, N., Dave, S., & Kumar, A. (2020). Introduction to nanofood. In U. Hebbar, S. Ranjan, N. Dasgupta & R. Kumar Mishra (Eds.), *Nano-food engineering*. Springer.

Moges, F. D., Patel, P., Parashar, S., & Das, B. (2020). Mechanistic insights into diverse nano-based strategies for aquaculture enhancement: A holistic review. *Aquaculture, 519*, 734770.

Moore, M. N. (2006). Do nanoparticles present ecotoxicological risk for the health of the aquatic environment. *Environment International, 32*, 967–976.

Munangandu, H. M., Fredriksen, B. N., Mutoloki, S., Brudeseth, B., Kuo, T. Y., Marjara, I. S., Dalmo, R. A., & Evensen, O. (2012). Comparison of vaccine efficacy for different antigen delivery systems for infectious pancreatic necrosis virus vaccines in Atlantic salmon (*Salm salar* L.) in a cohabitation challenge model. *Vaccine, 30*, 4007–4016.

Muruganandam, M., Chipps, S. R., & Ojasvi, P. R. (2019). On the advanced technologies to enhance fisheries production and management. *Acta Scientific Agriculture, 3*(8), 216–222.

Naguib, M., Mahmoud, U. M., Mekkawy, I. A., & Sayed, A. E. D. H. (2020). Hepatotoxic effects of silver nanoparticles on *Clarias gariepinus*; Biochemical, histopathological, and histochemical studies. *Toxicology Reports, 7*, 133–141.

Oberdörster, G., Oberdörster, E., & Oberdörster, J. (2005). Nanotoxicology: An emerging discipline evolving from studies of ultrafine particles. *Environmental Health Perspectives, 113*, 823–839.

Pandey, H., Rani, R., & Agarwal, V. (2016). Liposomes and their applications in cancer therapy. *Brazilian Archives of Biology and Technology, 59*, e16150477.

Park, M., Shin, S. K., Do, Y. H., Yarish, C., & Kim, J. K. (2018). Application of open water integrated multi-trophic aquaculture to intensive monoculture: A review of the current status and challenges in Korea. *Aquaculture, 497*, 174–183.

Patolsky, F., Zheng, G. F., Hayden, O., Lakadamyali, M., Zhuang, X. W., & Lieber, C. M. (2004). Electrical detection of single viruses. *Proceedings of the National Academy of Sciences, 101*, 14017–14022.

Pulkkinen, K., Suomalainen, L. R., Read, A. F., Ebert, D., Rintamäki, P., & Valtonen, E. T. (2010). Intensive fish farming and the evolution of pathogen virulence: The case of columnaris disease in Finland. *Proceedings of the Royal Society B: Biological Sciences, 277*(1681), 593–600.

Qiu, D. K., Jia, Y. J., Gong, Y. M., Zheng, Y. Y., Wang, G. X., & Zhu, B. (2020). Optimizing the immunization procedure of single-walled carbon nanotubes based vaccine against grass carp reovirus for grass carp. *Aquaculture, 533*(3), 736152.

Rajeshkumar, S., Venkatesan, C., Sarathi, M., Sarathbabu, V., Thomas, J., Anver Basha, K., & Hameed, A. S. S. (2009). Oral delivery of DNA construct using chitosan nanoparticles to protect the shrimp from white spot syndrome virus (WSSV). *Fish & Shellfish Immunology, 26*, 429–437.

Ramya, V. L., Sharma, R., Gireesh-Babu, P., Patchala, S. R., Rather, A., Nanda-Psawar, P. C., & Eswaran, S. (2014). Development of chitosan conjugated DNA vaccine against nodavirus in *Macrobrachium rosenbergii* (De Man, 1879). *Journal of Fish Diseases, 37*(9), 815–824.

Rather, M. A., Bhat, I. A., Sharma, N., Gora, A., Ganie, P. A., & Sharma, R. (2016). Synthesis and characterization of *Azadirachta indica* constructed silver nanoparticles and their immunomodulatory activity in fish. *Aquaculture Research, 48*, 3742–3754.

Rivas-Aravena, A., Fuentes, Y., Cartagena, J., Brito, T., Poggio, V., Torre, J. L., & Spencer, E. (2015). Development of a nanoparticle-based oral vaccine for Atlantic Salmon against ISAV using an alphavirus replicon as adjuvant. *Fish & Shellfish Immunology, 45*, 157–166.

Ross, K. A., Loyd, H., Wu, W., Huntimer, L., Wannemuehler, M. J., Carpenter, S., & Narasimhan, B. (2014). Structural and antigenic stability of H5N1 hemagglutinin trimer upon release from polyanhydride NPs. *Journal of Biomedical Material Research Part A, 102*(11), 4161–4168.

Roy, S., Kumar, V., Barman, D., & Mandal, S. (2012). Nanotechnology and its application in fisheries and aquaculture. *AQUA International, 20*(6), 2–3.

Saleh, M., & El-Matbouli, M. (2015). Rapid detection of cyprinid herpesvirus-3 (CyHV-3) using a gold nanoparticle-based hybridization assay. *Journal of Virological Methods, 217*, 50–54.

Saleh, M., Soliman, H., Schachner, O., & El-Matbouli, M. (2011). Antibody-coated gold NPs immunoassay for direct detection of *Aeromonas salmonicida* in fish tissues. *Journal of Fish Diseases, 34*(11), 845–852.

Saleh, M., Soliman, H., Schachner, O., & El-Matbouli, M. (2012). Direct detection of unamplified spring viraemia of carp virus RNA using unmodified gold nanoparticles. *Diseases of Aquatic Organisms, 100*, 3–10.

Sanginario, A., Miccoli, B., & Demarchi, D. (2017). Carbon nanotubes as an effective opportunity for cancer diagnosis and treatment. *Biosensors, 7*, 9.

Sayed, A. E. D. H., Mekkawy, I. A., Mahmoud, U. M., & Nagiub, M. (2020). Histopathological and histochemical effects of silver nanoparticles on the gills and muscles of African catfish (*Clarias garepinus*). *Scientific African, 7*, e00230.

Seetang-Nun, Y., Jaroenram, W., Sriurairatana, S., Suebsing, R., & Kiatpathomchai, W. (2013). Visual detection of white spot syndrome virus using DNA-functionalized gold NPs as probes combined with loop-mediated isothermal amplification. *Molecular and Cellular Probes, 27*(2), 71–79.

Sendra, M., Blasco, J., & Araujo, C. (2018a). Is the cell wall of marine phytoplankton a protective barrier or a nanoparticle interaction site? Toxicological responses of *Chlorella autotrophica* and *Dunaliella salina* to Ag and CeO2 nanoparticles. *Ecological Indicators, 95*, 1053–1067.

Sendra, M., Este, M. P., Garrido, I. M., Gatica, J. M., & Blasco, J. (2017a). CeO2 NPs, toxic or protective to phytoplankton? Charge of nanoparticles and cell wall as factors which cause changes in cell complexity. *The Science of the Total Environment, 590–591*, 304–315.

Sendra, M., Este, M. P., Gatica, J. M., Garrido, I. M., & Blasco, J. (2017b). Direct and indirect effects of silver nanoparticles on freshwater and marine microalgae (*Chlamydomonas reinhardtii* and *Phaeodactylum tricornutum*). *Chemosphere, 179*, 279–289.

Sendra, M., Volland, M., Balbi, T., Fabbri, R., Yeste, M. P., Gatica, J. M., Canesi, L., & Blasco, J. (2018b). Cytotoxicity of CeO2 nanoparticles using *in vitro* assay with *Mytilus galloprovincialis* hemocytes: Relevance of zeta potential, shape and biocorona formation. *Aquatic Toxicology, 200*, 13–20.

Sendra, M., Yeste, M. P., Garrido, I. M., Gatica, J. M., & Blasco, J. (2017c). Homoagglomeration and heteroagglomeration of TiO2, in nanoparticle and bulk form, onto freshwater and marine microalgae. *The Science of the Total Environment, 592*, 403–411.

Shaalan, M. I., El-Mahdy, M. M., Theiner, S., El-Matbouli. M., & Saleh, M. (2017). *In vitro* assessment of the antimicrobial activity of silver and zinc oxide NPs against fish pathogens. *Acta Veterinaria Scandinavica, 59*(1), 49.

Shaalan, M., Saleh, M., El-Mahdy, M., & El-Matbouli, M. (2016). Recent progress in applications of NPs in fish medicine: A review. *Nanomedicine: NBM, 12*, 701–710.

Shah, B. R., & Mraz, J. (2020). Advances in nanotechnology for sustainable aquaculture and fisheries. *Reviews in Aquaculture, 12*, 925–942.

Shin, G. H., Kim, J. T., & Park, H. J. (2015). Recent developments in nano-formulations of lipophilic functional foods. *Trends in Food Science Technology, 46*(1), 144–157.

Sibaja-Luis, A. I., Ramos-Campos, E. V., de Oliveira, J. L., & Fernandes, L. (2019). Trends in aquaculture sciences: From now to use of nanotechnology for disease control. *Reviews in Aquaculture, 11*(1), 119–132.

Singh, H., Thompson, A., Liu, W., & Corredig, M. (2012). Liposomes as food ingredients and nutraceutical delivery systems. *Encapsulation Technologies and Delivery Systems for Food Ingredients and Nutraceuticals,* 287–318.

Skjermo, J., & Bergh, Ø. (2004). High-M alginate immunostimulation of Atlantic halibut (*Hippoglossus hippoglossus* L.) larvae using Artemia for delivery, increases resistance against vibriosis. *Aquaculture, 238*, 107–113.

Skjolding, L. M., Sørensen, S. N., Hartmann, N. B., Hjorth, R., Hansen, S. F., & Baun, A. (2016). A critical review of aquatic ecotoxicity testing of nanoparticles - the quest for disclosing nanoparticle effects. *Angewandte Chemie, 55*, 15224–15239.

Stromme, M., Brohede, U., Atluri, R., & Garcia-Bennett, A. E. (2009). Mesoporous silica-based nanomaterials for drug delivery: Evaluation of structural properties associated with release rate. *Wiley Interdisciplinary Reviews Nanomedicine and Nanobiotechnology, 1*, 140–148.

Surve, D. H., Paul, A. T., & Jindal, A. B. (2018). Nanotechnology based delivery of nutraceuticals. In *Environmental nanotechnology* (pp. 63–107).

Swain, P., Nayak, S. K., Sasmal, A., Behera, T., Barik, S. K., Swain, S. K., & Jayasankar, P. (2014). Antimicrobial activity of metal-based nanoparticles against microbes associated with diseases in aquaculture. *World Journal of Microbiology and Biotechnology, 30*, 2491–2502.

Swain, P., Sasmal, A., Nayak, S. K., Barik, S. K., Mishra, S. S., Mohapatra, K. D., Swain, S. K., Saha, J. N., Sen, A. K., & Jayasankar, P. (2016). Evaluation of selected metal nanoparticles on hatching and survival of larvae and fry of Indian major carp, rohu (*Labeo rohita*). *Aquaculture Research, 47*, 498–511.

Tafaghodi, M., Sajadi Tabasi, S. A., & Payan, M. (2007). Alginate microsphere as a delivery system and adjuvant for autoclaved *Leishmania major* and *Quillaja saponin*: Preparation and characterization. *Iranian Journal of Pharmaceutical Sciences, 3*, 61–68.

Tandberg, J., Lagos, L., Ropstad, E., Smistad, G., Hiorth, M., & Winther-Larsen, H. C. (2018). The use of chitosan-coated membrane vesicles for immunization against salmonid rickettsial septicemia in an adult zebrafish model. *Zebrafish, 15*(4), 372–381.

Tian, J., Sun, X., Chen, X., Yu, J., Qu, L., & Wang, L. (2008). The formulation and immunisation of oral poly (DL-lactide-co-glycolide) microcapsules containing a plasmid vaccine against lymphocystis disease virus in Japanese flounder (*Paralichthys olivaceus*). *International Immunopharmacology, 8*(6), 900–908.

Tian, J., & Yu, J. (2011). Poly (lactic-co-glycolic acid) NPs as candidate DNA vaccine carrier for oral immunization of Japanese flounder (*Paralichthys olivaceus*) against lymphocystis disease virus. *Fish & Shellfish Immunology, 30*, 109–117.

Toubanaki, D. K., Margaroni, M., & Karagouni, E. (2015). Nanoparticle-based lateral flow biosensor for visual detection of fish nervous necrosis virus amplification products. *Molecular and Cellular Probes, 3*, 158–166.

Vaseeharan, B., Ramasamy, P., & Chen, J. C. (2009). Antibacterial activity of silver nanoparticles (AgNPs) synthesized by tea leaf extracts against pathogenic *Vibrio harveyi* and its protective efficacy on juvenile *Feneropenaeus indicus*. *Letters in Applied Microbiology, 50*(4), 352–356.

Vicari, T., Dagostim, A. C., Klingelfus, T., Limberger-Galvan, G., Sampaio-Monteiro, P., da Silva-Pereira, L., Silva-de Assis, H. C., & Cestari, M. M. (2018). Co-exposure to titanium dioxide nanoparticles (NpTiO2) and lead at environmentally relevant concentrations in the Neotropical fish species *Hoplias intermedius*. *Toxicology Reports, 5*, 1032–1043.

Vignardi, C. P., Hasue, F. M., Sartorio, P. V., Cardoso, C. M., Machado, A. S., Passos, M. J., Santos, T. C. A., Nucci, J. M., Hewer, T. L. R., Watanabe, I. S., Gomes, V., & Phan, N. V. (2015). Genotoxicity, potential cytotoxicity and cell uptake of titanium dioxide nanoparticles in the marine fish *Trachinotus carolinus* (Linnaeus, 1766). *Aquatic Toxicology, 158*, 218–229.

Volland, M., Hampel, M., Katsumiti, A., Yeste, M. P., Gatica, J. M., Cajaraville, M., & Blasco, J. (2018). Synthesis methods influence characteristics, behaviour and toxicity of bare CuO NPs compared to bulk CuO and ionic Cu after *in vitro* exposure of *Ruditapes philippinarum* hemocytes. *Aquatic Toxicology, 199*, 285–295.

World Bank. (2013). Fish to 2030: Prospects for fisheries and aquaculture. Agriculture and environmental services discussion paper 3. In World Bank Report Number 83177- GLB. The World Bank, Washington, DC.

Wu, H. Q., Wei, X. W., Shao, M. W., & Gu, J. S. (2004). Synthesis of zinc oxide nanorods using carbon nanotubes as templates. *Journal of Crystal Growth, 9*(1), 184–189.

Yang, S. Y., Wu, J. L., Tso, C. H., Ngou, F. H., Chou, H. Y., Nan, F. H., Horng, H. E., & Lu, M. W. (2012). A novel quantitative immunomagnetic reduction assay for Nervous necrosis virus. *Journal of Veterinary Diagnostic Investigation, 24*(5), 911–917.

Yasumoto, S., Yoshimura, T., & Miyazaki, T. (2006). Oral immunization of common carp with a liposome vaccine containing *Aeromonas hydrophila* antigen. *Fish Pathology, 41*, 45–49.

Yeh, S. P., Chang, C. A., Chang, C. Y., Liu, C. H., & Cheng, W. (2008). Dietary sodium alginate administration affects fingerling growth and resistance to Streptococcus Sp. and Iridovirus, and juvenile non-specific immune responses of the Orange-spotted grouper, *Epinephelus coioides*. *Fish & Shellfish Immunology, 25*, 19–27.

Yun, S., Jun, J. W., Giri, S. S., Kim, H. J., Chi, C., Kim, S. G., Kim, S. W., & Park, S. C. (2017). Efficacy of PLGA microparticle-encapsulated formalin-killed *Aeromonas hydrophila* cells as a single-shot vaccine against *A. hydrophila* infection. *Vaccine, 35*(32), 3959–3965.

Zhang, C., Wang, G. X., & Zhu, B. (2020). Application of antigen presenting cell-targeted nanovaccine delivery system in rhabdovirus disease prophylactics using fish as a model organism. *Journal of Nanobiotechnology, 18*(1), 1–13.

Zhang, L., Zeng, Z., Hu, C., Bellis, S. L., Yang, W., Su, Y., Zhang, X., & Wu Y. (2015). Controlled and targeted release of antigens by intelligent shell for improving applicability of oral vaccines. *Biomaterials, 77*, 307–319.

Zhao, L., Seth, A., Wibowo, N., Zhao, C. X., Mitter, N., Yu, C., & Middelber, A. P. J. (2014). Nanoparticle vaccines. *Vaccine, 32*, 327–337.

Zhao, Z., Xiong, Y., Zhang, C., Jia, Y. J., Qiu, D. K., Wang, G. X., & Zhu, B. (2020). Optimization of the efficacy of a SWCNTs-based subunit vaccine against infectious spleen and kidney necrosis virus in mandarin fish. *Fish & Shellfish Immunology, 106*, 190–196.

Zheng, F., Liu, H., Sun, X., Zhang, Y., Zhang, B., Teng, Z., & Wang, B. (2016). Development of Oral DNA vaccine based on chitosan nanoparticles for the immunization against reddish body Iridovirus in turbots (*Scophthalmus maximus*). *Aquaculture, 452*, 263–271.

Zhu, B., Liu, G. L., Gong, Y. X., Ling, F., & Wang, G. X. (2015). Protective immunity of grass carp immunized with DNA vaccine encoding vp7 gene of grass carp reovirus using carbon nanotubes as carrier molecule. *Fish & Shellfish Immunology, 42*, 325–334.

Zhu, B., Zhang, C., Zhao, Z., & Wang, G. X. (2020). Targeted delivery of mannosylated NPs improve prophylactic efficacy of immersion vaccine against fish viral disease. *Vaccines, 8*(1), 87.

CHAPTER 22

Gut Morphometry, Enzymes, and Digestive Physiology of Coldwater Fishes

ASIFA WALI[1,*], TASADUQ HUSSAIN SHAH[1], FAROOZ AHMAD BHAT[2], and ISHRAT MOHD[1]

[1]Division of Fisheries Resource Management, Faculty of Fisheries-Rangil, Sher-e-Kashmir University of Agricultural Sciences and Technology, Rangli, Jammu and Kashmir, India

[2]Division of Fisheries Resource Management, Faculty of Fisheries, Skuast-K, Rangil, Ganderbal, Jammu and Kashmir, India

*Corresponding author

ABSTRACT

In order to address nutritional issues in fish feeding, this chapter focuses on the nutritional physiology of coldwater fishes. The histological examination of alimentary canal structures along the digestive tube will assist in calculating all possible methods of feeding for the species and can be supplemented with an enzymatic study. Determining the evolution of the alimentary canal at the histomorphological level will contribute to a more accurate comprehension of dietary needs and nutritional demands at different developmental stages. According to the enzyme activities that may improve survival, the chapter confirms the stages at which acceptable food may be administered. It also specifies what food the fish can digest and absorb exogenous nutrients at each stage. This chapter can also be used to establish a baseline for the healthy fish gut, which will help with understanding changes to the digestive system at various stages of trout

Recent Advances in Sustainable Aquaculture and Fisheries. Shabir Ahmad Dar, Imtiyaz Qayoom, Gohar Bilal Wani, Adnan Abubakr, and Seemab Zehra (Eds.)

DOI: 10.1201/9781998511372-22

life cycles and pave the way for the creation of an informed aquaculture program that will further enhance the management of fish stocks.

22.1 INTRODUCTION

Aquacultured or wild-caught fish provides a significant amount of protein to many developing countries, whether from marine, coastal, off-shore, or freshwater sources. In 2017, fish made up almost 17% of all animal protein consumed worldwide. Furthermore, fish provide about 20% of the average animal protein consumed annually by 3.3 billion people. Taking into consideration the growing demand for quality food, we need to handle natural resources carefully, protect them from overexploitation, and ensure they are managed and governed in a sustainable way to fulfill the increasing need for high-quality food. The global fishing industry is approximated to have generated 179 million tons of fish in 2018. Over 156 million tons of the total were used to feed humans, exceeding the average annual consumption of 20.5 kg. Fisheries provide livelihoods and income for millions of people, especially in developing countries. Globally, it is estimated that more than 158 million people are directly dependent on fish-related activities (fishing, fish farming, fish processing, and trading), most of whom are small businesses in developing nations.

22.2 FISH AS FOOD

Fish is an especially nutrient-dense food, rich in a number of micronutrients that are often lacking in low-income diets. Fish contains a high amount of superior quality protein and essential nutrients, including iodine, vitamin D, long-chain PUFA and fish oil (EPA, DHA), calcium, iron, zinc, and vitamin A. Aquaculture and fisheries, taken together, play a crucial role for food security in the sense that they are a source of food (availability), livelihoods and incomes, particularly for marginalized populations. In spite of this, the importance of fisheries and aquaculture for food security and nutrition has often been underestimated by the food security community and by those involved in the specialized discussions of fish, which have often focused instead on management rather than its significance in human health.

A systematic review of international development and research agencies working on nutrition and food security identified fish as a major omission from policies that could help reduce micronutrient deficiency (Allison et al., 2013). Additionally, the sector is going through a great deal of change and difficulties. Coastal areas and waters are threatened by overfishing, pollution, and competition. Aquaculture's dramatic growth raises many questions about its environmental impact on land, water, and biodiversity, and it contends with competition from other users of both land and water. Fish demand globally is increasing, and the international fish trade is growing at the same time, forcing major economic changes in both sectors. In turn, they result in the emergence of large-scale actors and improved integration with food chains and international trade over traditional small-scale units, with important economic and social implications ranging from economic development to changes in the nature of the workplace and topics relating to fish trade, as well as the relative contribution to food security.

A coldwater ecosystem is the one that maintains oxygen and thermal levels to ensure the health of trouts, mahseers, and snow trouts. Generally, the temperature sits within the tolerance range of trout and salmon from the family *Salmonidae*. Electrochemical parameters, as well as physical, chemical, and geochemical characteristics of different water bodies, are also important for the distribution and abundance of various coldwater fish species. The Indian subcontinent is generally a home to coldwater fishes associated with the Himalayan and sub-Himalayan zones to the north, as well as watersheds draining the southern slope of the Deccan plateau. Coldwater fisheries occupy a significant position in India's fisheries sector. There are a variety of cold water and hill fisheries resources in the country, spread out over high- and low-altitude natural lakes, streams, and rivers in the Himalayan and peninsular regions. It has streams and rivers that extend over 10,000 km, natural lakes that cover 20,500 ha, and both natural and man-made reservoirs covering 50,000 ha and brackish water lakes covering an area of 2500 ha in high-altitude regions.

The teleosts, which form an important source of food supply for humans nowadays, are acclimatized to every type of aquatic habitat and show versatile feeding behavior. There exists a notable variety in both form and functional features that can be observed in fish digestive systems, which are related to diversity in feeding habits, classification, body structure, mass, dimensions, and sexuality. Understanding the morphology of the alimentary canal in teleosts is becoming increasingly crucial for comprehending fish digestive physiology and enhancing nutritional protocols for fish. To grasp the mechanics of histo-physiological and nutritional functions, it is essential to study digestive tract structure and identification. In fishes, there is less knowledge about digestive actions compared to mammals.

Testing the intestinal enzymes of a fish can reveal details about its dietary physiology. Feed caliber directly influences growth support, with nutritional value determined by the ease of digestion and absorption by the animal (Akintunde, 1985). Feeding patterns influence the distribution and functionality of digestive enzymes throughout the gastrointestinal tract (GIT) (Tengjaroenkul et al., 2000). Varied feeding habits at different developmental stages are attributed to digestive processes influenced by dimensions of fish and food category (Tramati et al., 2005). There is an apparent reciprocity between the digestive organs' physiological and anatomical development. Because their surface area is smaller and more susceptible to enzymatic activity, adults typically catch larger animals in natural settings. A stronger gastroenteric work is necessary for this. Understanding digestive catalyst could be helpful in creating enhanced diets and cultivation methods that are more effective (Uys and Hecht, 1987). The paramount factors affecting the energy expenditure of aquatic creatures are timing and weather. Aquatic life consists primarily of thermally responsive creatures. As such, external temperature affects their biological functions, including evolution, procreation, and lifespan. Ectothermic animals have the peculiar quality of being able to adjust to varying temperatures in their surroundings, with established minimum and maximum climatic tolerance thresholds. The influence of water temperature on fish metabolism and enzyme production has been found in a number of experiments conducted during various time periods. In a farmed fishery, a feed's aptitude to promote fish growth is determined not only by its nutrient profile but also by an animal's intrinsic capacity to intake, digest, absorb, and metabolize these nutriments. The way that resources are used by aquatic beings depends on how well their alimental tract enzymes work (Areekijseree et al., 2004; Day et al., 2011). There is less knowledge about fish digestion than about animals, even though the digestion enzymes studied so far by researchers in fish are similar to those observed in other invertebrates. Fish age and developmental stage affect digestive enzyme activity significantly (Klahan et al., 2009; Hidalgo et al., 2011). Additionally, for fish to be able to utilize ingested nutrients, natural enzymes must be present in the right places within the intestinal tract and along its lumen. Feeding habits and intestinal morphology along with the tract influence the dispersion gastrointestinal enzymatic activity and level of the magnitude (Cockson and Bourne, 1972; Tengjaroenkul et al., 2000).

Carbohydrates, protein, and lipids are the main components of fish diets. Regardless of the complexity of the nutrients, each of them is degraded by digestive enzymes that can break them down into simple units that can be absorbed and subsequently metabolized. Different animal groups have varying enzyme complement systems reflecting their adaptations to diets in a complex manner (Moitra and Das, 1967). The digestive processes of fish are poorly understood compared to mammals, despite the fact that the digestive enzymes studied in fish are qualitatively similar to those studied in other vertebrates. All physiological processes in lower vertebrates are governed by energy consumption. As a result of food digestion, they oxidize organic molecules such as carbohydrates, lipids, and proteins to produce their energy demands. Throughout fish metabolism, digestion is an elemental process since it determines whether nutrients are available for all biological processes (Solovyev and Gibsert, 2016). Being fundamental to the process of digestion, digestive enzymes play important roles since the overall efficiency of digestion hinges on their functional characteristics and types (Pujante et al., 2017). In the digestive tract of fish, enzymes are largely responsible for the degradation of nutrients. After examining the digestive physiology of fish, researchers have determined a correlation between enzymes and diet. The ability to measure digestive enzyme activity is, therefore, of potential value for obtaining complete and accurate information about fish digestive physiology.

22.3 ENZYMES AND GUT METRICS

Various enzymes involved in digestion are studied extensively in aquaculture. As a result, studies on digestive secretions have provided useful information in identifying the limits to dietary utilization of protein and carbohydrates and helped formulate nutritional strategies for fish feed and diet formulation (Debnath et al., 2007). Under natural conditions, feeding ecology and trophic niche are reflected by digestive enzyme activities in all aquatic species, similar findings emerge as in the case of piscesians, and are well correlated with their diet (German *et al.*, 2004). Different species have different digestive enzyme activities. Biotic factors including size, age, and origin, as well as abiotic factors like temperature, season, and diet influences enzyme levels or profiles can change (Hofer, 1979). Fish are classified based on their food preferences, which include carnivory, herbivory, or omnivory (Pujante *et al.,* 2017). In addition to the gut length and relative gut mass, fish can be classified according to Zihler's index (ZI), which quantifies the relationship between the length of the intestinal tract and tenfold the body weight squared (German and Horn, 2006; Zihler, 1981). The fish gut length has been assessed as an indicator of dietary strategy based on the relative length of gut (RGL) and ZI (German and Horn, 2006; Al-Hussaini, 1947). In addition to determining the feeding state of fish, RGM (also known as digestion somatic index) is also helpful in assessing the number of tissues to the digestive systems in fish (German and Horn, 2006). There are seven primary digestive enzymes identified in fish, encompassing trypsin, carboxypeptidase A- and B, maltase, amylase, lipase, and alkaline phosphatase (Bone and Moore, 2008). Compared to herbivorous and omnivorous fish, carnivorous species are generally reported to possess displaying elevated levels of proteolytic enzyme activity (Kuz'mina, 1996). Additionally, carbohydrases are not considered to significantly contribute to predatory fish digestion because their levels are entirely different from those in herbivorous and omnivorous fish (Hidalgo et al., 1999). Generally, the activity of proteases (such as trypsin) in carnivorous species increases as they age, whereas it decreases in phytophagous species. Carbohydrase activity progressively decreases in carnivorous fish as they age but increases in herbivorous and omnivorous fish (Kuz'mina, 1996).

Fishes and vertebrates employ an identical approach in their alimentary tube (Wilson and Castro, 2011). Variations in architecture are correlated with differences in fish alimentary systems, evolutionary relationships, ontogenies, food formulations, and habitats (Xiong et al., 2011). Unlike other fish, carnivorous teleosts have relatively large stomachs and short intestines. As they feed on a large number of animals as part of their diet, omnivore fish usually have a large stomach and long intestines. It is the characteristic of herbivore fishes with the longest complex intestine to have an intestinal bulb or a small gut or not have a stomach.

Under a variety of environmental conditions, the GIT of fish serves as a vital component for growth, nutrition, and survival. Prior to passing into the esophagus, food is captured in the mouth and processed there. Eventually, the latter connects with the stomach through a muscular and mucus-secreting channel. Digestion takes place during the presence of food in the stomach, which is the portion of the GIT with the most diversifications. Nutrition is consumed by the intestine, typically supported by the slender, finger-like appendages, and pyloric caeca, which are present between the stomach and large intestine. The anus, distal intestine, and rectum contain mucus-secreting cells that aid in the expulsion of undigested food. Fish rely on enzymes within the GIT wall and lumen to effectively utilize ingested nutrients. In addition to safeguarding the epithelium from mechanical damage, mucus plays a crucial role in swiftly eliminating pathogens and irritants. It also forms a diffusion barrier for ions between luminal contents and the epithelial lining, shielding them from irritation.

By histochemical approaches, most teleost fish have been described extensively as having mucus-secreting cells in the GIT. However, examining tiny, studied non-model organisms can be prudent when looking at specific environmental conditions or whenever scientific questions expand. A vital organ system of fishes that connects with the environment is the GIT, which plays a commendable role in growth, nutrition, and survival under stressful conditions. Moreover, endocrine cells regulate several functions of various segments of the GIT, whose frequency and distribution are affected by environmental conditions and the particular fish (Diaz *et al.,* 2003). It is, therefore, crucial to sample at different geographical settings to enhance comprehension of the mechanisms driving morphological diversification. Research on the intestinal morphology of fish is an effective instrument for understanding the processes involved in the intake, breakdown, and uptake of nutrients from food and has recently been done for various species (Xiong et al., 2011; Germano et al., 2013; Lokka et al., 2013). Due to the high diversity of teleost fish and different positions in the trophic chain, fish possess morphological diversity and functional diversity in their digestive systems (Diaz et al., 2003). Fish use their mouths and their oropharyngeal cavities to select food, seize it, and then transfer it to the esophagus (Al-Hussaini, 1949; Rodrigues et al., 2006, Wilson and Castro, 2011). Based on anatomical and histological studies, we are able to connect those cavities with the nutrient processing system (Rodrigues et al., 2006; Hanke et al., 2008) and eating habits (Al-Hussaini, 1946). A tubular organ designed for conveying, storing, digestion, and absorption of food is the food pipe, digestive sac, intestinal tract, and caecal structures (Seixas-Filho et al., 2000; Albrecht et al., 2001; Wilson and Castro, 2011). Histologically, the composition of these regions of the digestive tract varies among different species. These regions typically consist of four layers: the mucosal layer, submucosal layer, muscular layer, and serosal layer (Albrecht et al., 2001; Xiong et al., 2011; Wilson and Castro, 2011).

Research shows that the main differences between the layers are observed in the epithelium, concerning feeding (Albrecht et al., 2001), ingested food type (Germano et al., 2013), and mucosal absorbing and producing factors (Albrecht et al., 2001). External characteristics (such as the size and shape of the mouth, the shape of the caudal fin) and internal traits (e.g., teeth present or absent, stomach size and shape, gut length) are important indicators of feeding ecology in fish. (Wooton, 1998). Exploration of the relations between various feeding-related morphological characteristics with body length, such as body girth (Stergiou and Karpouzi, 2003), mouth dimensions, and gut length (Karachle and Stergiou, 2007) is of great importance for understanding the biology and ecology of fishes (Froese and Pauly, 2008); as well as pinpointing the ecological role of a species in the aquatic food webs (Karpouzy and Stergiou, 2003).

The length of the gut serves as a crucial indicator for understanding the feeding behavior across various invertebrate classes, including fishes (Kramer and Bryant, 1995), reptiles (O'Grady et al., 2005), birds (Ricklefs, 1996), and mammals (Chivers and Hladik, 1980). Furthermore, it provides a basis for interspecific comparisons in fishes. The GIT of fish exhibits significant morphological and functional diversity, aiding in taxonomy, determination of feeding habits, and habitat preferences (Hellberg and Bjerkas, 2000; Murray et al., 1996). Specialized mucosal structures are present throughout the digestive tract to optimize secretion, absorption, and digestive processes (Ezeasor and Stokoe, 1981). Thus, the fish intestine's main function is to finish digestion initiated in the stomach and facilitate nutrient absorption (Wilson and Castro, 2011). Additionally, the gross histological characteristics of fish GIT vary according to taxonomy, feeding habits, and body morphology (Al Abdulhadi, 2005). The morphological characteristics of the GIT of carnivorous fish diverge from that of herbivorous fish. The predatory fish tend to have shorter intestines than the herbivorous fish due to the diet's lower percentage of plant materials (El-Bakary and El-Gammal, 2010; Buddington and Doroshov, 1986). In terms of histology, carnivorous fish typically exhibit a higher density of mucus cells in the esophagus compared to herbivorous fish (Abuzinadah, 1990).

Fish have a unique intestine structure that depends upon their extraordinary environment and feeding habits, and their body shape and weight. Fish's gastrointestinal systems exhibit an astonishing range in appearance and operation, which is related to differences in consumption patterns and classification (Al-Abdulhadi, 2005). Similar to other vertebrates, fish have an alimentary tract that is lengthwise separated into the esophagus, stomach, intestines, and rectum. The alimentary channel varies in diameter (Treer et al., 1995). In fish, there are several organs attached to the alimentary tube, including the tongue and teeth in the mouth and digestive glands beyond the skin, specifically the liver, gall bladder, and exocrine pancreas; however, the salivary glands are usually absent in the mouth (Treer et al., 1995). A histological examination shows the alimentary canal wall to consist of four distinct layers beginning at the cranial end and decreasing to the caudal end. The membrane's lumen is four layers: the mucus, the submucus, the muscularis externa, and the serosa or adventitia (Kierszenbaum and Tres, 2012). Understanding the structural characteristics and design of the digestive tract is crucial for identifying an appropriate feeding regime for trout, a fish of increasing aquaculture importance. Considering how rapidly the aquaculture industry is growing, there is an increasing need for specialized feeds tailored to specific production situations. Nutritionists and feed manufacturers have centered most of their research on determining which feedstuffs from the wide array of feedstuffs available to the feed industry may be used to produce lower-cost aquaculture feeds. In contrast to this, the successful mass rearing of fish and shrimp larvae has a large economic importance in marine aquaculture. Therefore, current research in fish nutrition and feeding is largely devoted to developing artificial diets for larvae of more commonly raised fish. In both cases, a detailed knowledge concerning the digestive physiology of aquatic animals is a prerequisite when formulating feeds; it follows, therefore, that studies aimed at determining the main features of the digestive system of different species, as well as evaluating the possible effect of anti-nutritive compounds present in many feeds are needed. Furthermore, some workers recommend that research efforts be focused on the development of simple *in vitro* digestion techniques for quick estimation of nutrient digestibility (Tacon, 1995).

Sport and food fish of the rainbow sp. (*Oncorhynchus mykiss*) are into extensive commercialization. The largest possible dimensions of rainbow trout are 120 cm in length overall, 25.4 kg weight, and 11 years old. Mature rainbow trout normally weigh 2–3 kg. The upper, colder water regions of the rivers and seas are home to rainbow trout. As in the instance of other trout, the environment and diet of rainbow trout typically dictate their color and morphology. Different freshwater sources provide shelter to many local stocks and strains of rainbow trout, of which the high commercial types have been used for reproductive intent. The holotype rainbow trout breed yields multiple enhanced varieties with outstanding features such as roughness, immunity to diseases, greater growth rate, and the capacity to procreate in captivity. Rainbow trout's native habitats are freshwater sources along the Pacific regions of Asia and North America.

More than 82 countries have received rainbow trout introductions, including virtually all those with favorable conditions for its culture. This is due to the species' greater tolerance for a variety of environmental variables than those of various kinds of trout. The quantity and quality of the dietary product, the chronological stage of the fish, and the ecosystem in which rainbow trout reside all affect what the fish is able to eat naturally. Rainbow trout, as opposed to other species of fish, are combative and voracious feeders. When terrestrial insects fall into the water, they eat almost anything, like opportunistic feeders. Pertaining to creatures such as insects, these are the developmental phases of moths and butterflies (*Lepidoptera*), adult beetles (*Coleoptera*), flies (*Diptera*), and ants (*Formicidae*). Despite rainbow trout having minimal physical traits for obtaining and processing prey, they are primarily carnivores. Apart from their diminutive size and basic teeth, they lack complex systems for ensnaring, retaining, or ingesting prey. The salmonid has a broad esophagus and a stomach that is formed like a Y. It consumes its meal whole. The total number of pyloric caeca, which typically extend into the midsection at the pyloric end, is significant from a taxonomic standpoint in salmonid populations. Because the pancreas is dispersed throughout the layers of fat and connective tissue, it is difficult to see in the pyloric caeca. A bile canal that connects the upper midsection of the gut to the liver's mid lobe is present in the bigger specimens. Among the midgut and the hindgut, there is no clear distinction. The kidney comprises not only a thin, almost translucent swim bladder but also a significant portion of the abdominal chamber, just above the air bladder. All things taken into account, rainbow trout are often indicative of salmonids. To sum up, this fish is a prime instance of a comparatively fundamental (unspecialized) carnivore. It has an extensively extendable belly chamber, a short gut for digesting meals, which includes little to no unprocessed stuff, and good swimming skills for collecting the prey. The overall length of the gut (from the esophagus to the anus) is typically 0.6–0.8 times the length of the entire body, making it not much longer compared to that of teleosts.

This chapter focuses on the nutritional physiology of *O. mykiss* and associated facets; therefore, it could also help to solve dietary problems in fish feeding. Evaluating the progression of the digestive tube at the histo-morphological level will help in the precise understanding of feed and nutritional requirements at different developmental stages, histological examination of alimentary canal structures along the digestive tube helps in calculating all the possible ways of feeding for the species which cannot be fully justified by direct observations and can be complemented with the enzymatic study. To determine the different phases at which suitable nutrition could be given based on the activity of the enzymes, more research is required, that could lead to better survival, and exactly what food the fish can digest and absorb exogenous nutrients at what stage. Both histological information and enzymatic study will provide a better understanding than the single method approach. Assessing the development of the digestive tract through the straightforward measurements of different parameters of the gut helps in the precise knowledge of feed and nutritional requirements at different developmental stages of *O. mykiss* to offer improved feeding and nurturing methods to boost fish production and, in turn, give a fillip to the economy. Further, the knowledge gained under the research works can be utilized for creating a foundation for the trout gut's health, which is beneficial in understanding changes to the digestive system in different life cycle stages of trout and bring about the establishment of a knowledgeable farming program and further improve the management of trout stocks.

22.4 CONCLUSION

Protein-rich diet should be provided to fry and the early stage of *O. mykiss* as reflected by higher trypsin values for this stage. It can be concluded that protein is more digestible by fish at the early development of life than advanced stages, which show a definitive profile of histology regarding lipids and carbohydrate metabolism amplified by various cells like enterocytes, supra nuclear vacuoles in combination with lipase and amylase activity. By formulating feeds based on the enzyme profile studies like the present work, digestibility of the feed will be increased resulting in the faster growth and higher profits. Various manipulation to trout feeds is now the need of the hour as feeding of trout is highly cost-intensive. So, by feed manipulation according to digestive ability of various life stages of *O. mykiss* keeping in view the enzyme activity of that stage, will result in enhanced growth and development through appropriate nutrient utilization, which ultimately lead to improved productions of trout fish ultimately contributing not only to the economy but also to per capita consumption adding to health perspective. We now have an understanding that developing artificial diets for fish larvae is more complicated than just finding the right combination of nutrients. In the formulation of microdiets for fish larvae, a "holistic" strategy is needed.

KEYWORDS

- **enzyme**
- **feeding**
- **gut**
- **histomorphology**
- **saliva**
- **stocks**
- **trout**

REFERENCES

Abdulhadi, H. (2005). Some comparative histological studies on alimentary tract of tilapia fish (*Tilapia spilurus*) and sea bream (*Mylio cuvieri*). *Egyptian Journal of Aquatic Research, 31*, 387–397.

Abuzinadah, O. A. (1990). *Studies on red sea fish* (Ph.D. thesis, Department of Zoology, School of Biological Science, University College of Swansea).

Akintunde, E. A. (1985). Digestive enzymes in the gut of *Sarotherodon galilaeus* (Syn. *Tilapia galilea*) (Family Cichlidae) of Lake Kainji. *Journal of Science, 18*, 22–25.

Albrecht, M. P., Ferreira, M. F. N., & Caramaschi, E. P. (2001). Anatomical features and histology of the digestive tract of two related neotropical omnivorous fishes (Characiformes; Anastomidae). *Journal of Fish Biology, 58*, 419–430.

Al-Hussaini, A. H. (1946). The anatomy and histology of the alimentary tract of the coral feeding, *Mulloides auriflamma* (Forsk.). *Journal of Morphology, 78*, 121–154.

Al-Hussaini, A. H. (1947). *The feeding habits and the morphology of the alimentary tract of some teleosts: Living in the neighbourhood of the marine biological station, Ghardaqa, Red Sea*. Fouad I University Press.

Al-Hussaini, A. H. (1949). On the functional morphology of the alimentary tract of some fishes in relation to differences in their feeding habits; anatomy and histology. *Quarterly Journal of Microscopical Science, 90*, 109–139.

Allison, E. H., Delaporte, A., & Hellebrandt de Silva, D. (2013). Integrating fisheries management and aquaculture development with food security and livelihoods for the poor. Report submitted to the Rockefeller Foundation, School of International Development, University of East Anglia Norwich, UK. p. 124.

Areekijseree, M., Engkagul, A., Kovitvadhi, U., Thongpan, A., & Mingmuang, M. (2004) Temperature and pH characteristics of amylase and proteinase of adult freshwater pearl mussel, *Hyriopsis* (*Hyriopsis*) *bialatus*, Simpson 1900. *Aquaculture, 234*(1–4), 575–587.

Bone, Q., & Moore, R. H. (2008). Biology of fishes (Third Edition). Taylor and Francis Group Edition.

Buddington, R. K., & Doroshov, S. I. (1986). Structural and functional relations of white sturgeon alimentary canal (*Acipenser transmontanus*). *Journal of Morphology, 190*(2), 201–213.

Chivers, D. J., & Hladik, C. M. (1980). Morphology of the gastrointestinal tract in primates; Comparisons with other animals in relation to diet. *Journal of Morphology, 166*(3), 337–386.

Cockson, A., & Bourne, D. (1972). Enzymes in the digestive tract of two species of euryhaline in fish. *Comparative Biochemistry and Physiology. Part A, Computer Physiology, 41*(4), 715–718.

Day, R. D., German, D. P., & Tibbetts, I. R. (2011). Why can't young fish eat plants? Neither digestive enzymes nor gut development preclude herbivory in the young of a stomach less marine herbivorous fish. *Comparative Biochemistry and Physiology, 158*(1), 23–29.

Debnath, D., Pal, A. K., Sahu, N. P., Yengkokpam, S., Baruah, K., & Choudhury, D. (2007). Digestive enzymes and metabolic profile of *Labeo rohita* fingerlings fed diets with different crude protein levels. *Biochemistry and Molecular Biology, 146*(1), 107–114.

Diaz, A. O., García A. M., Devincenti, C. Y., & Goldemberg, A. L. (2003). Morphological and histochemical characterization of the mucosa of the digestive tract in *Engraulis anchoita*. *Anatomia, Histologia, Embryologia, 32*, 341–346.

El-Bakary, N. E. R., & El-Gammal, H. L. (2010). Comparative histological, histochemical and ultrastructural studies on the proximal intestinal of flathead greymullet (*Mugil cephalus*) and seabream (*Sparus aurata*). *World Applied Science Journal, 8*(4), 477–485.

Ezeasor, D. N., & Stokoe, W. M. (1981). Light and electron microscopic studies on the absorptive cells of the intestine, caeca and rectum of the adult rainbow trout, *Salmo gairdneri*, Rich. *Journal of Fish Biology, 18*(5), 527–544.

Froese, R., & Pauly, D. (2008). Fishbase (version 07/2008). http://www.fishbase.org.

German, D. P., & Horn, M. H. (2006). Gut length and mass in herbivorous and carnivorous prickleback fishes (Teleostei: Stichaeidae): Ontogenetic, dietary, and phylogenetic effects. *Marine Biology, 148*(5), 1123–1134.

German, D. P., Horn, M. H., & Gawlicka, A., (2004). Digestive enzyme activities in herbivorous and carnivorous prickleback fishes (Teleostei: Stichaeidae): Ontogenetic, dietary, and phylogenetic effects. *Physiological and Biochemical Zoology, 77*, 789–804.

Germano, R. M., Stabille, S. R., Mari, R. B., Pereira, J. N. B., Faglioni, J. R. S., & Miranda Neto, H. (2013). Morphological characteristics of the *Pterodoras g]-ranulosus* digestive tube (Valenciennes, 1821) (Osteichthyes, Doradidae). *Acta Zoologica, 95*(2), 166–175.

Hanke, B. V., Meyer, A. A., & Oliveira, E. (2008). Análise histológica de estruturas sensoriais de *Centropomus parallelus* (Poey, 1896) (Centropomidae) relacionadas ao hábito alimentar. *RUBS, 1*, 16–23.

Hellberg, H., & Bjerkås, I. (2000). The anatomy of the oesophagus, stomach and intestine of the common wolffish (*Anarhichas lupus* L.): A basis for diagnostic work and research. *Acta Veterinaria Scandinavica, 41*(3), 283–297.

Hidalgo, F. B., Barrios, A. F., Farnés, O. C., & Hernández, K. U. (2011). Digestive enzymes of two freshwater fishes (Limia vittata and Gambusia punctata) with different dietary preferences at three developmental stages. *Comparative Biochemistry and Physiology Part B: Biochemistry and Molecular Biology, 158*(2), 136–141.

Hidalgo, M. C., Urea, E., & Sanz, A. (1999). Comparative study of digestive enzymes in fish with different nutritional habits. Proteolytic and amylase activities. *Aquaculture, 170*, 267–283.

Hofer, R. (1979). The adaptation of digestive enzymes to temperature, season and diet in roach, *Rutilus rutilus* L., and rudd *Scardinius erythrophthalmus* L. 1. Amylase. *Journal of Fish Biology, 14*(6), 565–572.

Karachle, P. K., & Stergiou, I. K. (2007, September 5). "Uncoiling" information on fish gut. In J. Snoeks, G. Boden, & T. Musschoot (Eds.), *5rd Fishbase Mini Symposium, Tervuren, Belgium* (pp. 17–18). Royal Museum for Central Africa (RMCA).

Karpouzy, V. S., & Stergiou, K. I. (2003). The relationships between mouth size and shape and body length for 18 species of marine fishes and their trophic implications. *Journal of Fish Biology, 62*(6), 1353–1365.

Kierszenbaum, A. L., & Tres, L. L. (2012). Histology and cell biology. *An introduction to pathology* (p. 701). Philadelphia: Elsevier.

Klahan, R., Areechon, N., Yoonpundh, R., & Engkagul A. (2009). Characterization and activity of digestive enzymes in different sizes of Nile Tilapia (*Oreochromis niloticus* L.). *Kasetsart Journal (Natural Science), 43*, 143–153.

Kramer, D. L., & Bryant, M. J. (1995). Intestine length in the fishes of a tropical stream: 2. Relationships to diet-the long and short of a convoluted issue. *Environmental Biology of Fishes, 42*(2), 129–141.

Kuz'mina, V. V. (1996). Influence of age on digestive enzyme activity in some freshwater teleosts. *Aquaculture, 148*(1), 25–37.

Løkka, G., Austbo, L., Falk, K., Bjerkås, I., & Koppang, E. O. (2013). Intestinal morphology of the wild Atlantic salmon (*Salmo salar*). *Journal of Morphology, 274*, 859–876.

Moitra, S. K., & Das, K. M. (1967). Influences of environment on the digestive enzyme of a fish, *Tilapia mossambica* (Peters). *Current Science, 36*, 381–382.

Murray, H. M., Wright, G. M., & Goff, G. P. (1996). A comparative histological and histochemical study of the post-gastric alimentary canal from three species of pleuronectid, the Atlantic halibut, the yellowtail flounder and the winter flounder. *The Journal of Fish Biology, 482*, 187–206.

O'Grady, S. P., Morando, M., Avila, L., & Dearing, D. M. (2005). Correlating diet and digestive tract specialization: Examples from the Lizard family Liolaemidae. *Zoology, 108*(3), 201–210.

Pujante, I. M., DõÂaz-LoÂpez, M., Mancera, J. M., & Moyano, F. J. (2017). Characterization of digestive enzymes protease and alpha-amylase activities in the thick-lipped grey mullet (*Chelon labrosus,* Risso 1827). *Aquaculture Research, 48*(2), 367–376.

Ricklefs, R. E. (1996). Morphometry of the digestive tracts of some passerine birds. *The Condor 98*(2), 279 –292.

Rodrigues, S. S., Navarro, R. D., & Menin, E. (2006). Adaptações anatômicas da cavidade bucofaringiana de *Leporinus macrocephalus* Garavello e Britski, 1988 (Characiformes, Anostomidae) em relação ao hábito alimentar. *Revista Biotemas, 19*(1), 51–58.

Seixas-Filho, J. T., Brás, J. M., Gomide, A. T. M., Oliveira, M. G. A., Donzele, J. L., & Menin, E. (2000). Anatomia funcional emorfometria dos intestinos e cecos pilóricos do teleostei (Pisces) de Água Doce Piau (*Leporinus friderici*, Bloch, 1794). *Revista Brasileira de Zoologia, 29*, 2181–2192.

Solovyev, M., & Gisbert, E. (2016). Influence of time, storage temperature and freeze/thaw cycles on the activity of digestive enzymes from gilthead sea bream (*Sparus aurata*). *Fish Physiology and Biochemistry, 42*(5), 1383–1394.

Stergiou, K. I., & Karpouzi, S. V. (2003). Length–girth relationships for several marine fishes. *Fisheries Research, 60*(1), 161–168.

Tacon, A. G. J. (1995). Application of intensive and semi-intensive fish. *The Journal of Applied Ichthyology, 11*, 205–221.

Tengjaroenkul, B., Smith, B. J., Caceci, T., & Smith, S. A. (2000). Distribution of intestinal enzyme activities along the intestinal tract of cultured Nile tilapia, *Oreochromis niloticus* L. *Aquaculture, 182*, 317–327.

Tramati, C., Savona, B., & Mazzola, A. (2005). A study of the pattern of digestive enzymes in *Diplodus puntazzo* (Cetti 1777) (Osteichthyes, Sparidae): Evidence for the definition of nutritional protocols. *Aquaculture International, 13*, 89–95

Treer, T., Safner, R., Aničić, I., & Lovrinov, M. (1995). Ribarstvo. Zagreb: Nakladni zavod Globus. p. 463.

Uys, W., & Hecht T. (1987). Assays on the digestive enzymes of sharp tooth catfish *Clarias gariepinus* (Pisces: Clariidae). *Aquaculture, 63*, 301–313.

Wilson, J. M., & Castro, L. F. C. (2011). Morphological diversity of the gastrointestinal tract in fishes. In M. Grosell, A. P. Farrell, & C. J. Brauner (Eds.), *The multifunctional gut of fish* (pp. 42–44). Boston, Mass, USA: Academic Press.

Wooton, R. J. (1998). Ecology of teleost fishes. 2nd Edn. *Fish and Fisheries Series No. 24*. London: Kluwer Academic Publishers.

Xiong, D., Zhang, L., Yu, H., Xie, C., Kong, Y., Zeng, Y., Huo, B., & Liu, Z. (2011). A study of morphology and histology of the alimentary tract of *Glyptosternum maculatum* (Sisoridae, Siluriformes). *Acta Zoologica, 92*, 161–169.

Zihler, F. (1981). Gross morphology and configuration of digestive tracts of Cichlidae (Teleostei, Perciformes): Phylogenetic and functional, significance. *Netherlands Journal of Zoology, 32*(4), 544–571.

CHAPTER 23

Seaweed: An Ecosystem Service for Aquaculture Sustainability

ABOOBUCKER SIDDIK ABUL KASIM* and ASAAD H. MOHAMED

Beacon Development, King Abdullah University of Science and Technology, Thuwal, Jeddah, Saudi Arabia

*Corresponding author

ABSTRACT

The majority of aquaculture operations, either land-based or offshore-based, discharge varying levels of contaminants, including nitrogen and phosphorus, to the aquatic environment. The discharged nutrients accumulate in the marine system causing coastal eutrophication and ecosystem imbalance. The sustainable treatment required to reduce the impact of these contaminants on the coastal environment. The configuration of seaweeds in the aquaculture system is constantly increasing to reduce the pollution impact by the nutrients uptake. The seaweeds are proving their effectiveness in removing nutrients from land-based and offshore aquaculture due to significant developments in their cultivation over the past few years. Incorporating seaweeds in the aquaculture system can contribute significantly to the economic well-being of aquaculture producers by harvesting the biomass. In this chapter, we discussed the ecosystem service offered by seaweeds, the integration of seaweeds in the aquaculture operations to remediate the excess nutrient discharged in the coastal waters. Additionally, the cultivation of economically valuable species provides incentives for the aquaculture producers.

Recent Advances in Sustainable Aquaculture and Fisheries. Shabir Ahmad Dar, Imtiyaz Qayoom, Gohar Bilal Wani, Adnan Abubakr, and Seemab Zehra (Eds.)

DOI: 10.1201/9781998511372-23

23.1 INTRODUCTION

Aquaculture industry growth relies on innovative research to improve the sustainability and diversification of the aquaculture sector (Chopin et al., 2012). A huge effort is made during recent years to build more sustainable systems for the growth of aquaculture sector (Mangano et al., 2019). The expansion of the aquaculture sector and the establishment of sustainable practices could alleviate an increasing demand for aquaculture products in intensive aquaculture systems. Intensive aquaculture is associated with considerable organic and inorganic waste release from uneaten feeds into the environment (Wang et al., 2012). The release of organic and inorganic matters into the marine environment has a considerable impact on the quality of the marine water (Fang et al., 2016). A due consideration of the ecological and environmental impacts of intensive aquaculture systems has resulted in increased interest in alternative sustainable practices, mainly integrated multi-trophic aquaculture (IMTA). In addition, the sustainable practice helps in managing the environmental impacts of aquaculture, while also providing additional aquaculture product within one system. Hence, the sustainability of aquaculture is greatly dependent on extractive species-based aquaculture (Neori et al., 2004).

Incorporation of fin-fish aquaculture with extractive species like seaweeds may provide a management opportunity for mitigating climate change impacts and addressing coastal acidification (Chopin, 2015; Nederlof et al., 2022). Seaweeds are important components in marine benthic ecosystems and play a major role in the ecosystem by providing habitat for marine invertebrates, nutrient assimilation and improving marine water quality (Al-Hafedh et al., 2015; El-Sherbiny et al., 2021). Due to their multi-functions in ecosystems and sustainability, seaweeds are a critical part of the sustainable aquaculture system. Seaweeds are recognized as a contributor to mitigate environmental contaminants, reduce ecological footprints and public opposition toward intensive aquaculture (Troell et al., 2009). The seaweeds are used as nutrient strippers in inland aquaculture and coastal mariculture, further, it has been demonstrated as an excellent ecotechnology approach (Neori et al., 2004). Aquaculture's sustainability and success are determined by selecting seaweed species capable of rapid growth, high nutrient uptake, and high biomass conversion rates.

23.2 GLOBAL PRODUCTION AND ECONOMIC VALUE OF SEAWEEDS

Seaweeds have a wide variety of environmental and ecological applications and the demand for the biomass is rapidly rising. In various countries, seaweeds are produced commercially due to an increased interest in the product. An estimated US $10 billion is generated annually by the seaweed industry in products and services for direct and indirect human uses (Rebours et al., 2014). The production of seaweed for human consumption accounts for about 83% of the total, while the rest is used for fertilizers and animal feed additives, pharmaceutical applications, and environmental and biotechnological purposes (Rebours et al., 2014). Five genera of seaweeds including *Saccharina*, *Undaria*, *Porphyra*, *Kappaphycus,* and *Gracilaria* represented around 98 % of the globally cultured seaweed. *Saccharina* and *Kappaphycus* are an important source of raw materials for the food and food-based polymer industries. Laminaria, Palmaria, and Wakame seaweeds are highly valuable, and their economical values were estimated between 18 and 160 USD per kg (van den Burg et al., 2016). Further, developing a successful marketing strategy for seaweed as human food, as well as value-added concepts can greatly enhance seaweed production.

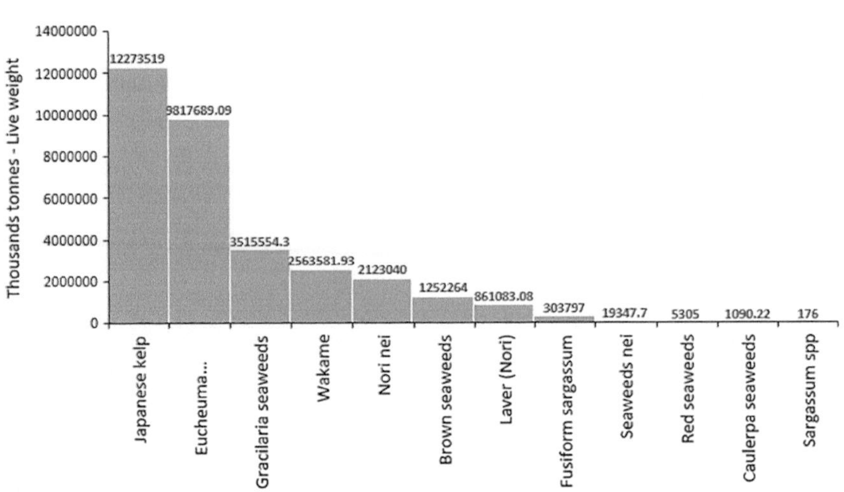

FIGURE 23.1 Global mariculture production of marine seaweeds (FAO, 2021).

Source: Adapted from FAO, 2021

TABLE 23.1 Economical Value of Seaweeds and Seaweed Products (van den Burg et al., 2016).

S. No.	Seaweeds	Price in USD
1	Laminaria	18– 3.5/kg
2	Palmaria	27–35.25/kg
3	Wakame	160/kg
	Seaweed products	
4	**Animal feed (Dry matter)**	
	• *Saccharina latissima*	61.60/metric ton
	• *Palmaria palmata*	168.22/metric ton
	• *Ulva lactuca*	67.29/metric ton
5	**Hydrocolloids**	
	• Alginates • Carrageenan • Agar	3500–12,000/metric ton

TABLE 23.2 Global Aquaculture Production of Seaweeds (FAO, 2020).

Seaweeds	Produced in 2018 (Thousand Ton, Live Weight)
Japanese kelp (***Laminaria japonica***)	11,448.30
Eucheuma seaweeds nei (***Eucheuma spp***.)	9237.50
Gracilaria seaweeds (***Gracilaria spp.***)	3454.80
Wakame (***Undaria pinnatifida***)	2320.40
Norinei (***Porphyra*** spp.)	2017.80
Elkhorn sea moss (***Kappaphycus alvarezi***i)	1597.30
Brown seaweeds (Phaeophyceae)	891.50
Laver (***Porphyra tenera***)	855.00
Fusiform sargassum (***Sargassum fusiforme***)	268.70
Spiny eucheuma (***Eucheuma denticulatum)***	174.90
Spirulina nei (***Spirulina spp***.)	69.60
Seaweeds nei (algae)	22.50
Other algae	27.80
Total	32,386.20

23.3 NUTRIENT PHYSIOLOGY OF SEAWEEDS

Like terrestrial plants, seaweeds utilize the light source as an energy source through the photosynthesis reaction. Seaweeds assimilate the carbon, nitrogen, and phosphorus present in the water column and it is a fundamental part of the biochemical process (Pajares and Ramos, 2019; White and Dyhrman, 2013). Furthermore, nitrogen and phosphorus are key elements in the genetic material, energy transfer, and structural support of organisms (Aczel, 2019; Bringslid, 2019). A basic understanding of seaweed nutrient physiology and nutrient requirements of seaweeds is important to identify the potential seaweeds for the bioremediation process. The acquisition of nutrients in seaweeds is through inorganic nutrient uptake, assimilation of inorganic nutrients to organic molecules and storage in cellular vacuoles (Roleda and Hurd, 2019). The acquisition of dissolved organic nitrogen (nitrate) and dissolved inorganic phosphorus (phosphates) in seaweeds is through active diffusion (Harrison and Hurd, 2001). The nitrate and phosphate in the seaweeds are assimilated to ammonium (NH_4^+) and polyphosphates, respectively, and it is an energy-dependent process (Boyd and Hurd, 2009; Yang et al., 2017). The dissolved inorganic carbon (CO_2) requires passive diffusion through the cell membrane (Raven et al., 2011) and passive transport is an energy-independent process that requires no photosynthesis reaction (Roleda and Hurd, 2019). During active transport of nitrates to the cellular membrane, the nitrates can be stored in vacuoles or reduced to nitrites and then to ammonium (Forbord et al., 2021). Therefore, the knowledge of seaweed's nutrient physiology, nutrient uptake rate and assimilation capacity is important before incorporating the species in an aquaculture system.

23.4 ECOSYSTEM SERVICE OF SEAWEEDS

The expansion of offshore aquaculture operation contributed to the vast release of inorganic nutrients (nitrogen and phosphorus) to the surrounding water column (Kautsky et al., 1999). The offshore cage aquaculture operations face major challenges to minimize the environmental impacts or eutrophication in the marine ecosystem. Hence, the aquaculture industries have been forced to develop sustainable strategies that could efficiently remove biological wastes or nutrients posing threats to the environment (Marinho-Soriano et al., 2009; Neori et al., 2004; Troell et al., 2003; Wurts, 2000). Seaweed offers a wide range of ecosystem services including bioremediation for coastal eutrophication, mitigation of ocean acidification, and habitat formation (Langton et al., 2019).

Bioremediation with seaweeds can be accomplished with a variety of aquaculture systems. Artificial ponds, indoor tanks, and marine offshore cages are some of the examples of aquaculture systems. There are a multitude of seaweed species that can be used in bioremediation, but it depends on the type of nutrients discharged from the aquaculture system and environmental conditions. Numerous research has been suggested that seaweed has nutrient assimilation capability that can efficiently fix nitrogen rather than being cause for eutrophication (Resende et al., 2022). In this regard, integration of seaweeds culture with fed fish aquaculture has been suggested as bioremediation strategy to improve marine water quality and increase economic benefits. Seaweeds shown to reduce total ammonia nitrogen, nitrite, and nitrate generated by intensive aquaculture operations (Marinho-Soriano et al., 2011; Troell et al., 1997; Wu et al., 2015). Depending on the species, seaweeds are capable of assimilating 29–100% of the total ammonia nitrogen in the water column (Ashkenazi et al., 2019; Wu et al., 2015).

23.4.1 ULVA SPECIES

The seaweed *Ulva lactuca* has been widely tested for the bioremediation efficacy against the coastal pollution (Al-Hafedh et al., 2015; Al-Hafedh et al., 2012; Elizondo-González et al., 2018; Msuya and Neori, 2008). The seaweed Ulva is an effective nutrient pump for removing dissolved inorganic nitrogen and phosphorus from the aquatic ecosystem. Due to the rapid growth rate and high photosynthetic activity, *Ulva lactuca* has been used widely in the aquaculture effluent treatment (Al-Hafedh et al., 2015; Al-Hafedh et al., 2012; Elizondo-González et al., 2018; Msuya and Neori, 2008; Sode et al., 2013). *Ulva lactuca* is a proven species that could efficiently assimilate the nitrogen and phosphorus in the aquaculture effluents. *Ulva lactuca* has a high nitrogen removal efficiency of 83% and phosphorus of 41% (Al-Hafedh et al., 2012). In another study by Chung et al. (2002), reported that *Ulva lactuca* removes 40–90% of NH_4 from aquaculture discharge. *Ulva lactuca* incorporated in the fish culture could efficiently assimilate 85–90% of NH_3-N (Neori et al., 2003). Overall, the reports suggested that the incorporation of *Ulva lactuca* with finfish or shrimp aquaculture could efficiently remove 80–90% of total ammonia nitrogen from the discharged effluents.

Similarly, *Ulva reticulata* reported to have a removal efficiency of 94% NH_3-N from shrimp culture effluent (Seema and Jayasankar, 2005) and in finfish effluent the removal efficiency for ammonia, phosphate, and nitrate in finfish waste water are at the rate of 63, 58, and 54%, respectively (Mwandya et al., 2001). *Ulva reticulata* removed nitrite, nitrate and phosphate, and nitrate from shrimp effluent with average efficiencies of 98.6, 33.3, and 89.4%, respectively (Rabiei et al.). *Ulva rotundatan* has a removal efficiency of 97.7% NH_3-N from aquaculture effluent (Martinez-Aragón et al., 2002). Kang et al. (2011) studied the bioremediation potential of *Ulva pertusa* and found that *U. pertusa* exhibited the highest total ammonia nitrogen uptake rate from the Korean rockfish waste water. Generally, the uptake of nitrogen by seaweed has been shown to be high, in addition to the nitrogen stored by seaweeds for growth during nitrogen shortages (Abreu et al., 2011). Hence, the results suggest that *Ulva* species could be the potential species of interest in the sustainable aquaculture system.

23.4.2 GRACILARIA SPECIES

Gracilaria, a seaweed that has been introduced as a biofilter in aquaculture systems to reduce dissolved nutrient pollution (Buschmann et al., 1996; Wu et al., 2015). *Gracilaria* species can play a vital role in bio-assimilation of excess nutrients when co-cultured with finfish species. Similar to the *Ulva* species, the seaweed species of *Gracilaria* found to have bioremediation and bio-assimilation capacity of nitrogen and phosphorus. *Gracilaria arcuata* has the removal efficiency of 80% nitrogen and 41% phosphorus from the aquaculture effluents (Al-Hafedh et al., 2012). In a study by Zhou et al. (2006), an extrapolation of data showed that cultivation of *Gracilaria lemaneiformis* in one hectare would yield an annual harvest of more than 70 kg live weight of the seaweed. Further, it would sequester 220 kg of nitrogen and 30 kg of phosphorus from seawater. *Gracilaria verrucosa* was co-cultured with *Pseudosciaena crocea* and the nutrient assimilation capacity of ammonium, inorganic nitrogen, and inorganic phosphate was determined (Huo et al., 2012). The maximum reduction efficiency of nitrite, nitrate, phosphate and ammonia nitrogen was 48, 47, 58, and 61%, respectively. Similarly, co-cultivation of *Gracilaria chouae* with the *Sparus microcephalus* and the maximum reduction efficiency of NO_3-N, NO_2-N, NH_3-N, PO_4-P was 37.76, 36.99, 29.27 and 40.64%, respectively (Wu et al., 2015). Huo et al. (2011) studied the bioremediation potential of *Gracilaria verrucosa* in Hangzhou Bay of East China Sea. The Hangzhou Bay was subjected to severe eutrophication by anthropogenic activities and industrial pollution. The large-scale cultivation of Gracilaria verrucosa improved the coastal water quality and the dissolved inorganic nitrogen and phosphorus were significantly lowered. The concentration of ammonia nitrogen, nitrite, nitrate, and phosphate was decreased by 54.12, 49.81, 75.54, and 49.00%, respectively. In recirculatory aquaculture system, the seaweeds *Gracilaria edulis* and *Gracilaria changii* have been demonstrated their potential as an effective biofilter (Mawi et al., 2020). The removal efficiency of ammonium, nitrate, and phosphate by *Gracilaria edulis* was found to be 72.5, 58.8, and 45.9%, respectively. Similarly, the removal efficiency of ammonium, nitrate, and phosphate by *Gracilaria changii* was found to be 71.0, 56.8, and 43.5%, respectively.

23.4.3 CAULERPA SPECIES

Caulerpa spp. have been widely studied for the biotechnological, pharmaceutical, nutraceutical, and cosmetic applications (Castiglia et al., 2021). In recent years, *Caulrepa* has been investigated for the bioremediation process for nutrient removal from aquaculture discharge (Bambaranda et al., 2019). *Caulerpa* spp. showed effective performance in the bioremediation of industrial dyes from waste water, biosorption of heavy metals and reduction of nutrients in aquaculture effluents (Apiratikul and Pavasant, 2008; Marungrueng and Pavasant, 2006; Mithra et al., 2012). The incorporation of *Caulerpa* spp. in recirculatory aquaculture system efficiently reduced the inorganic nutrients including nitrite, nitrate, ammonia, and phosphate. Bambaranda et al. (2019) reported that *Caulerpa lentillifera* decreased the inorganic nutrients like ammonia, nitrite, nitrate and phosphate up to 48.03, 19.23, 31.33, and 25%, respectively. In another study by Tanaka et al. (2020) showed that *Caulerpa lentillifera*, *Caulerpa racemosa*, and *Caulerpa sertularioides* efficiently absorb phosphorus from the seawater effluents. *Caulerpa lentillifera* significantly reduced ammonia, nitrite, nitrate, and phosphate concentrations in a polyculture system with *Litopenaeus vannamei* (Ly et al., 2021). In addition, the incorporation of *Caulerpa lentillifera* with white leg shrimp improved the growth performance and feed conversion ratio (Ly et al., 2021). *Caulerpa taxifolia* tested for the biofiltration capacity of ammonia–nitrogen, nitrite, nitrate, and phosphate (Mithra et al., 2012).

23.4.4 KAPPAPHYCUS SPECIES

Kappaphycus spp. are the carrageenophytes, and cultivated extensively in South East Asian countries for its economically valuable carrageenans.

*Kappaphycus alva*rezii and *Kappaphycus striatum* has been investigated to remove ammonia nitrogen from the milk fish culture tanks. The results showed that *Kappaphycus* sp. significantly reduced the ammonia nitrogen in the fish farm effluent (Rodrigueza and Montaño, 2007a). The integration of *Kappaphycus alvarezii* in a floating fish cage reduced the inorganic ammonia nitrogen and phosphate on average of 22.5–34% and 5–30%, respectively (Kambey et al., 2020). *Kappaphycus alvarezii* has been investigated for the biofiltration capacity of ammonia–nitrogen, nitrite, nitrate, and phosphate. The results showed that *Kappaphycus alvarezii* would effectively reduce ammonia nitrogen, nitrite, nitrate, and phosphate from waste water (Mithra et al., 2012). *Kappaphycus alvarezii* has a high affinity for metal sorption. Praveen and Vijayaraghavan (2015) studied the kinetics of *Kappaphycus alvarezii,* during the 45 min process, 90% of the heavy metals like lead, cadmium, copper, and nickel was efficiently removed. Despite the bioremediation capacity of *Kappaphycus* spp., the economic contribution to the farmers is significantly high.

23.4.5 SACCHARINA LATISSIMA

S. latissima is an economically valuable kelp species and *S. latissima* has often been considered a useful species for scalable processes. In the bioremediation standup point, *S. latissima* cultivated in the vicinity of salmon farms eliminates 5% of the total nitrogen for two years. (Sanderson et al., 2012). In IMTA system, the incorporated *S. latissima* was found to have higher tendency for nitrogen assimilation and removal. The maximum removal of nitrogen and phosphorus was 5.02–7.2 g Nm^{-1} and 0.86–1.23 g Nm^{-1}, respectively (Marinho et al., 2015). Grebe et al. (2021) studied the bio extraction potential of *S. latissima* along the Eastern coast of USA. As a result, *S. latissima* is known to remove nitrogen in a density-dependent manner. In addition, the removal capacity of *S. latissima* has been estimated to range between 19.2 and 176.0 kg N ha^{-1}.

23.4.6 SARGASSUM SPECIES

Sargassum species is a brown seaweed with complex cell wall rich in carotenoids, phenols, polysaccharides, and alginates (Balboa et al., 2013; Holdt and Kraan, 2011). *Sargassum* species has found to have biosorption capacity of absorbing the pollutants and inorganic nutrients

TABLE 23.3 Bioremediation Capability of Seaweeds Removing Nitrogen and Phosphorus from Different Sources.

Seaweeds	Nutrient Source	Nutrient Type	Nutrient Removal Rate	Nutrient Removal Efficiency	References
Ulva lactuca	Aquaculture	Nitrogen	6.4 ± 1.1 (g N m^{-2} day^{-1})	17±3 (% d^{-1})	Msuya and Neori (2008)
	Aquaculture	Nitrogen	0.42 ± 0.34 (g N m^{-2} day^{-1})	83.06%	(Al-Hafedh et al., 2012)
		Phosphate (PO_4-P)	1.07 ± 0.34 (g P m^{-2} day^{-1})	41.11%	
	Reject water from anaerobic sludge	Nitrogen	17.74 ± 1.17 (mg N d^{-1})	12.36 ± 1.91 (% d^{-1})	(Sode et al., 2013)
		Phosphorus	1.31 ± 0.22 (mg P d^{-1})	4.75 ± 0.91 (% d^{-1})	
	Aquaculture	Nitrogen	0.26–0.31 (g N m^{-2} day^{-1})	48.82–57.74%	(Al-Hafedh et al., 2015)
		Phosphate (PO_4-P)	0.32–0.41 (g P m^{-2} day^{-1})	4.04–19.65%	
	Aquaculture	Nitrogen		25.9%	(Elizondo-González et al., 2018)
		Phosphate (PO_4-P)		24.6%	
	Aquaculture	Total Ammonia nitrogen		85–90%	(Neori et al., 2003)
	Aquaculture	Ammonium	1.9 μmoles g dw^{-1} min^{-1}	40–90%	(Chung et al., 2002)
	Aquaculture	Nitrogen	0.68 ± 0.47 mg N g DW^{-1} D^{-1}		(Tremblay-Gratton et al., 2018)
		Phosphorus	0.07 ± 0.02 mg P g DW^{-1} D^{-1}		
Ulva reticulata	Aquaculture	Ammonia nitrogen		94%	(Seema and Jayasankar, 2005)
		Nitrate	28.39–24.21 μmol nitrogen L^{-1}	5%	
	Aquaculture	Nitrate		58%	(Mwandya et al., 2001)
		Phosphate (PO_4-P)		54%	
Ulva rotundatan	Aquaculture	Total ammonia nitrogen		97.7%	(Martınez-Aragon et al., 2002)
Ulva pertusa	Aquaculture	Dissolved inorganic nitrogen (DIN)		54%	(Kang et al., 2011)
		Ammonium		83.9%	
		Phosphate		30.6%	

TABLE 23.3 *(Continued)*

Seaweeds	Nutrient Source	Nutrient Type	Nutrient Removal Rate	Nutrient Removal Efficiency	References
Ulva rigida	Aquaculture	Total ammonia nitrogen (TAN)	1.6 (g TAN m^{-2} day^{-1})	100%	(Ashkenazi et al., 2019)
Gracilaria arcuata	Aquaculture	Nitrogen	0.45 (g N m^{-2} day^{-1})	80.15%	(Al-Hafedh et al., 2012)
		Phosphate (PO_4-P)	1.03 (g P m^{-2} day^{-1})	41.06%	
Gracilaria lemaneiformis	Aquaculture	Dissolved inorganic nitrogen (DIN)	5.24 kg	20.98%	(Wei et al., 2017)
		Dissolved inorganic phosphorus (DIP)	0.81 kg	28.61%	
		Nitrogen	10.64 (μmol g^{-1} DW h^{-1})	80%	(Zhou et al., 2006)
		Phosphate (PO_4-P)	0.38 (μmol g^{-1} dry weight h^{-1})	60%	
		Nitrogen		54.34–67.70%	(Li et al., 2015)
Gracilaria verrucosa	Fish processing industry	Nitrogen		67.6%	(Sari and Adharini, 2020)
		Phosphate (PO_4-P)		96%	
	Aquaculture	Ammonia nitrogen		54.12%	(Huo et al., 2011)
		Nitrate		75.54%	
		Nitrite		49.81%	
		Phosphate		49%	
Gracilaria tikvahiae	Aquaculture	Nitrogen	0.17–1.54 (g N m^{-2} day^{-1})	35%	(Samocha et al., 2015)
Gracilaria chouae	Aquaculture	Ammonia nitrogen	0.062 mg L^{-1}	29.27%	(Wu et al., 2015)
		Nitrate nitrogen		37.76%	
		Nitrite nitrogen		36.99%	
		Phosphate (PO_4-P)	0.033 mg L^{-1}	40.64%	
Gracilaria vermiculophylla	Aquaculture	Nitrogen	40.54 ± 2.02 g m^{-2} $month^{-1}$		(Abreu et al., 2011)

TABLE 23.3 *(Continued)*

Seaweeds	Nutrient Source	Nutrient Type	Nutrient Removal Rate	Nutrient Removal Efficiency	References
Gracilaria caudata	Aquaculture	Dissolved inorganic nitrogen	8 μmol N L^{-1}	44.5%	(Marinho-Soriano et al., 2011)
		Phosphate (PO_4-P)	0.17–1.70 μmol L^{-1}		
Gracilaria birdiae	Aquaculture	Ammonium (NH_4^+)	7.34 (μmol L^{-1})	34%	(Marinho-Soriano et al., 2009)
		Phosphate (PO_4^{3-})	0.75 (μmol L^{-1})	93.5%	
Gracilaria edulis	Recirculatory aquaculture system	Ammonium (NH_4^+)		72.5%	(Mawi et al., 2020)
		Nitrate (NO_3-N)		58.8%	
		Phosphate (PO_4^{3-})		45.9%	
Gracilaria changii		Ammonium (NH_4^+)		71%	
		Nitrate (NO_3-N)		56.8%	
		Phosphate (PO_4^{3-})		43.5%	
Gracilaria conferta		Total ammonia nitrogen (TAN)	0.51 (g TAN m^{-2} day^{-1})	50%	(Ashkenazi et al., 2019)
Hypneamusciformis			0.52 (g TAN m^{-2} day^{-1})	60%	
Dictyotadichotoma	Aquaculture	Ammonium (NH_4^+)		69.2–99.9%	(Madkour et al., 2020)
		Phosphate (PO_4^{3-})		90.7–100%	
Kappaphycus alvarezii, Kappaphycus sp., Kappaphycus striatum	Aquaculture	Total Ammonia Nitrogen (TAN)		41–66%	(Rodrigueza and Montaño, 2007b)
Kappaphycus alvarezii	Aquaculture	Ammonium (NH_4^+)		22.5–34%	(Kambey et al., 2020)
		Phosphate (PO_4^{3-})		5–30%	
Kappaphycus alvarezii	Aquaculture	Nitrate (NO_3-N)		54.54%	(Mithra et al., 2012)
		Nitrite (NO_2-N)		92.3%	
		Phosphate (PO_4^{3-})		94.5%	

TABLE 23.3 *(Continued)*

Seaweeds	Nutrient Source	Nutrient Type	Nutrient Removal Rate	Nutrient Removal Efficiency	References
Caulerpa lentillifera	Effluent from RAS	Ammonia nitrogen	0.66 mg L^{-1}	48.03%	(Bambaranda et al., 2019)
		Nitrate (NO_3-N)		31.33%	
		Nitrite (NO_2-N)		19.23%	
		Phosphate (PO_4-P)	0.3 mg L^{-1}	25%	
Chondrus crispus	Aquaculture	Ammonia nitrogen	4 kg m^{-2} at 13°C 425 (μmol N L^{-1} d^{-1})		(Kim et al., 2013)
Palmaria palmata			4 kg m^{-2} at 16°C 211 (μmol N L^{-1} d^{-1})		
Laminaria saccharina	Aquaculture	Dissolved inorganic nitrogen (DIN)	5.4–8.3 (μmol g^{-1} DW h^{-1})	26–40%	(Subandar et al., 1993)
Palmaria palmata	Aquaculture	Total nitrogen (TN)		12% of TN for two years	(Sanderson et al., 2012)
Palmaria palmata	Aquaculture	Nitrogen	0.42 ± 0.26 mg N g DW^{-1} D^{-1}		(Tremblay-Gratton et al., 2018)
		Phosphorus	0.03 ± 0.07 mg P g DW^{-1} D^{-1}		
Saccharina latissima	Aquaculture	Total nitrogen (TN)		5% of TN for two years	(Sanderson et al., 2012)
Saccharina latissima	Aquaculture	Nitrogen	5.02–7.2 g N m^{-1}		(Marinho et al., 2015)
		Phosphorus	0.86–1.23 g P m^{-1}		
Saccharina latissima	Aquaculture	Total nitrogen	19.2 (±4.8)–176.0 (±7.7) kg N ha^{-1}		(Grebe et al., 2021)
Sargassum polycystum	Aquaculture	Nitrate		51.79–80%	(Yassir and Adharini, 2021)
		Phosphate		18.46–86.30%	

in the coastal environment (Davis et al., 2000; Saldarriaga-Hernandez et al., 2020; Santos et al., 2018). Yu et al. (2014) demonstrated the bioremediation potential of *Sargassum hemiphyllum* and *Sargassum henslowianum*. The study results indicated that the *Sargassum* has high potential to remove nitrogen and phosphorus from the fish farm effluents. *Sargassum polycystum* would be the potential candidate species for IMTA system and bioremediation process (Yassir and Adharini, 2021). The results showed that *Sargassum polycystum* improved the water quality by absorbing the nitrates and phosphates in the shrimp effluent.

23.4.7 PALMARIA PALMATA

The impacts of offshore cage culture in the marine ecosystem need an amelioration process for marine water quality improvement. The incorporation of seaweed cultivation in the sea cage culture of fish has been implicated as an alternative process for bioremediation while providing an extra incentive for the farmer (Troell et al., 2009). *Palmaria palmata was* investigated for their potential to filter out phosphate, ammonium, and nitrate from the aquaculture effluents (Sanderson et al., 2012; Tremblay-Gratton et al., 2018). The study's results suggested that *Palmaria palmata* has significantly reduced the ammonia, nitrate, and phosphate from the fish farm effluents, meanwhile, the daily growth rate of *Palmaria palmata* was found to be 8.9% (Corey et al., 2013; Sanderson et al., 2012).

KEYWORDS

- **bioremediation**
- **ecosystem service**
- **integrated multi-trophic aquaculture**
- **producers**
- **seaweeds**
- **sustainability**

REFERENCES

Abreu, M. H., Pereira, R., Yarish, C., Buschmann, A. H., & Sousa-Pinto, I. (2011, February 25). IMTA with *Gracilaria vermiculophylla*: Productivity and nutrient removal performance of the seaweed in a land-based pilot scale system. *Aquaculture, 312*(1), 77–87. https://doi.org/10.1016/j.aquaculture.2010.12.036.

Aczel, M. R. (2019). What is the nitrogen cycle and why is it key to life? *Frontiers for Young Minds, 7*, 41.

Al-Hafedh, Y. S., Alam, A., & Buschmann, A. H. (2015). Bioremediation potential, growth and biomass yield of the green seaweed, *Ulva lactuca* in an integrated marine aquaculture system at the Red Sea coast of Saudi Arabia at different stocking densities and effluent flow rates. *Reviews in Aquaculture, 7*(3), 161–171. https://doi.org/10.1111/raq.12060.

Al-Hafedh, Y. S., Alam, A., Buschmann, A. H., & Fitzsimmons, K. M. (2012). Experiments on an integrated aquaculture system (seaweeds and marine fish) on the Red Sea coast of Saudi Arabia: Efficiency comparison of two local seaweed species for nutrient biofiltration and production. *Reviews in Aquaculture, 4*(1), 21–31. https://doi.org/10.1111/j.1753-5131.2012.01057.x.

Apiratikul, R., & Pavasant, P. (2008). Batch and column studies of biosorption of heavy metals by *Caulerpa lentillifera*. *Bioresource Technology, 99*(8), 2766–2777.

Ashkenazi, D. Y., Israel, A., & Abelson, A. (2019). A novel two-stage seaweed integrated multi-trophic aquaculture. *Reviews in Aquaculture, 11*(1), 246–262. https://doi.org/10.1111/raq.12238.

Balboa, E. M., Conde, E., Moure, A., Falqué, E., & Domínguez, H. (2013). *In vitro* antioxidant properties of crude extracts and compounds from brown algae. *Food chemistry, 138*(2–3), 1764–1785.

Bambaranda, B., Tsusaka, T. W., Chirapart, A., Salin, K. R., & Sasaki, N. (2019). Capacity of *Caulerpa lentillifera* in the removal of fish culture effluent in a recirculating aquaculture system. *Processes, 7*(7), 440.

Boyd, P. W., & Hurd, C. L. (2009). Ocean nutrients. *Washington DC American Geophysical Union Geophysical Monograph Series, 187*, 139–160.

Bringslid, I. H. (2019). *Phosphorus in Saccharina latissima: Initial uptake, storage, and possibilities for recycling in IMTA-systems* (Master thesis, Bringslid, Ingeborg Hollekim).

Buschmann, A. H., Troell, M., Kautsky, N., & Kautsky, L. (1996). Integrated tank cultivation of salmonids and *Gracilaria chilensis* (Graciariales, Rhodophyta). *Hydrobiologia*, 326–327.

Castiglia, D., Landi, S., & Esposito, S. (2021). Advanced applications for protein and compounds from microalgae. *Plants, 10*(8), 1686.

Chopin, T. (2015). Marine aquaculture in Canada: Well-established monocultures of finfish and shellfish and an emerging integrated multi-trophic aquaculture (IMTA) approach including seaweeds, other invertebrates, and microbial communities. *Fisheries, 40*(1), 28–31.

Chopin, T., Cooper, J. A., Reid, G., Cross, S., & Moore, C. (2012). Open-water integrated multi-trophic aquaculture: Environmental biomitigation and economic diversification of fed aquaculture by extractive aquaculture. *Reviews in Aquaculture, 4*(4), 209–220.

Chung, I. K., Kang, Y. H., Charles, Y., George, P. K., & Lee, J. (2002). Application of seaweed cultivation to the bioremediation of nutrient-rich effluent. *Algae, 17*(3), 187–194.

Corey, P., Kim, J. K., Duston, J., Garbary, D. J., & Prithiviraj, B. (2013). Bioremediation potential of *Palmaria palmata* and *Chondrus crispus* (Basin Head): Effect of nitrate and ammonium ratio as nitrogen source on nutrient removal. *Journal of Applied Phycology, 25*(5), 1349–1358.

Davis, T., Volesky, B., & Vieira, R. (2000). Sargassum seaweed as biosorbent for heavy metals. *Water Research, 34*(17), 4270–4278.

El-Sherbiny, M., Satheesh, S., & Ba-Akdah, M. A. (2021). Physiological responses of marine macroalgae to chlorine dioxide treatment. *Thalassas: An International Journal of Marine Sciences, 37*(1), 291–302.

Elizondo-González, R., Quiroz-Guzmán, E., Escobedo-Fregoso, C., Magallón-Servín, P., & Peña-Rodríguez, A. (2018). Use of seaweed *Ulva lactuca* for water bioremediation and as feed additive for white shrimp *Litopenaeus vannamei*. *PeerJ, 6*, e4459.

Fang, J., Zhang, J., Xiao, T., Huang, D., & Liu, S. (2016). Integrated multi-trophic aquaculture (IMTA) in Sanggou Bay, China. *Aquaculture Environment Interactions, 8*, 201–205.

FAO. (2020). *The state of world fisheries and aquaculture: Sustainability in action.* Rome.

FAO. (2021). Fishery and Aquaculture Statistics. Global aquaculture production 1950–2019 (FishstatJ), version updated 202. www.fao.org/fishery/statistics/software/fishstatj/en.

Forbord, S., Etter, S. A., Broch, O. J., Dahlen, V. R., & Olsen, Y. (2021, January 1). Initial short-term nitrate uptake in juvenile, cultivated *Saccharina latissima* (Phaeophyceae) of variable nutritional state. *Aquatic Botany, 168*, 103306. https://doi.org/10.1016/j.aquabot.2020.103306.

Grebe, G. S., Byron, C. J., Brady, D. C., Geisser, A. H., & Brennan, K. D. (2021, June 1). The nitrogen bioextraction potential of nearshore *Saccharina latissima* cultivation and harvest in the Western Gulf of Maine. *Journal of Applied Phycology, 33*(3), 1741–1757. https://doi.org/10.1007/s10811-021-02367-6.

Harrison, P. J., & Hurd, C. L. (2001). Nutrient physiology of seaweeds: Application of concepts to aquaculture. *Cahiers de biologie marine, 42*(1–2), 71–82.

Holdt, S. L., & Kraan, S. (2011). Bioactive compounds in seaweed: Functional food applications and legislation. *Journal of Applied Phycology, 23*(3), 543–597.

Huo, Y., Wu, H., Chai, Z., Xu, S., Han, F., Dong, L., & He, P. (2012, January 25). Bioremediation efficiency of *Gracilaria verrucosa* for an integrated multi-trophic aquaculture system with *Pseudosciaena crocea* in Xiangshan harbor, China. *Aquaculture, 326–329*, 99–105. https://doi.org/10.1016/j.aquaculture.2011.11.002.

Huo, Y. Z., Xu, S. N., Wang, Y. Y., Zhang, J. H., Zhang, Y. J., Wu, W. N., Chen, Y. Q., & He, P. M. (2011, April 1). Bioremediation efficiencies of *Gracilaria verrucosa* cultivated in an enclosed sea area of Hangzhou Bay, China. *Journal of Applied Phycology, 23*(2), 173–182. https://doi.org/10.1007/s10811-010-9584-9.

Kambey, C. S. B., Sondak, C. F. A., & Chung, I. K. (2020). Potential growth and nutrient removal of *Kappaphycus alvarezii* in a fish floating-net cage system in Sekotong Bay, Lombok, Indonesia. *Journal of the World Aquaculture Society, 51*(4), 944–959. https://doi.org/10.1111/jwas.12683.

Kang, Y. H., Park, S. R., & Chung, I. K. (2011). Biofiltration efficiency and biochemical composition of three seaweed species cultivated in a fish-seaweed integrated culture. *Algae, 26*(1), 97–108.

Kautsky, N., Troell, M., & Folke, C. (1999). Ecological engineering for increased production and environmental improvement in open sea aquaculture. *Ecological Engineering for Wastewater Treatment, 11*, 89–97.

Kim, J. K., Duston, J., Corey, P., & Garbary, D. J. (2013, November 15). Marine finfish effluent bioremediation: Effects of stocking density and temperature on nitrogen removal capacity of *Chondrus crispus* and *Palmaria palmata* (Rhodophyta). *Aquaculture, 414–415*, 210–216. https://doi.org/10.1016/j.aquaculture.2013.08.008.

Langton, R., Augyte, S., Price, N., Forster, J., Noji, T., Grebe, G., St Gelais, A., & Byron, C. J. (2019). An ecosystem approach to the culture of seaweed. *OAA Technical Memorandum NMFS-F/SPO* (p. 195). https://repository.library.noaa.gov/view/noaa/35734.

Li, M., Song, X., Sun, G., Ma, X., & Liu, Y. (2015). Optimal ratios of oyster to Gracilaria improving removal efficiency of nitrogen nutrients in aquaculture wastewater. *Transactions of the Chinese Society of Agricultural Engineering, 31*(11), 243–248.

Ly, K. V., Murungu, D. K., Nguyen, D. P., & Nguyen, N. A. T. (2021). Effects of different densities of sea grape *Caulerpa lentillifera* on water quality, growth and survival of the whiteleg shrimp *Litopenaeus vannamei* in polyculture system. *Fishes, 6*(2), 19. https://www.mdpi.com/2410-3888/6/2/19.

Madkour, A., Rashedy, S., & Touliabah, H. (2020). Bioremediation of ammonium and phosphates and their effect on the macroalgae *Dictyota dichotoma* (Hudson) *Lamouroux Native* to the Red Sea Coast, Egypt. *Egyptian Journal of Botany, 60*(3), 671–680. https://doi.org/10.21608/ejbo.2020.21997.1430.

Mangano, M. C., Giacoletti, A., & Sarà, G. (2019). Dynamic energy budget provides mechanistic derived quantities to implement the ecosystem based management approach. *Journal of Sea Research, 143*, 272–279.

Marinho-Soriano, E., Azevedo, C. A. A., Trigueiro, T. G., Pereira, D. C., Carneiro, M. A. A., & Camara, M. R. (2011, January 1). Bioremediation of aquaculture wastewater using macroalgae and Artemia. *International Biodeterioration and Biodegradation, 65*(1), 253–257. https://doi.org/10.1016/j.ibiod.2010.10.001.

Marinho-Soriano, E., Nunes, S. O., Carneiro, M. A. A., & Pereira, D. C. (2009, February 1). Nutrients' removal from aquaculture wastewater using the macroalgae *Gracilaria birdiae. Biomass and Bioenergy, 33*(2), 327–331. https://doi.org/10.1016/j.biombioe.2008.07.002.

Marinho, G. S., Holdt, S. L., Birkeland, M. J., & Angelidaki, I. (2015, October 1). Commercial cultivation and bioremediation potential of sugar kelp, *Saccharina latissima*, in Danish waters. *Journal of Applied Phycology, 27*(5), 1963–1973. https://doi.org/10.1007/s10811-014-0519-8.

Martınez-Aragon, J. F., Hernandez, I., Perez-Llorens, J. L., Vasquez, R., & Vergara, J. J. (2002). Biofiltering efficiency in removal of dissolved nutrients by three species of estuarine macroalgae cultivated with sea bass (*Dicentrarchus labrax*) waste waters 1. *Phosphate Journal of Applied Phycology, 14*, 365–374.

Marungrueng, K., & Pavasant, P. (2006). Removal of basic dye (Astrazon Blue FGRL) using macroalga *Caulerpa lentillifera. Journal of Environmental Management, 78*(3), 268–274.

Mawi, S., Krishnan, S., Din, M. F. M. D., Arumugam, N., & Chelliapan, S. (2020, February 1). Bioremediation potential of macroalgae *Gracilaria edulis* and *Gracilaria changii* co-cultured with shrimp wastewater in an outdoor water recirculation system. *Environmental Technology and Innovation, 17*, 100571. https://doi.org/10.1016/j.eti.2019.100571.

Mithra, R., Sivaramakrishnan, S., Santhanam, P., Kumar, D. S. D., & Kumar, N. (2012, January 1). Investigation on nutrients and heavy metal removal efficacy of seaweeds, *Caulerpa taxifolia* and *Kappaphycus alvarezii* for wastewater remediation. *Journal of Algal Biomass Utilization, 3*, 21–27.

Msuya, F. E., & Neori, A. (2008). Effect of water aeration and nutrient load level on biomass yield, N uptake and protein content of the seaweed *Ulva lactuca* cultured in seawater tanks. *Journal of Applied Phycology, 20*(6), 1021–1031.

Mwandya, A. W., Mtolera, M., Pratap, H. B., & Jiddawi, N. S. (2001). Macroalgae as biofilters of dissolved inorganic nutrients in an integrated mariculture tank system in Zanzibar. In M. D. Richmond, & J. Francis (Eds.), *Marine science development in tanzania and Eastern Africa* (pp. 159–170).

Nederlof, M. A., Verdegem, M. C., Smaal, A. C., & Jansen, H. M. (2022). Nutrient retention efficiencies in integrated multi-trophic aquaculture. *Reviews in Aquaculture, 14*(3), 1194–1212.

Neori, A., Chopin, T., Troell, M., Buschmann, A. H., Kraemer, G. P., Halling, C., Shpigel, M., & Yarish, C. (2004). Integrated aquaculture: Rationale, evolution and state of the art emphasizing seaweed biofiltration in modern mariculture. *Aquaculture, 231*(1–4), 361–391.

Neori, A., Msuya, F. E., Shauli, L., Schuenhoff, A., Kopel, F., & Shpigel, M. (2003, November 1). A novel three-stage seaweed (*Ulva lactuca*) biofilter design for integrated mariculture. *Journal of Applied Phycology, 15*(6), 543–553. https://doi.org/10.1023/B:JAPH.0000004382.89142.2d.

Pajares, S., & Ramos, R. (2019, November 29). Processes and microorganisms involved in the marine nitrogen cycle: Knowledge and gaps [Review]. *Frontiers in Marine Science, 6*, 739. https://doi.org/10.3389/fmars.2019.00739.

Praveen, R. S., & Vijayaraghavan, K. (2015, August 14). Optimization of Cu(II), Ni(II), Cd(II) and Pb(II) biosorption by red marine alga *Kappaphycus alvarezii*. *Desalination and Water Treatment, 55*(7), 1816–1824. https://doi.org/10.1080/19443994.2014.927334.

Raven, J. A., Giordano, M., Beardall, J., & Maberly, S. C. (2011). Algal and aquatic plant carbon concentrating mechanisms in relation to environmental change. *Photosynthesis Research, 109*(1), 281–296.

Rebours, C., Marinho-Soriano, E., Zertuche-González, J. A., Hayashi, L., Vásquez, J. A., Kradolfer, P., Soriano, G., Ugarte, R., Abreu, M. H., & Bay-Larsen, I. (2014). Seaweeds: An opportunity for wealth and sustainable livelihood for coastal communities. *Journal of Applied Phycology, 26*(5), 1939–1951.

Resende, L., Flores, J., Moreira, C., Pacheco, D., Baeta, A., Garcia, A. C., & Rocha, A. C. S. (2022, January). Effective and low-maintenance IMTA system as effluent treatment unit for promoting sustainability in coastal aquaculture. *Applied Sciences-Basel, 12*(1). <Go to ISI>://WOS:000751360600001.

Rodrigueza, M. R. C., & Montaño, M. N. E. (2007a, December 1). Bioremediation potential of three carrageenophytes cultivated in tanks with seawater from fish farms. *Journal of Applied Phycology, 19*(6), 755–762. https://doi.org/10.1007/s10811-007-9217-0.

Rodrigueza, M. R. C., & Montaño, M. N. E. (2007b, August 7). Bioremediation potential of three carrageenophytes cultivated in tanks with seawater from fish farms. *Journal of Applied Phycology, 19*(6), 755. https://doi.org/10.1007/s10811–007-9217-0.

Roleda, M. Y., & Hurd, C. L. (2019, September 3). Seaweed nutrient physiology: Application of concepts to aquaculture and bioremediation. *Phycologia, 58*(5), 552–562. https://doi.org/10.1080/00318884.2019.1622920.

Saldarriaga-Hernandez, S., Hernandez-Vargas, G., Iqbal, H. M. N., Barceló, D., & Parra-Saldívar, R. (2020, May 1). Bioremediation potential of *Sargassum* sp. biomass to tackle pollution in coastal ecosystems: Circular economy approach. *Science of the Total Environment, 715*, 136978. https://doi.org/10.1016/j.scitotenv.2020.136978.

Samocha, T. M., Fricker, J., Ali, A. M., Shpigel, M., & Neori, A. (2015, September 1). Growth and nutrient uptake of the macroalga *Gracilaria tikvahiae* cultured with the shrimp *Litopenaeus vannamei* in an integrated multi-trophic aquaculture (IMTA) system. *Aquaculture, 446*, 263–271. https://doi.org/10.1016/j.aquaculture.2015.05.008.

Sanderson, J. C., Dring, M. J., Davidson, K., & Kelly, M. S. (2012, July 2). Culture, yield and bioremediation potential of *Palmaria palmata* (Linnaeus) Weber and Mohr and *Saccharina latissima* (Linnaeus) C.E. Lane, C. Mayes, Druehl & G. W. Saunders adjacent to fish farm cages in northwest Scotland. *Aquaculture, 354–355*, 128–135. https://doi.org/10.1016/j.aquaculture.2012.03.019.

Santos, S. C., Ungureanu, G., Volf, I., Boaventura, R. A., & Botelho, C. M. (2018). Macroalgae biomass as sorbent for metal ions. In *Biomass as renewable raw material to obtain bioproducts of high-tech value* (pp. 69–112). Elsevier.

Sari, D. R., & Adharini, R. I. (2020). The utilization of *Gracilaria verrucosa* as fish processing wastewater biofilter. *E3S Web of Conferences, 147*, 02022. https://doi.org/10.1051/e3sconf/202014702022.

Seema, C., & Jayasankar, R. (2005). Removal of nitrogen load in the experimental culture system of seaweed and shrimp. *Journal of the Marine Biological Association of India, 47*(2), 150–153.

Sode, S., Bruhn, A., Balsby, T. J. S., Larsen, M. M., Gotfredsen, A., & Rasmussen, M. B. (2013, October 1). Bioremediation of reject water from anaerobically digested waste water sludge with macroalgae (*Ulva lactuca*, Chlorophyta). *Bioresource Technology, 146*, 426–435. https://doi.org/10.1016/j.biortech.2013.06.062.

Subandar, A., Petrell, R. J., & Harrison, P. J. (1993, August 1). Laminaria culture for reduction of dissolved inorganic nitrogen in salmon farm effluent. *Journal of Applied Phycology, 5*(4), 455–463. https://doi.org/10.1007/BF02182738.

Tanaka, Y., Ashaari, A., Mohamad, F. S., & Lamit, N. (2020, March 15). Bioremediation potential of tropical seaweeds in aquaculture: Low-salinity tolerance, phosphorus content, and production of UV-absorbing compounds. *Aquaculture, 518*, 734853. https://doi.org/10.1016/j.aquaculture.2019.734853.

Tremblay-Gratton, A., Boussin, J. C., Tamigneaux, É., Vandenberg, G. W., & Le François, N. R. (2018, April 1). Bioremediation efficiency of *Palmaria palmata* and *Ulva lactuca* for use in a fully recirculated cold-seawater naturalistic exhibit: Effect of high NO_3 and

PO_4 concentrations and temperature on growth and nutrient uptake. *Journal of Applied Phycology, 30*(2), 1295–1304. https://doi.org/10.1007/s10811-017-1333-x.

Troell, M., Halling, C., Neori, A., Chopin, T., Buschmann, A. H., & Kautsky, N. (2003). Integrated mariculture: Asking the right questions. *Aquaculture, 226*, 69–90.

Troell, M., Halling, C., Nilsson, A., Buschmann, A., Kautsky, N., & Kautsky, L. (1997). Integrated marine cultivation of *Gracilaria chilensis* (Gracilariales, Rhodophyta) and salmon cages for reduced environmental impact and increased economic output. *Aquaculture, 156*(1–2), 45–61.

Troell, M., Joyce, A., Chopin, T., Neori, A., Buschmann, A. H., & Fang, J. G. (2009). Ecological engineering in aquaculture—potential for integrated multi-trophic aquaculture (IMTA) in marine offshore systems. *Aquaculture, 297*(1–4), 1–9.

van den Burg, S. W., van Duijn, A. P., Bartelings, H., van Krimpen, M. M., & Poelman, M. (2016). The economic feasibility of seaweed production in the North Sea. *Aquaculture Economics and Management, 20*(3), 235–252.

Wang, X., Olsen, L. M., Reitan, K. I., & Olsen, Y. (2012). Discharge of nutrient wastes from salmon farms: Environmental effects, and potential for integrated multi-trophic aquaculture. *Aquaculture Environment Interactions, 2*(3), 267–283.

Wei, Z., You, J., Wu, H., Yang, F., Long, L., Liu, Q., Huo, Y., & He, P. (2017, August 15). Bioremediation using *Gracilaria lemaneiformis* to manage the nitrogen and phosphorous balance in an integrated multi-trophic aquaculture system in Yantian Bay, China. *Marine Pollution Bulletin, 121*(1), 313–319. https://doi.org/10.1016/j.marpolbul.2017.04.034.

White, A., & Dyhrman, S. (2013, May 21). The marine phosphorus cycle [Editorial]. *Frontiers in Microbiology, 4*. https://doi.org/10.3389/fmicb.2013.00105.

Wu, H., Huo, Y., Han, F., Liu, Y., & He, P. (2015, February 15). Bioremediation using *Gracilaria chouae* co-cultured with *Sparus macrocephalus* to manage the nitrogen and phosphorous balance in an IMTA system in Xiangshan Bay, China. *Marine Pollution Bulletin, 91*(1), 272–279. https://doi.org/10.1016/j.marpolbul.2014.11.032.

Wurts, W. A. (2000). Sustainable aquaculture in the twenty-first century. *Reviews in Fisheries Science, 8*(2), 141–150.

Yang, S. Y., Huang, T. K., Kuo, H. F., & Chiou, T. J. (2017, June 1). Role of vacuoles in phosphorus storage and remobilization. *Journal of Experimental Botany, 68*(12), 3045–3055. https://doi.org/10.1093/jxb/erw481.

Yassir, M., & Adharini, R. (28–29 July 2021). The effectiveness of *Sargassum polycystum* C. Agardh (1824) density to reduce nitrate and phosphate in vannamei shrimp aquaculture. *IOP Conference Series: Earth and Environmental Science, 2021 4th International Symposium on Marine and Fisheries Research, Yogyakarta, Indonesia (Virtual)* (Vol. 919, p. 012009). IOP Publishing Ltd.

Yu, Z., Zhu, X., Jiang, Y., Luo, P., & Hu, C. (2014, August 30). Bioremediation and fodder potentials of two *Sargassum* spp. in coastal waters of Shenzhen, South China. *Marine Pollution Bulletin, 85*(2), 797–802. https://doi.org/10.1016/j.marpolbul.2013.11.018.

Zhou, Y., Yang, H., Hu, H., Liu, Y., Mao, Y., Zhou, H., Xu, X., & Zhang, F. (2006, March 10). Bioremediation potential of the macroalga *Gracilaria lemaneiformis* (Rhodophyta) integrated into fed fish culture in coastal waters of north China. *Aquaculture, 252*(2), 264–276. https://doi.org/10.1016/j.aquaculture.2005.06.046.

CHAPTER 24

Biofloc: A Modern Way to Sustainable Aquaculture Production

EDIGA ARUN GOUD[1,*], AITWAR VAIJNATH[2], and SURYAWANSHI UMESH[3]

[1]Dr. Balasaheb Sawant Konkan Krishi Vidyapeeth, Dapoli, Department of Aquaculture, College of Fisheries, Shirgaon, Ratnagiri, Maharashtra, India

[2]Karnataka Veterinary, Animal and Fisheries Sciences University, Bidar, Department of Aquatic Environment Management, College of Fisheries, Mangaluru, Karnataka, India

[3]Karnataka Veterinary, Animal and Fisheries Sciences University, Bidar, Department of Fisheries Resource Management, College of Fisheries, Mangaluru, Karnataka, India

*Corresponding author

ABSTRACT

Biofloc is a new sustainable technology that has great potential for fish production. The demand for aquatic food is increasing with growing global population, so to fulfil this demand the aquaculture is intensifying at a much faster rate which leads to more waste generation. Managing the wastes generated is an emerging issue for environmental safeguard. Biofloc technology works on the principle of nutrient recycling, which recycles the waste generated to natural food for fish using heterotrophic bacteria via complex bio-pathways. By this, the feed costs can be reduced by up to 30%, resulting in higher profitability. It is also a water quality management tool that uses microbes to convert detritus waste into consumable flocs. This system also supports higher stocking densities,

Recent Advances in Sustainable Aquaculture and Fisheries. Shabir Ahmad Dar, Imtiyaz Qayoom, Gohar Bilal Wani, Adnan Abubakr, and Seemab Zehra (Eds.)

DOI: 10.1201/9781998511372-24

improves growth, immunity, and digestive activity of digestive enzymes. This technology encompassed sustainable aquaculture production with low cost and technical demands. Biofoc also helps in achieving some of the sustainable development goals (SDGs).

24.1 INTRODUCTION

With increasing global population, there is a continuous increase in demand for aquatic food. It is projected that by 2050 the human population may reach 9.6 billion. Poverty is indicated by food and nutrition insecurity (FAO, 2018). Greater presence of poverty in developing countries is associated with the inadequate adoption of technology and innovation management practices (TIMPs) and by the climatic changes in agricultural productive areas (FAO, 2018; Brander, 2010). The required amount of food fish for human consumption was supplied by the capture fisheries sector but today it is not the case, as the capture fisheries sector cannot supply enough fish food required for the increasing population, so there is a need for development of sustainable aquaculture techniques to cope up the demand and supply gap (FAO, 2018).

In many countries, aquaculture has originated as an ideal food production tool. And also, it plays a very crucial role in alleviating poverty, hunger, malnutrition, promoting health, generating livelihood, economic growth, and also assures the better use of natural resources (FAO, 2017). Nevertheless, the sustainable aquaculture development is mainly challenged by the restricted availability of natural resources and by the effect of industry on the environment (Costa-Pierce et al., 2012). So, keeping these challenges in mind, one should focus on developing a system that uses lesser resources like land, water, energy, and capital and also should have a minimal impact on the environment despite high productivity and profitability (FAO, 2017). Sustainable aquaculture development also contributes to the fulfilment of various sustainable development goals (SDGs) of FAO mainly SDG 1 (no poverty), SDG 2 (zero hunger, achieve food security, improve nutrition, and promote sustainable agriculture), SDG 8 (promoting inclusive and sustainable economic growth, decent work for all), and SDG 14 (conservation and sustainable use of water bodies for sustainable development) (FAO, 2017).

Global aquaculture intensification with higher stocking densities and feed inputs resulted in greater productivity but also created some challenges like higher or excess nitrogen loadings and greater danger of illness that might put farmers at greater risk as well as the possibility to contaminate or introduce pathogens to natural water sources (Boyd, 2003). Due to the discharge of these excess effluents with high amount of organic load, nitrogenous waste, harmful metabolites with greater BOD and COD makes aquaculture unsustainable industry (Martinez-Porchas and Martinez-Cordova, 2012).

There are also some other challenges like conflict for land, water and for other limited natural resources, the introduction of exotic species globally which may impact native species, the overexploitation of oceanic fish catches for fishmeal and fish oil production, the spreading of pathogens and the development of antibiotic-resistant bacteria (Naylor et al., 2001). So, it is necessary to develop certain strategies to deal with these problems. In consideration to these challenges, the culture system's physicochemical variables have been altered to encourage the growth of specific biotic communities, helping to improve both the recirculation of nutrients and the use of the biomass from these biotic communities as a primary source of food for the cultured organisms (Martinez-Cordova et al., 2015). Biofloc is one of the sustainable aquaculture systems that solves the above challenges. Since nutrients are continually recycled and reused in the culture medium with little to no water exchange, biofloc technology is advised as the next "blue revolution."

Moreover, with the ability of biofloc-based aquaculture systems to provide relatively more biosecure, eco-friendly, and economically viable aquaculture production, the interest has increased in these systems during the past thirty years. Biofloc also helps to deliver high-quality fish juveniles, which are one of the most crucial production inputs that directly help to increase output and productivity. Biofloc technology assists in supplying high-quality seeds by strengthening the immunity and robustness of the larvae and also helps in improving the reproductive efficiency of aquaculture animals (Ekasari et al., 2015a). Some other recent advances in sustainable aquaculture are recirculating aquaculture systems (RAS) and integrated multi-trophic aquaculture (IMTA).

24.2 BIOFLOC TECHNOLOGY

The fundamental idea behind biofloc technology is the recycling of waste nutrients, particularly nitrogen, into microbial biomass that may either be consumed directly by the cultured animals or harvested and converted into feed components (Avnimelech, 2009). National Agricultural Library

Glossary (United States Department of Agriculture) defined biofloc as the use of aggregates of bacteria, algae, or protozoa which are held together in a matrix along with particulate organic matter to improve water quality, waste treatment, and disease prevention in intensive aquaculture systems. Hargreaves (2006) defined biofloc as a suspended growth system that relies on live and dead organic matter aggregates, filamentous and floc-forming bacteria, microbial grazers, and active phytoplankton biomass to maintain water quality.

Several researchers have referred biofloc technology under a variety of names over time, some of them are suspended bacteria-based systems or active sludge (Rakocy et al., 2004), microbial floc system (Avnimelech, 2007), single cell protein production system (Avnimelech et al., 1989), and zero exchange autotrophic-heterotrophic system (Burford et al., 2003). Physicochemical forces of attraction and a polymer matrix made of substances like polysaccharides, proteins, and humic complexes keep these flocs together. Biofloc also aids in improving water quality by microbially degrading harmful nitrogenous waste (ammonia and nitrite), either by converting it to less hazardous nitrate or by turning it into microbiological biomass. The nitrogen absorption by bacterial growth lowers the ammonium content, which is much faster than the nitrification process, as the heterotrophs grow ten times faster and produce more microbial biomass per unit substrate than nitrifying bacteria. (Hargreaves, 2006). This practically eliminates the need for expensive equipment or installation, making it appealing to farmers of all economic levels.

A well-balanced carbon and nitrogen ratio is essential for the conversion of ammonia and other organic nitrogenous wastes into bacterial biomass. Carbon and nitrogen ratio is maintained by the addition of carbon source (starch, molasses, cellulose, wheat flour, rice flour, glucose, etc.) externally. Heterotrophic bacterial growth is promoted by using carbon as energy source, which they use to break down all the extra feeds and inorganic waste in the system and turn them into natural feeds for the cultured animals. About 0.5 g of microbial biomass is produced per unit substrate of heterotrophic bacteria for every gram of substrate C used (Eding et al., 2006).

24.3 BACTERIAL COMMUNITIES AND THEIR ROLE

Biofloc is a microbial pool and has many beneficial bacteria that help to eliminate nitrogen waste by their conversion into less toxic forms, that is, nitrite and nitrate, and these bacteria act as feed, help in maintaining water quality, and improve the immune response of the culture organisms. The nitrogen conversion happens in three pathways with the help of photo-autotrophic, chemoautotrophic, and heterotrophic bacteria. Algae uptake nitrogen by photoautotrophic process, ammonia nitrogen is converted into nitrite and finally to less toxic nitrate by chemoautotrophic bacteria, and the heterotrophic bacteria help in assimilating toxic ammonia into bacterial biomass (Avnimelech, 1999). These conversion techniques do not exist independently in biofloc, there is a constant interaction between all three paths, with the dominant one relying on system management. Algae (photoautotrophic) and heterotrophic dominated biofloc system was observed at C/N 9 and C/N 18 (Xu et al., 2016). Other characteristics of their bacteria are shown in Table 24.1.

TABLE 24.1 Characteristics of Bacteria Found in Biofloc and Their Function.

Heterotrophic Bacteria	Photoautotrophic Bacteria	Chemoautotrophic Bacteria
Require a greater amount of dissolved oxygen and dominate at higher oxygen rates	Require low oxygen and prevail at lower oxygen rates	Require a moderate amount of oxygen
Bacterial activity is higher at limited light conditions	Bacterial activity is higher during daytime when the photosynthesis is at its peak	Bacterial activity is not dependent on light and photosynthesis
Predominated by heterotrophic bacteria	Predominated by algae	Predominated by nitrifying bacteria (*Nitrobacter*, *Nitrosomonas*, *Nitrosococcus*, etc.)
High C/N ratio favors these bacteria	Low C/N ratio favors these bacteria	Moderate C/N ratio favors these bacteria
Help in converting ammonia–nitrogen to consumable bacterial biomass	Help in decreasing ammonia–nitrogen through its uptake by phytoplankton	Help in converting ammonia–nitrogen to nitrite and then to less toxic nitrate

Biofloc consists of suspended particles and organic matter (60–70%), heterogenous microorganisms (protozoa, rotifer, fungi, nematode, algae, and bacteria) and other inorganic matter (dead cells, binary ions, colloids, salts, and polymers) (30–40%) (Khanjani and Sharifini, 2020). As bacteria proliferate and move from algae to bacteria in a biofloc system, the color of the biofloc changes from green to brown. According to Avnimelech (2007), the concentration of bacteria in biofloc can range from 10^6 to 10^9

colonies per millilitre, with each millilitre having 10–30 mg of dry matter. A microbial number of 3.36×10^7 heterotrophic bacteria per ml signifies that a biofloc system is fully developed (Khanjani, 2015).

24.4 C/N RATIO

Carbon–nitrogen ratio (C/N) is crucial in the biofloc system for the conversion of hazardous inorganic nitrogen compounds into beneficial microbial biomass which serve as a direct source of food for the cultured aquatic species. Carbonaceous materials are added to the water to maintain C/N ratio. At a C/N ratio 10:1, around 10 mg of ammonium N/l could be completely taken up while glucose was provided as a carbon source (Avnimelech, 2007). According to Lancelot (1985), as long as the C/N ratio is larger than 10, the inorganic nitrogen will get fully disabled. If the C/N ratio in the biofloc water was kept higher, ammonium absorption by heterotrophic microorganisms would prevent the deposition of harmful inorganic substances such as NH_4^+ and NO_2. A change from photoautotrophic or chemoautotrophic communities to heterotrophic communities, which have a notable impact on water quality and floc production, can occur when carbon–nitrogen ratios are improved directly from feed or other carbon additions (Avnimelech, 1999). An inverse relationship was observed between the protein content and C/N ratio of feed (Avnimelech, 2015). Because of this, it is essential to choose feed components that will promote the growth of dense microbial flocs that can swiftly absorb inorganic nitrogenous compounds to satisfy the needs of the target fish species. Another crucial indicator of water quality is total suspended solids (TSS), whose concentration in aquatic habitats is influenced by the C/N ratio. The amount of carbohydrate addition required to lower the total ammonia nitrogen (TAN) concentration by 1 ppm N is 20 g/m^3, assuming that the additional carbohydrate comprises 50% C (Avnimelech, 1999).

C/N ratio also affects the turbidity. TSS, alkalinity, and settleable solids (SS), which in turn affect the growth performance of culture organisms. The species and origin of carbon sources will determine the ideal amount of C/N ratios. A C/N ratio of 10:1 and 20:1 is required for the best growth of *Oreochromis niloticus* and carps, respectively (Yu et al., 2020). Ideal C/N ratio is unknown but a ratio of 15:1 has shown a positive effect on most of the species (Dauda, 2019). The floc volume, survival rate, FCR, and weight gain were all favorably impacted by the 15 and 20 C/N ratios for shrimp (Panigrahi et al., 2018). Exogenous carbon sources should be lowered or halted altogether after nitrifying bacteria have stabilized to prevent excess production of bioflocs. A quick rise in SS, TSS, volatile suspended solids (VSS), and water turbidity might result from an excessive addition of organic carbon (Xu et al., 2016).

24.5 BIOFLOC TECHNOLOGY IN AQUACULTURE

Biofloc has various applications in aquaculture. It has been used in the nursery phase, grow-out phase for shrimp, carp, catfish, and tilapia, the culture of freshwater prawn, formation and broodstock maturation of fish and shrimp, and also it has been used as a feed ingredient, that is, as biofloc meal. Apart from this, biofloc helps in maintaining water quality, acts as feed, improves growth performance, the immune system, and digestive enzymes activity.

24.5.1 USE OF BIOFLOC IN NURSERY PHASE

Nursing phase is a very crucial and intermediate phase between the hatchery and the grow-out phase. Nursing helps to enhance the growth and survival rate of larvae. Biofloc has been used successfully for the nursery phase of fish and shrimp. Numerous researches asserted that microbial floc serves as a source of essential nutrients including amino acids, vitamins, and mineral supplements to clarify the impact of the biofloc system on growth and survival (Decamp et al., 2002).

24.5.2 USE OF BIOFLOC FOR DEVELOPMENT OF BROODSTOCK

The need for high-quality fish seed is rising as aquaculture is rapidly intensifying and also the vertical transfer of pathogens from wild broodstock in the shrimp sector is currently a hot topic. So, to overcome these problems, researchers started using biofloc as a way to biosecurity along with using microbial flocs as a nutrient meal. Use of biofloc for *O. niloticus* broodstock improved the fecundity, egg diameter, and fry production rate (Ekasari et al., 2015b). Microbial floc also reduces the spawner's latency period. The latency period serves as both a criterion to determine the cut-off period for non-spawning females and an indicator of the multiple spawning capabilities of females. Lower latency period was observed in floc spawners of *Litopenaeus stylirostris* and *F. duorarum* than that of control (Emerenciano et al., 2013).

24.5.3 USE OF BIOFLOC AS FOOD

Aquaculture intensification has increased the feed demand, and moreover, the increasing price of feed is another major concern for farmers. These problems can be solved by using biofloc technology as it provides an uninterrupted supply of feed all day. This microbial organic soup significantly improves the feed utilization efficiency and lowers feed conversion ratio with little to no retention of ejected nutrients, thereby reducing the feed cost. In addition to providing a complete aquatic food source, bioflocs have a dynamic nutritional value with bioactive substances. Ballester et al. (2010) stated that bioflocs contain about 29.1% nitrogen-free extract, 4.7% crude fat, 8.3% fibre, 30.4% crude protein, and 39.2% ash based on dry matter. Protein and fat concentrations were greatest in particles larger than 100 μm, whereas essential amino acids were concentrated in flocs smaller than 48 μm. Vitamins are a crucial element for the development and survival of aquaculture species and biofloc contains vitamin C in the range 0–54 μg/g dry matter (Crab, 2010) along with other essential vitamins. Biofloc also acts as probiotics, which enhances the digestive enzymes activity and immunity of many aquatic species. Culture of common carp in biofloc showed 75% improvement in the activity of digestive enzymes and growth performance. Bioflocs can also supply immunostimulatory and bioactive substances like carotenoids, and soluble vitamins that may help farmed fish develop their immune systems.

24.5.4 USE OF BIOFLOC AS A TOOL TO MAINTAIN WATER QUALITY

Biofloc plays a very prominent role in maintaining water quality. Maintaining good water quality and a healthy environment is very essential and is of primary concern in aquaculture. Aquaculture intensification led to the use of a huge quantity of feeds rich in protein, which contain about 65% of nitrogen. It is known that all the feed given is not consumed by the culture organisms. The leftover feed will ultimately deteriorate the water quality along with the release of toxic ammonia. It is stated earlier that the biofloc system helps in reducing the N (ammonia and nitrite) content and organic load with the help of heterotrophic bacteria. These bacteria convert toxic N compounds into consumable microbial protein. Along with heterotrophic bacterial N conversion, other mechanisms like nitrification, autotrophic N conversion, and denitrification help in N conversion. Nutrient recycling of inorganic phosphorous is also done by heterotrophic bacteria and makes them available to the culture species.

24.6 ECONOMICAL ASPECTS

The key factor that impacts the development of aquaculture is the cost of feed. So, lowering the production costs and increasing profitability are essential objectives of aquaculture. Aquaculture expenses are significantly reduced in the biofloc system as compared to the traditional system, and profitability is also higher in the biofloc treatments as the biofloc system significantly improves growth rate and also has a lower feed conversion ratio. Also, with supporting higher stocking rates, biofloc also improves the survival rate of the culture species, which ultimately contributes to higher production. The cost of producing one kilogram of tilapia and green tiger shrimp (*Penaeus semisulcatus*) with biofloc was reduced by 10% (De Schryver and Verstraete, 2009) and 33%, respectively (Megahed, 2010). Organic and inorganic fertilizers are no longer necessary for biofloc culture system, so thereby reducing the capital required for filters and their installation. Additionally, biofloc systems provide more cost-effective advantages, such as a 30% reduction in water treatment costs and a twofold increase in protein usage efficiency when compared to traditional water treatment methods.

24.7 LIMITATIONS

The biofoc has some drawbacks, as does any other technology. Keeping the C:N ratio steady during the culture phase is a significant problem for this novel technology. Another barrier to this idea is that it needs a steady source of electricity. Monitoring of floc volumes, oxygen levels, TSS, and ammonia concentration should be done repeatedly and should be maintained in optimal ranges. Therefore, to operate this microbial system, in-depth knowledge, much greater skills, and better-run laboratories are needed.

24.8 CONCLUSION

Biofloc is an emerging sustainable aquaculture production system, which has a number of advantages over conventional clean water systems like the ability to control water quality, providing proteinaceous microbial feed, zero or minimal water exchange, biosecurity, enhancing the immunity, growth, digestibility, and survival rate of culture animals. All these attributes help in accomplishing SDGs. Since biofloc technology goes against the conventional idea that pond water must be clear, convincing farmers to adopt the method is a significant challenge. Despite all these benefits, biofloc is a complex system which requires better understanding and technical skills. Biofloc may not only be utilized by herbivorous or detritivorous organisms. More species, particularly carnivorous species, can be tested with various carbon sources and ratios to learn whether BFT might benefit them in some way. Therefore, it is advised that additional study be done to create the best possible conditions for various species in the system.

KEYWORDS

- **aquaculture**
- **biofloc**
- **C/N ratio**
- **enzyme**
- **heterotrophic bacteria**
- **sustainable**

REFERENCES

Avnimelech, Y. (1999). Carbon/nitrogen ratio as a control element in aquaculture systems. *Aquaculture, 176*(3–4), 227–235.

Avnimelech, Y. (2007). Feeding with microbial flocs by tilapia in minimal discharge bioflocs technology ponds. *Aquaculture, 264*, 140–147.

Avnimelech, Y., Mokady, S., & Schoroder, G. L. (1989). Circulated ponds as efficient bioreactors for single cell protein production. *The Israeli Journal of Aquaculture, 41*(2), 58–66

Avnimelech, Y. (2009). *Biofloc technology - A practical guide book.* Baton Rouge, LA: The World Aquaculture Society, p. 182.

Avnimelech, Y. (2015). *Biofoc technology, a practical guidebook.* 3rd Edition. Baton Rouge, LA: World Aquaculture Society. p. 258.

Ballester, E. L. C., Abreu, P. C., Cavalli, R. O., Emerenciano, M., De Abreu, L., & Wasielesky Jr, W. (2010). Effect of practical diets with different protein levels on the performance of *Farfantepenaeus paulensis* juveniles nursed in a zero exchange suspended microbial flocs intensive system. *Aquaculture Nutrition, 16*(2), 163–172.

Boyd, C. E. (2003). Guidelines for aquaculture effluent management at the farm-level. *Aquaculture, 226*, 101–112.

Brander, K. (2010). Impacts of climate change on fisheries. *Journal of Marine Systems, 79*(3–4), 389–402.

Burford, M. A., Thompson, P. J., McIntosh, R. P., Bauman, R. H., & Pearson, D. C. (2003). Nutrient and microbial dynamics in high-intensity, zero-exchange shrimp ponds in Belize. *Aquaculture, 219*, 393–411.

Costa-Pierce, B. A., Bartley, D. M., Hasan, M., Yusoff, F., Kaushik, S. J., Rana, K., Lemos, D., Bueno, P., & Yakupitiyage, A. (2012). Responsible use of resources for sustainable aquaculture. In *Farming the waters for people and food. Proceedings of the global conference on aquaculture* (pp. 113–147).

Crab, R. (2010). *Bioflocs technology: An integrated system for the removal of nutrients and simultaneous production of feed in aquaculture* (Doctoral dissertation, Ghent University).

Dauda, A. B. (2019). Biofloc technology: A review on the microbial interactions, operational parameters and implications to disease and health management of cultured aquatic animals. *Reviews in Aquaculture, 12*(2), 1193–1210.

De Schryver, P., & Verstraete, W. (2009). Nitrogen removal from aquaculture pond water by heterotrophic nitrogen assimilation in lab-scale sequencing batch reactors. *Bioresource Technology, 100*(3), 1162–1167.

Decamp, O., Conquest, L., Forster, I., & Tacon, A. G. J. (2002). The nutrition and feeding of marine shrimp within zero-water exchange aquaculture production systems: Role of eukaryotic microorganisms. In *World aquaculture society: Microbial approaches to aquatic nutrition within environmentally sound aquaculture production systems* (pp. 79–86).

Eding, E. H., Kamstra, A., Verreth, J. A. J., Huisman, E. A., & Klapwijk, A. (2006). Design and operation of nitrifying trickling filters in recirculating aquaculture: A review. *Aquacultural Engineering, 34*(3), 234–260.

Ekasari, J., Rivandi, D. R., Firdausi, A. P., Surawidjaja, E. H., Zairin, M., Bossier, P., & De Schryver, P. (2015a). Biofloc technology positively affects Nile tilapia (*Oreochromis niloticus*) larvae performance. *Aquaculture,* 441, 72–77.

Ekasari, J., Zairin Jr, M., Putri, D. U., Sari, N. P., Surawidjaja, E. H., & Bossier, P. (2015b). Biofloc-based reproductive performance of Nile tilapia *Oreochromis niloticus* L. broodstock. *Aquaculture Research*, *46*(2), 509–512.

Emerenciano, M., Cuzon, G., Paredes, A., & Gaxiola, G. (2013). Evaluation of biofloc technology in pink shrimp *Farfantepenaeus duorarum* culture: Growth performance, water quality, microorganisms profile and proximate analysis of biofloc. *Aquaculture International*, *21*, 1381–1394.

FAO. (2017). *Indicators: Measuring up to the 2030 agenda for sustainable development.* Rome: FAO. p. 39. http://www.fao.org/3/a-i6919e.pdf.

FAO. (2018). *The state of world fisheries and aquaculture: Meeting the sustainable development goals*. Rome: Contributing to Food Security and Nutrition for All, Food and Agriculture Organization of the United Nations. p. 200.

Hargreaves, J. A. (2006). Photosynthetic suspended-growth systems in aquaculture. *Aquacultural Engineering, 34*, 344–363.

Khanjani, M. (2015). *The effect of different feeding levels in biofloc system on water quality, growth performance and carcass composition of Pacific white shrimp (Litopenaeus vannamei Boone, 1931)* (Doctoral dissertation, Ph. D. thesis, Hormozgan University, Hormozgan, Iran).

Khanjani, M. H., & Sharifinia, M. (2020). Biofloc technology as a promising tool to improve aquaculture production. *Reviews in Aquaculture*, *12*(3), 1836–1850.

Lancelot, C. (1985). Carbon-nitrogen relationship in nutrient metabolism of coastal marine ecosystems. *Advances in Aquatic Microbiology*, *3*, 263–321.

Martínez-Cordova, L. R., Emerenciano, M., Miranda-Baeza, A., & Martínez-Porchas, M. (2015). Microbial-based systems for aquaculture of fish and shrimp: An updated review. *Reviews in Aquaculture*, *7*(2), 131–148.

Martinez-Porchas, M., & Martinez-Cordova, L. R. (2012). World aquaculture: Environmental impacts and troubleshooting alternatives. *The Scientific World Journal, 2012,* 389623.

Megahed, M. E. (2010). The effect of microbial biofloc on water quality, survival and growth of the green tiger shrimp (*Penaeus semisulcatus*) fed with different crude protein levels. *Journal of the Arabian Aquaculture Society*, *5*(2), 119–142.

Naylor, R. L., Williams, S. L., & Strong, D. R. (2001). Aquaculture--A gateway for exotic species. *Science*, *294*(5547), 1655–1656.

Panigrahi, A., Saranya, C., Sundaram, M., Kannan, S. V., Das, R. R., Kumar, R. S., Rajesh, P., & Otta, S. K. (2018). Carbon: Nitrogen (C:N) ratio level variation influences microbial community of the system and growth as well as immunity of shrimp (*Litopenaeus vannamei*) in biofloc based culture system. *Fish and Shellfish Immunology*, *81*, 329–337.

Rakocy, J. E., Bailey, D. S., Thoman, E. S., & Shultz, R. C. (2004). Intensive tank culture of tilapia with a suspended, bacterial-based, treatment process. In *New Dimensions on Farmed Tilapia: Proceedings of the Sixth International Symposium on Tilapia in Aquaculture* (pp. 584–596).

Xu, W. J., Morris, T. C., & Samocha, T. M. (2016). Effects of C/N ratio on biofloc development, water quality, and performance of *Litopenaeus vannamei* juveniles in a biofloc-based, high-density, zero-exchange, outdoor tank system. *Aquaculture*, *453*, 169–175.

Yu, Z., Li, L., Zhu, R., Li, M., Duan, J., Wang, J. Y., Liu, Y. H., & Wu, L. F. (2020). Monitoring of growth, digestive enzyme activity, immune response and water quality parameters of Golden crucian carp (*Carassius auratus*) in zero-water exchange tanks of biofloc systems. *Aquaculture Reports*, *16*, 100283.

Index

B

C

D

E

F

K

L

M

N

R

S

T

For Product Safety Concerns and Information please contact our EU representative GPSR@taylorandfrancis.com Taylor & Francis Verlag GmbH, Kaufingerstraße 24, 80331 München, Germany